Use R!

Series editors

Robert Gentleman Kurt Hornik Giovanni Parmigiani

More information about this series at http://www.springer.com/series/6991

Marius Hofert • Ivan Kojadinovic •
Martin Mächler • Jun Yan

Elements of Copula Modeling with R

Springer

Marius Hofert
Department of Statistics and Actuarial Science
University of Waterloo
Waterloo, Ontario, Canada

Ivan Kojadinovic
Laboratory of Mathematics and its Applications
University of Pau and Pays de l'Adour
Pau, France

Martin Mächler
Seminar for Statistics
ETH Zurich
Zurich, Switzerland

Jun Yan
Department of Statistics
University of Connecticut
Storrs, Connecticut, USA

ISSN 2197-5736 ISSN 2197-5744 (electronic)
Use R!
ISBN 978-3-319-89634-2 ISBN 978-3-319-89635-9 (eBook)
https://doi.org/10.1007/978-3-319-89635-9

Library of Congress Control Number: 2018940269

Mathematics Subject Classification (2010): 62H05, 65C10, 62H12, 62H15, 62P05, 62P12, 65C60

This Springer imprint is published by the registered company Springer Nature Switzerland AG
The registered company address is: Gewerbestrasse 11, 6330 Cham, Switzerland

To Saisai

Nire emazteari eta haurrei, maitasunez

To Liselotte, with thanks for many hours of waiting

To Jiafeng, Bohan, and Jolin, with love

Preface

The aim of this book is to show how some of the main steps involved in the statistical modeling of continuous multivariate distributions using copulas can be carried out in the R statistical environment (R Core Team , 2017) with the R package copula. The R package copula originally emerged from the authors' research interests; it has been available on the Comprehensive R Archive Network (CRAN) since 2005 and has been under constant development ever since. Judging from users' feedback, the package is applied in areas such as hydrology, environmental sciences, quantitative risk management, insurance, and finance.

Compared to existing monographs on copulas, the originality of this book is to illustrate how theoretical concepts can be applied in practice with the R package copula. To this end, numerous stand-alone examples are provided. This is both of pedagogical and of practical interest as the R source code and reproduced outputs such as figures not only allow one to better understand the subtleties of the theoretical notions but also enable one to solve real-world problems in the various fields of interest.

The book targets (possibly future) statisticians and (financial, hydrological, and other) engineers alike who would like to understand how theoretical notions and practical computations around copula modeling can be applied in R without an overwhelming amount of mathematics. Readers are nonetheless expected to have a basic knowledge of (multivariate) probability and statistics, in particular of random vectors, simulation algorithms, estimation methods, and statistical tests.

Although the book addresses most of the practical issues arising when modeling multivariate data with copulas, it only covers a modest part of the field. In particular, important aspects such as dynamic copula models, vine copulas, and copula modeling for censored or discontinuous data are not dealt with. This is partly due to a selection bias: Concepts that are not implemented in the R package copula at the time of writing are not presented in the book. Many of these more specialized notions are, however, available in other R packages. For an up-to-date description of the functionalities of other packages dealing with copulas, see, for example, the CRAN View "Distributions" available at https://cran.r-project.org/web/views/Distributions.html.

As the R statistical environment, and most free software, the R package `copula` comes with no warranty. It is distributed with the hope that it can be useful to others.

Waterloo, ON, Canada — Marius Hofert
Pau, France — Ivan Kojadinovic
Zürich, Switzerland — Martin Mächler
Storrs, CT, USA — Jun Yan
November 2017

Reference

R Core Team. (2017). R: A Language and Environment for Statistical Computing. R Foundation for Statistical Computing, Vienna, Austria. https://www.R-project.org.

Contents

Chapter 1
Introduction

1.1 A Motivating Example

Assume that one is given the two bivariate data sets displayed in Fig. 1.1 and asked to compare them in terms of the "dependence" between the two underlying variables. The first (respectively, second) data set, denoted by (x_{i1}, x_{i2}), $i \in \{1, \dots, n\}$ (respectively, (y_{i1}, y_{i2}), $i \in \{1, \dots, n\}$), is assumed to consist of $n = 1000$ *independent observations* (that is, a realization of independent copies) of a bivariate random vector (X_1, X_2) (respectively, (Y_1, Y_2)). Roughly speaking, comparing the two data sets in terms of dependence means comparing the way X_1 and X_2 are related with the way Y_1 and Y_2 are related.

The scatter plots do not reveal the presence of ties in the *component samples* $x_{1j}, \dots, x_{nj}$ and $y_{1j}, \dots, y_{nj}$, $j \in \{1, 2\}$, of the data sets, suggesting that (X_1, X_2) and (Y_1, Y_2) are *continuous random vectors*. Saying that (X_1, X_2) is continuous means that its *distribution function (df)* H defined by $H(\boldsymbol{x}) = \mathbb{P}(X_1 \leq x_1, X_2 \leq x_2)$, $\boldsymbol{x} \in \mathbb{R}^2$ (where $\mathbb{P}$ denotes the underlying probability measure) is continuous or, equivalently, that the univariate dfs F_1 and F_2 defined by $F_1(x) = \mathbb{P}(X_1 \leq x)$ and $F_2(x) = \mathbb{P}(X_2 \leq x)$, $x \in \mathbb{R}$, respectively, are continuous. Note in passing that F_1 and F_2 can be recovered from H by $F_1(x) = H(x, \infty) = \lim_{x_2 \to \infty} H(x, x_2)$ and $F_2(x) = H(\infty, x) = \lim_{x_1 \to \infty} H(x_1, x)$, $x \in \mathbb{R}$, and are thus also called the *margins* of H or *marginal dfs* of (X_1, X_2). In order to compare the two data sets in terms of the bespoke dependence, one may decide to estimate the linear correlation coefficient between X_1 and X_2, and compare it with the one between Y_1 and Y_2. A standard computation gives approximately 0.77 (respectively, 0.56) for the sample linear correlation coefficient between X_1 and X_2 (respectively, Y_1 and Y_2). This seems to indicate that the dependence between X_1 and X_2 is stronger than the dependence between Y_1 and Y_2. This conclusion does not appear unrealistic in view of Fig. 1.1.

M. Hofert et al., *Elements of Copula Modeling with R*, Use R!,
https://doi.org/10.1007/978-3-319-89635-9_1

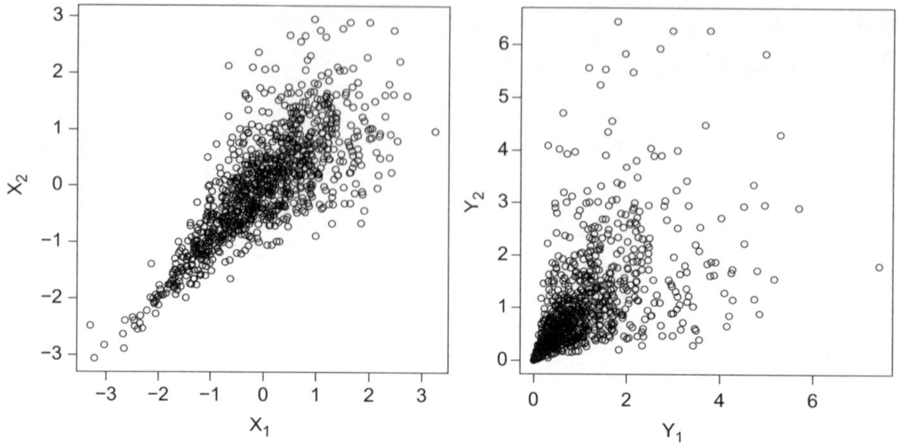

Fig. 1.1 Scatter plots of $n = 1000$ independent observations of (X_1, X_2) (left) and of (Y_1, Y_2) (right)

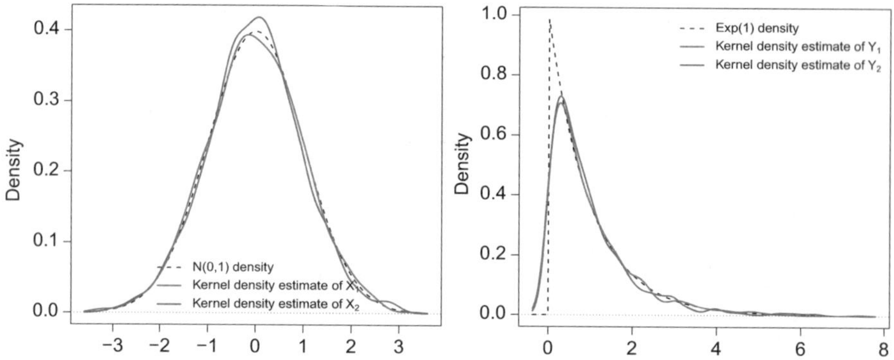

Fig. 1.2 Kernel density estimates of the densities of X_1, X_2 (left) and Y_1, Y_2 (right). The dashed curves represent the N(0, 1) (left) and Exp(1) (right) densities

As a normalized covariance, the correlation coefficient captures only one particular aspect of dependence: The strength of linear dependence between the underlying random variables. Given the shapes of the scatter plots in the left and right panels of Fig. 1.1, what was done thus far may therefore look at least incomplete. After some further thinking, one might decide to assess what the marginal distributions of (X_1, X_2) and (Y_1, Y_2) (that is, the distributions of X_1, X_2, Y_1, Y_2 separately) look like. A quick inspection of the component samples reveals that X_1 and X_2 seem to follow a standard normal distribution N(0, 1), while Y_1 and Y_2 seem to follow a standard exponential distribution Exp(1); see the kernel density estimates compared to the densities of these distributions in Fig. 1.2.

The fact that the two data sets in Fig. 1.1 differ at least marginally clearly affects one's perception of the possible difference in dependence. Because (Y_1, Y_2) appears to be marginally Exp(1) distributed, many of its realizations fall in the lower left corner of the plot on the right-hand side of Fig. 1.1, resulting in an area of high density in which it is more difficult to assess the relationship between Y_1 and Y_2. On the contrary, as (X_1, X_2) seems to be marginally N(0, 1) distributed, its bivariate observations in the plot on the left-hand side of Fig. 1.1 are much more spread out.

If we could transform the two data sets so that they become similar in terms of the underlying marginal dfs, their comparison in terms of dependence would be made on much fairer grounds.

1.2 Probability and Quantile Transformations

One transformation we could use is that to a standard uniform distribution, known as the *probability (integral) transformation*.

Lemma 1.2.1 (Probability Transformation) *Let F be a continuous df and let X have df F, that is, $X \sim F$. Then $F(X)$ is a standard uniform random variable, that is, $F(X) \sim \mathrm{U}(0, 1)$.* □

The probability transformation transforms a random variable with continuous df F to a standard uniform random variable. The continuity of F is crucial as otherwise the range of F would not contain $(0, 1)$.

Provided that the marginal dfs of the underlying random vectors (X_1, X_2) and (Y_1, Y_2) were known, the two data sets represented in Fig. 1.1 could be transformed into data sets consisting of observations of random vectors with standard uniform univariate marginal dfs, thereby making their comparison in terms of dependence fairer. Specifically, let F_1, F_2, G_1, and G_2 denote the dfs of X_1, X_2, Y_1, and Y_2, respectively, and let us assume for the moment that $F_1 = F_2$ is the N(0, 1) df and $G_1 = G_2$ is the Exp(1) df. Then the corresponding realizations of $(F_1(X_1), F_2(X_2))$ and $(G_1(Y_1), G_2(Y_2))$ are simply obtained by applying F_1 to the first component sample of the first data set (so to the realizations of X_1) and F_2 to the second component sample of the first data set (so to the realizations of X_2); similarly for the second data set. Scatter plots of the transformed data sets are displayed in Fig. 1.3.

The conclusion to be drawn from Fig. 1.3 is clear: The transformed data sets look similar (in fact, it follows from how the two data sets were constructed that they are actually identical). In other words, the distributions of $(F_1(X_1), F_2(X_2))$ and $(G_1(Y_1), G_2(Y_2))$ seem to be identical. If one is ready to accept the premise that the (for the moment vague) notion of dependence between the components of a continuous random vector should not be affected by its marginal distributions (a special case of an invariance principle; see Sect. 2.4), the conclusion is that the two data sets in Fig. 1.1 are indistinguishable in terms of dependence and only differ in terms of the underlying marginal dfs.

An alternative solution, for instance, would have been to transform the second data set so that it consists of observations of a bivariate random vector with standard normal margins. To this end, one needs the "converse" transformation of

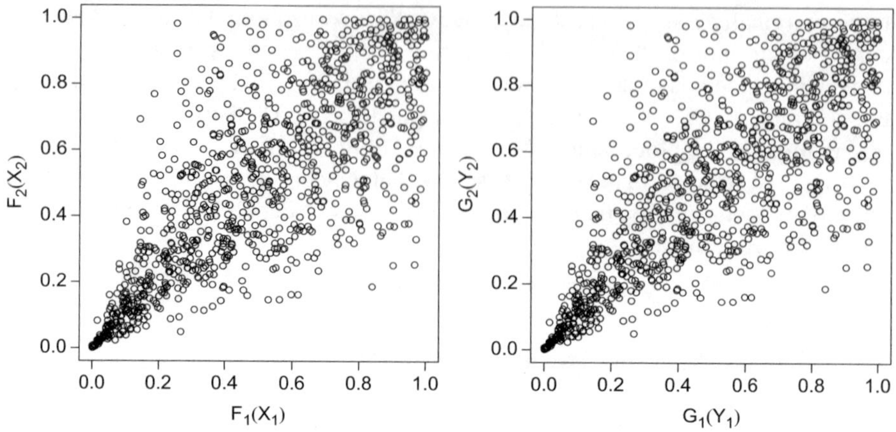

Fig. 1.3 Scatter plots of $n = 1000$ independent observations of the bivariate random vectors $(F_1(X_1), F_2(X_2))$ (left) and $(G_1(Y_1), G_2(Y_2))$ (right)

the probability transformation. It is called the *quantile transformation* and is given in Lemma 1.2.2 below. Note in passing that this transformation lies at the heart of the *inversion method* for pseudo-random number generation from nonuniform distributions; see Devroye (1986). The term "quantile" here refers to the following key concept: To any df F is associated a *quantile function* $F^{\leftarrow}$ defined by

$$F^{\leftarrow}(y) = \inf\{x \in \mathbb{R} : F(x) \geq y\}, \quad y \in [0, 1], \tag{1.1}$$

where inf stands for the infimum with the convention that $\inf \emptyset = \infty$. Note that this definition extends to increasing functions in general, providing a notion of an inverse for such functions, called *generalized inverse*. For continuous and strictly increasing dfs F, $F^{\leftarrow}$ equals the ordinary inverse F^{-1}. In many cases, one can work with $F^{\leftarrow}$ as if it were the ordinary inverse; see, for instance, Embrechts and Hofert (2013) for the conditions under which this can be done.

Lemma 1.2.2 (Quantile Transformation) *Let $U \sim \mathrm{U}(0, 1)$ and let F be any df. Then, the random variable $F^{\leftarrow}(U)$ has df F, that is, $F^{\leftarrow}(U) \sim F$.* □

The quantile transformation transforms U(0, 1)-variates into variates from a distribution with df F. Note that in contrast to F in Lemma 1.2.1, F in Lemma 1.2.2 does not need to be continuous.

To apply this result to transform (Y_1, Y_2) to a bivariate random vector with standard normal margins, one would first form $(G_1(Y_1), G_2(Y_2))$ (thus using Lemma 1.2.1) and then $\big(F_1^{\leftarrow}(G_1(Y_1)), F_2^{\leftarrow}(G_2(Y_2))\big)$ (thus using Lemma 1.2.2). To obtain the corresponding realizations, one simply needs to apply $F_1^{\leftarrow}(G_1(\cdot))$ and $F_2^{\leftarrow}(G_2(\cdot))$ to the first and second component samples, respectively, of the second data set. The resulting scatter plot is shown on the right-hand side of Fig. 1.4.

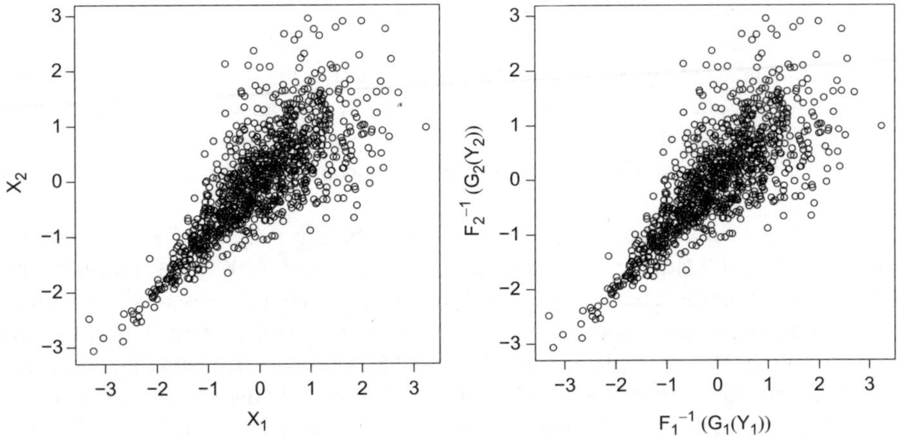

Fig. 1.4 Scatter plots of $n = 1000$ independent observations of the bivariate random vectors (X_1, X_2) (left) and $\big(F_1^{\leftarrow}(G_1(Y_1)), F_2^{\leftarrow}(G_2(Y_2))\big)$ (right)

As in Fig. 1.3, we see that the samples in Fig. 1.4 (which consist of independent observations of (X_1, X_2) and $\big(F_1^{\leftarrow}(G_1(Y_1)), F_2^{\leftarrow}(G_2(Y_2))\big)$, respectively) are equal. As before, we can thus conclude that the two data sets in Fig. 1.1 only differ in terms of their marginal dfs but are equal in terms of dependence.

1.3 Copulas

In this book, the informal notion of dependence will coincide with that of a *copula*, that is, a multivariate df with standard uniform univariate margins. The copula of (X_1, X_2) and the copula of (Y_1, Y_2) are simply the dfs of $(F_1(X_1), F_2(X_2))$ and $(G_1(Y_1), G_2(Y_2))$, respectively; this follows formally from Lemma 1.2.1 and the aforementioned invariance principle. The statement that "(X_1, X_2) and (Y_1, Y_2) have the same dependence" can then be rephrased as "(X_1, X_2) and (Y_1, Y_2) have the same copula."

In the above developments, the marginal distributions of the components of the two random vectors were assumed to be known. In practice, such an assumption is unrealistic and various sample analogues of the probability and quantile transformations can be applied alternatively.

The requirement that the margins of a copula be standard uniform is somewhat arbitrary. From the previous derivations, we see that the same conclusion for the two data sets is reached when the "standardization" is made to standard normal margins instead of standard uniform ones. For instance, Hoeffding (1940) originally considered multivariate dfs whose margins were $\mathrm{U}(-1/2, 1/2)$. Another example arises when studying dependence between componentwise block maxima in multivariate extreme-value theory: In that case, it is often more natural to standardize to unit Fréchet margins with df $F(x) = \exp(-1/x)$, $x > 0$. The important message is that no matter what continuous univariate distribution the margins are transformed to, it does not alter the philosophy behind the "copula approach" to the study of dependence. Due to the probability and quantile transformations (see Lemmas 1.2.1 and 1.2.2, respectively), the choice of U(0, 1) margins turns out to be a natural and convenient one.

In recent years, copulas have turned out to be the subject of a large number of scientific publications; see, for instance, Genest et al. (2009) for a bibliometric overview in finance. They were applied in a wide variety of areas such as quantitative risk management, econometric modeling, or environmental modeling, to name a very few; see, for example McNeil et al. (2015), Patton (2013), and Salvadori et al. (2007). The reason for what could be inelegantly called "the copula craze" lies in Sklar's Theorem (which will be precisely stated in Sect. 2.3). Let $(X_1, \dots, X_d)$ be a d-dimensional random vector and let $H(\boldsymbol{x}) = \mathbb{P}(X_1 \le x_1, \dots, X_d \le x_d)$, $\boldsymbol{x} \in \mathbb{R}^d$, be its df. The first part of Sklar's Theorem asserts that every d-dimensional df H can be expressed as

$$H(\boldsymbol{x}) = C\big(F_1(x_1), \dots, F_d(x_d)\big), \quad \boldsymbol{x} \in \mathbb{R}^d, \tag{1.2}$$

in terms of a d-dimensional copula C and the univariate marginal dfs $F_1, \dots, F_d$ obtained from H by $F_j(x_j) = H(\infty, \dots, \infty, x_j, \infty, \dots, \infty)$, $x_j \in \mathbb{R}$. The copula C is thus the function which connects or couples the marginal dfs $F_1, \dots, F_d$ to the multivariate df H, hence the name "copula." For estimation of H from data, this offers a great deal of flexibility as it allows one to model the marginal dfs $F_1, \dots, F_d$ separately from the dependence represented by the copula C, which is often of interest from a statistical and numerical point of view. The second part of Sklar's Theorem provides a converse. Given any copula C and univariate dfs $F_1, \dots, F_d$, a multivariate df H can be composed via (1.2) which then has univariate margins $F_1, \dots, F_d$ and "dependence structure" C. This is of interest in applications such as stress testing in finance, insurance and quantitative risk management, and in model building in general.

1.4 Structure and Philosophy of the Book

The book is organized as follows. Chapter 2 offers a basic introduction to copulas and presents their main properties along with the most important theoretical results such as the Fréchet–Hoeffding bounds, Sklar's Theorem, and the invariance principle. Chapter 3 introduces the main copula classes and the corresponding sampling procedures, along with some copula transformations that are important for practical purposes. The estimation of copulas is addressed in Chap. 4 from a parametric, semi-parametric, and nonparametric perspective. Chapter 5 presents graphical diagnostics, statistical tests and discusses model selection. Chapter 6 is concerned with more advanced topics such as the handling of ties, time series, and covariates (in a regression-like setting). The appendix contains information about the R session and package versions used throughout the book.

Overall, the theoretical concepts introduced in this book are illustrated by numerous R examples using functions predominantly from the R package `copula`. The examples and figures are stand-alone and reproducible. Most involve either synthetic or real data.

The book was written according to our best knowledge. Errors are, however, unavoidable in such a large project combining theoretical concepts and R code. In particular, it may happen that some of the code in the book does not work as expected anymore due to changes in one of the used R packages. An errata list is available at http://copula.r-forge.r-project.org/book/ along with other information on the book.

1.5 Additional Reading

As the title of the book indicates, only a small proportion of the existing research on copulas is touched upon in this work. Besides the literature cited in the book and on the help pages of the R package `copula`, the interested reader may consult one of the following references on copula theory and applications; note that this list is non-exhaustive and that some of the references are also cited later in this book at appropriate places.

Early monographs on copulas are Joe (1997) (with focus on novel probabilistic notions around copulas) and Nelsen (2006) (a well-known, readable introduction). An interesting historical perspective and introduction can be found in Durante and Sempi (2010). A more advanced probabilistic treatment of copulas is the recent Durante and Sempi (2015).

An overview paper of the theory and practice with financial applications in mind is given by Embrechts (2009). An important article well known for warning probabilists, statisticians, and financial data analysts alike about fallacies related to the use of linear correlation to measure dependence is Embrechts et al. (2002). For a more in-depth introductory treatment of copulas also stressing the latter topics in the context of quantitative risk management, see McNeil et al. (2015, Chapter 7).

Monographs on copulas mainly targeted towards the realm of finance are Cherubini et al. (2004, 2011) and Mai and Scherer (2014); see also Rémillard (2013, Chapter 8).

An easy-to-digest and statistically oriented introduction to copulas is given by Genest and Favre (2007); see also Genest et al. (2009) for additional material on goodness-of-fit testing. A recent statistically oriented monograph is Joe (2014).

A currently active area of research not covered in this book are vine copulas; see http://vine-copula.org for more details including publications and research projects.

References

Cherubini, U., Luciano, E., & Vecchiato, W. (2004). *Copula methods in finance*. Chichester: Wiley.

Cherubini, U., Mulinacci, S., Gobbi, F., & Romagnoli, S. (2011). *Dynamic copula methods in finance*. Chichester: Wiley.

Devroye, L. (1986). *Non-uniform random variate generation*. New York: Springer.

Durante, F., & Sempi, C. (2010). Copula theory: An introduction. In P. Jaworski, F. Durante, W. K. Härdle & W. Rychlik (Eds.), *Copula theory and its applications (Warsaw, 2009). Lecture notes in statistics* (pp. 3–32). Berlin: Springer.

Durante, F., & Sempi, C. (2015). *Principles of copula theory*. Boca Raton, FL: CRC Press.

Embrechts, P. (2009). Copulas: A personal view. *Journal of Risk and Insurance, 76*, 639–650.

Embrechts, P., & Hofert, M. (2013). A note on generalized inverses. *Mathematical Methods of Operations Research, 77*(3), 423–432.

Embrechts, P., McNeil, A. J., & Straumann, D. (2002). Correlation and dependency in risk management: Properties and pitfalls. In M. Dempster (Ed.), *Risk Management: Value at Risk and Beyond* (pp. 176–223). London: Cambridge University Press.

Genest, C., & Favre, A.-C. (2007). Everything you always wanted to know about copula modeling but were afraid to ask. *Journal of Hydrological Engineering, 12*, 347–368.

Genest, C., Gendron, M., & Bourdeau-Brien, M. (2009). The advent of copulas in finance, *European Journal of Finance, 15*, 609–618.

Genest, C., Rémillard, B., & Beaudoin, D. (2009). Goodness-of-fit tests for copulas: A review and a power study. *Insurance: Mathematics and Economics, 44*, 199–213.

Hoeffding, W. (1940). Massstabinvariante Korrelationstheorie, *Schriften des mathematischen Seminars und des Instituts für Angewandte Mathematik der Universität Berlin, 5*, 181–233.

Joe, H. (1997). *Multivariate models and dependence concepts*. London: Chapman & Hall.

Joe, H. (2014). *Dependence modeling with copulas*. Boca Raton, FL: Chapman & Hall/CRC.

Mai, J.-F., & Scherer, M. (2014). *Financial engineering with copulas explained*. London: Palgrave Macmillan.

McNeil, A. J., Frey, R., & Embrechts, P. (2015). *Quantitative risk management: Concepts, techniques and tools* (2nd ed.). Princeton, NJ: Princeton University Press.

Nelsen, R. B. (2006). *An introduction to copulas*. New York: Springer.

Patton, A. J. (2013). Copula methods for forecasting multivariate time series. In G. Elliott & A. Timmermann (Eds.), *Handbook of economic forecasting* (pp. 899–960). New York: Springer.

Rémillard, B. (2013). *Statistical methods for financial engineering*. Boca Baton, FL: Chapman & Hall/CRC.

Salvadori, G., De Michele, C., Kottegoda, N. T., & Rosso, R. (2007). *Extremes in nature: An approach using copulas. Water science and technology library* (Vol. 56). Berlin: Springer.

Chapter 2
Copulas

2.1 Definition and Characterization

Copulas are particular multivariate dfs. Recall that the df H of a d-dimensional random vector $\boldsymbol{X} = (X_1, \dots, X_d)$ is the function defined by

$$H(\boldsymbol{x}) = \mathbb{P}(\boldsymbol{X} \le \boldsymbol{x}) = \mathbb{P}(X_1 \le x_1, \dots, X_d \le x_d), \quad \boldsymbol{x} = (x_1, \dots, x_d) \in \mathbb{R}^d.$$

The df F_j of X_j, $j \in \{1, \dots, d\}$, can be recovered from the multivariate df H by $F_j(x_j) = H(\infty, \dots, \infty, x_j, \infty, \dots, \infty)$, $x_j \in \mathbb{R}$. This is why $F_1, \dots, F_d$ are also called the *univariate margins* of H or the *marginal dfs* of $\boldsymbol{X}$.

Definition 2.1.1 (Copula) A *copula* is a multivariate df with standard uniform univariate margins, that is, U(0, 1) margins. □

As mentioned in Chap. 1, the requirement that the margins be standard uniform can be regarded as arbitrary. For instance, Hoeffding (1940) worked with multivariate dfs whose margins are $\mathrm{U}(-1/2, 1/2)$, and, in multivariate extreme-value theory, when studying dependence between multivariate maxima, it is more natural to consider multivariate dfs whose margins are all unit Fréchet. The important message is that the way a multivariate distribution is "standardized" from the point of view of its margins does not alter the philosophy behind the concept of a copula. However, due to the probability and quantile transformations (see Lemmas 1.2.1 and 1.2.2), the choice of U(0, 1) margins turns out to be a rather sensible one.

One of the simplest copulas is the *independence copula*

$$\Pi(\boldsymbol{u}) = \prod_{j=1}^{d} u_j, \quad \boldsymbol{u} \in [0, 1]^d, \tag{2.1}$$

M. Hofert et al., *Elements of Copula Modeling with R*, Use R!,
https://doi.org/10.1007/978-3-319-89635-9_2

which is the df of a random vector $\boldsymbol{U} = (U_1, \dots, U_d)$ with $U_1, \dots, U_d \overset{\text{ind.}}{\sim} \mathrm{U}(0, 1)$, that is, with independent U(0, 1)-distributed components; this is easy to see from the fact that, for any $\boldsymbol{u} \in [0, 1]^d$, $\mathbb{P}(\boldsymbol{U} \le \boldsymbol{u}) = \mathbb{P}(U_1 \le u_1, \dots, U_d \le u_d) = \prod_{j=1}^d \mathbb{P}(U_j \le u_j) = \prod_{j=1}^d u_j = \Pi(\boldsymbol{u})$.

Example 2.1.2 (Independence Copula) In the R package `copula`, the copula Π is represented by an R object created using the function `indepCopula()`:

```
> library(copula)
> d <- 2
> ic <- indepCopula(dim = d)
```

To evaluate the copula Π, the function `pCopula()` can be used:

```
> set.seed(2008)
> u <- runif(d) # a random point in the unit hypercube
> (Pi <- pCopula(u, copula = ic)) # the value of the independence copula at u
```

```
[1] 0.3668219
```

To check the numerical equality between computations carried out with `pCopula()` on the one hand and directly with (2.1) on the other hand, we can use the function `all.equal()`. Also conveniently used in what follows is the function `stopifnot()`, which, as the name suggests, produces an error in case at least one of its arguments evaluates to `FALSE`:

```
> stopifnot(all.equal(Pi, prod(u))) # check numerical equality of the samples
```

When $d = 2$, surface (or perspective) plots and contour (or level) plots of the function Π can be easily produced with `wireframe2()` and `contourplot2()`, respectively, as follows:

```
> wireframe2  (ic, FUN = pCopula, # surface plot of the independence copula
               col.4 = adjustcolor("black", alpha.f = 0.25))
> contourplot2(ic, FUN = pCopula) # contour plot of the independence copula
```

The resulting plots are shown in Fig. 2.1. Note that, in this book, we shall use the expression *wireframe plot* for a surface or perspective plot by reference to the name of the corresponding function in the R package `lattice` of Sarkar (2008). □

Let $d = 2$ and notice from the wireframe plot in Example 2.1.2 that Π is zero on all edges of the unit square which start at $(0, 0)$, and that $\Pi(u_1, 1) = u_1$ and $\Pi(1, u_2) = u_2$ for all $u_1, u_2 \in [0, 1]$; these two properties are highlighted by dashed lines in the wireframe plot of Fig. 2.1. In fact, any d-dimensional copula C, $d \ge 2$, is *grounded* ($C(\boldsymbol{u}) = 0$ if $u_j = 0$ for at least one $j \in \{1, \dots, d\}$) and has *standard uniform univariate margins* (for any $j \in \{1, \dots, d\}$, $C(1, \dots, 1, u_j, 1, \dots, 1) = u_j$ for all $u_j \in [0, 1]$). For this reason `wireframe2(, FUN = pCopula)` by default visually emphasizes these four line segments for all bivariate copulas. Note that being grounded and having standard uniform univariate margins alone is not sufficient for a function $C : [0, 1]^d \to [0, 1]$ to be a df and thus a copula; for

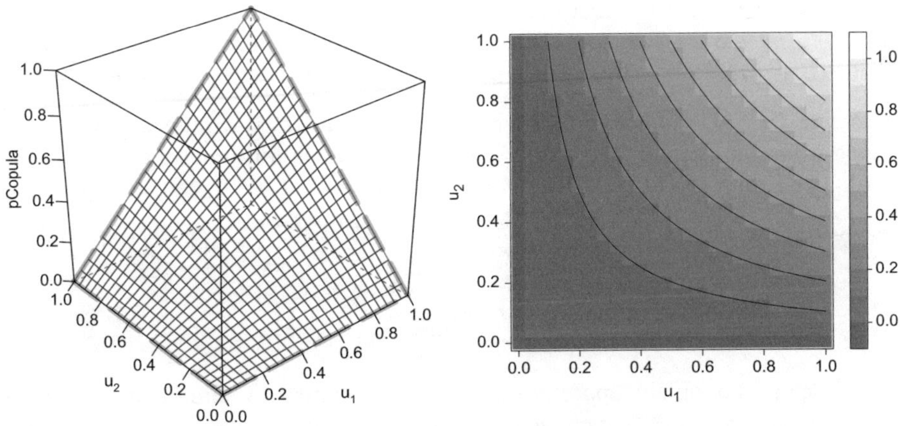

Fig. 2.1 Wireframe plot (left) and contour plot (right) of the independence copula

example, the function W in (2.6) for $d \geq 3$ is grounded and has standard uniform univariate margins but can be shown not to be a df.

In order for a function $C : [0, 1]^d \to [0, 1]$ to be a copula (and thus to obtain a characterization of copulas), one needs one additional property to guarantee that C is a multivariate df. To introduce it, we need the following additional definitions. For any $\boldsymbol{a}, \boldsymbol{b} \in [0, 1]^d$, $\boldsymbol{a} \leq \boldsymbol{b}$, let $(\boldsymbol{a}, \boldsymbol{b}]$ denote the hyperrectangle with lower end point $\boldsymbol{a}$ and upper end point $\boldsymbol{b}$ defined by $\{\boldsymbol{u} \in [0, 1]^d : \boldsymbol{a} < \boldsymbol{u} \leq \boldsymbol{b}\}$. Then, for any hyperrectangle $(\boldsymbol{a}, \boldsymbol{b}]$ in $[0, 1]^d$, define its *C-volume* as

$$\Delta_{(\boldsymbol{a},\boldsymbol{b}]} C = \sum_{\boldsymbol{i} \in \{0,1\}^d} (-1)^{\sum_{j=1}^d i_j} C\big(a_1^{i_1} b_1^{1-i_1}, \ldots, a_d^{i_d} b_d^{1-i_d}\big), \tag{2.2}$$

where the summation is taken over all 2^d-many vectors $\boldsymbol{i} = (i_1, \ldots, i_d)$ for $i_1, \ldots, i_d \in \{0, 1\}$. If the C-volumes of all hyperrectangles $(\boldsymbol{a}, \boldsymbol{b}]$ in $[0, 1]^d$ are nonnegative, that is,

$$\Delta_{(\boldsymbol{a},\boldsymbol{b}]} C \geq 0 \quad \text{for all } \boldsymbol{a}, \boldsymbol{b} \in [0, 1]^d,\ \boldsymbol{a} \leq \boldsymbol{b},$$

C is called *d-increasing*.

Notice that C-volumes have a natural interpretation. For $\boldsymbol{U} \sim C$, it can be verified that

$$\Delta_{(\boldsymbol{a},\boldsymbol{b}]} C = \mathbb{P}(\boldsymbol{U} \in (\boldsymbol{a}, \boldsymbol{b}]) \quad \text{for all hyperrectangles } (\boldsymbol{a}, \boldsymbol{b}] \text{ in } [0, 1]^d. \tag{2.3}$$

When $d = 2$, (2.2) becomes

$$\Delta_{(\boldsymbol{a},\boldsymbol{b}]} C = C(b_1, b_2) - C(b_1, a_2) - C(a_1, b_2) + C(a_1, a_2).$$

Let $C = \Pi$ and let us verify (2.3). On the one hand, we have that $\Delta_{(\boldsymbol{a},\boldsymbol{b}]}\Pi = b_1b_2 - b_1a_2 - a_1b_2 + a_1a_2 = (b_1 - a_1)(b_2 - a_2)$ and, on the other hand, we know that $\mathbb{P}(\boldsymbol{U} \in (\boldsymbol{a}, \boldsymbol{b}]) = \mathbb{P}(a_1 < U_1 \le b_1)\mathbb{P}(a_2 < U_2 \le b_2) = (b_1 - a_1)(b_2 - a_2)$.

Example 2.1.3 (C-volumes) In the R package copula, C-volumes can be computed using the function prob(). The following code illustrates the computation of $\Delta_{(\boldsymbol{a},\boldsymbol{b}]}\Pi$ for $\boldsymbol{a} = (1/4, 1/2)$ and $\boldsymbol{b} = (1/3, 1)$:

```
> a <- c(1/4, 1/2) # lower left end point
> b <- c(1/3, 1) # upper right end point
> stopifnot(0 <= a, a <= 1, 0 <= b, b <= 1, a <= b) # check
> p <- (b[1] - a[1]) * (b[2] - a[2]) # manual computation
> stopifnot(all.equal(prob(ic, l = a, u = b), p)) # check
```

We can also obtain approximations of C-volumes by simulation using the function rCopula() which generates independent observations from a given copula. We shall illustrate this in the case of the previous Π-volume below. But first, let us generate and plot a sample from Π (see Fig. 2.2):

```
> n <- 1000 # sample size
> set.seed(271) # set a seed (for reproducibility)
> U <- rCopula(n, copula = ic) # generate a sample of the independence copula
> plot(U, xlab = quote(U[1]), ylab = quote(U[2]))
```

Note that rCopula(n, ic) is merely a wrapper for matrix(runif(n*d), nrow = n):

```
> set.seed(271)
> stopifnot(all.equal(U, matrix(runif(n * d), nrow = n)))
```

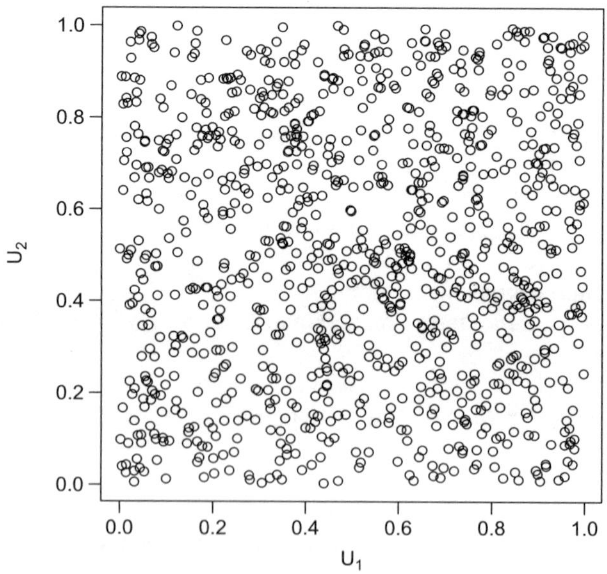

Fig. 2.2 $n = 1000$ independent observations of $\boldsymbol{U} \sim \Pi$

The Π-volume of $(1/4, 1/3] \times (1/2, 1]$ can then be approximated by the proportion of realizations of $\boldsymbol{U} \sim \Pi$ falling in this hyperrectangle:

```
> set.seed(314)
> U <- rCopula(1e6, copula = ic) # large sample size for good approximation
> ## Approximate the Pi-volume by the aforementioned proportion
> p.sim <- mean(a[1] < U[,1] & U[,1] <= b[1] & a[2] < U[,2] & U[,2] <= b[2])
> stopifnot(all.equal(p.sim, p, tol = 1e-2)) # note: may depend on seed
```

□

A copula C is called *absolutely continuous* if it admits a density. In this book, for practical purposes, we shall simply say that a copula C admits a *density* c if

$$c(\boldsymbol{u}) = \frac{\partial^d}{\partial u_d \dots \partial u_1} C(u_1, \dots, u_d), \quad \boldsymbol{u} \in (0,1)^d, \tag{2.4}$$

exists and is integrable. In the R package `copula`, copula densities can be evaluated with the function `dCopula()`. Note in passing that, if, for a function $C : [0,1]^d \to [0,1]$, c as defined in (2.4) is nonnegative for all $\boldsymbol{u} \in (0,1)^d$, then C is d-increasing; see, for instance, Example 2.6.3 where this result is used. This is due to the fact that any C-volume $\Delta_{(\boldsymbol{a},\boldsymbol{b}]}C$ (that is, the probability that $\boldsymbol{U} \sim C$ falls into $(\boldsymbol{a}, \boldsymbol{b}]$) can be expressed as the integral of c over $(\boldsymbol{a}, \boldsymbol{b}]$.

As an example, the independence copula Π is absolutely continuous with constant density $c(\boldsymbol{u}) = 1$, $\boldsymbol{u} \in (0,1)^d$; note the similarity to the univariate case ($\mathrm{U}(0,1)$ has the constant density 1 on $(0,1)$).

In the same way that parametric univariate distributions are crucial for univariate statistics, parametric copula families play a key role in the applications of copulas. In the remainder of this section, we introduce some parametric copula families.

Example 2.1.4 (Frank Copula) As a first example, consider the *Frank* family; see Genest (1987). Its members, parametrized by $\theta \in \mathbb{R} \setminus \{0\}$ are, in the bivariate case, the copulas defined by

$$C_\theta^{\mathrm{F}}(\boldsymbol{u}) = -\frac{1}{\theta} \log\left(1 + \frac{(\exp(-\theta u_1) - 1)(\exp(-\theta u_2) - 1)}{\exp(-\theta) - 1}\right), \quad \boldsymbol{u} \in [0,1]^2, \tag{2.5}$$

with the convention that $C_0^{\mathrm{F}} = \Pi$, as a consequence of the fact that the function in (2.5) converges to Π as $\theta \to 0$.

Two immediate consequences from (2.5) are groundedness (for example, if $u_1 = 0$, then $C_\theta^{\mathrm{F}}(\boldsymbol{u}) = -\log(1+0)/\theta = 0$) and standard uniform univariate margins (for example, if $u_2 = 1$, then $C_\theta^{\mathrm{F}}(\boldsymbol{u}) = -\log(1 + (\exp(-\theta u_1) - 1))/\theta = u_1$). The property of 2-increasingness can also be shown (for example, by deriving the density c_θ^{F} and showing that it is indeed nonnegative). The higher-dimensional version of C_θ^{F} will be defined in Chap. 3.

The parameter θ in the expression of C_θ^{F} controls the dependence between the components of $\boldsymbol{U} \sim C_\theta^{\mathrm{F}}$ as will be illustrated later in this example.

In the R package copula, the Frank copula C_θ^F is represented by an R object created using the function frankCopula():

```
> d <- 2 # dimension
> theta <- -9 # copula parameter
> fc <- frankCopula(theta, dim = d) # define a Frank copula
```

The created object fc thus represents a Frank copula with parameter $\theta = -9$. The copula C_θ^F and its corresponding density c_θ^F can be computed with the functions pCopula() and dCopula(), respectively:

```
> set.seed(2010)
> n <- 5 # number of evaluation points
> u <- matrix(runif(n * d), nrow = n) # n random points in [0,1]^d
> pCopula(u, copula = fc) # copula values at u
```

```
[1] 0.2515825004 0.1438027127 0.0005267892 0.0053807993 0.0075792044
```

```
> dCopula(u, copula = fc) # density values at u
```

```
[1] 0.9012539 1.8161995 0.1400732 0.4590708 2.4612347
```

Wireframe and contour plots of the functions C_θ^F and the corresponding density c_θ^F (for $d = 2$ and $\theta = -9$; see Fig. 2.3) can be easily produced as follows:

```
> wireframe2(fc, FUN = pCopula, # wireframe plot (copula)
             draw.4.pCoplines = FALSE)
> wireframe2(fc, FUN = dCopula, delta = 0.001) # wireframe plot (density)
> contourplot2(fc, FUN = pCopula) # contour plot (copula)
> contourplot2(fc, FUN = dCopula, n.grid = 72, # contour plot (density)
               lwd = 1/2)
```

Next, the function rCopula() is used to generate $n = 1000$ independent observations from C_θ^F for $\theta \in \{-9, 0, 9\}$ (the setting $\theta = 0$ generates a sample from the independence copula):

```
> set.seed(1946)
> n <- 1000
> U  <- rCopula(n, copula = fc)
> U0 <- rCopula(n, copula = setTheta(fc, value = 0))
> U9 <- rCopula(n, copula = setTheta(fc, value = 9))
> plot(U,  xlab = quote(U[1]), ylab = quote(U[2]))
> plot(U0, xlab = quote(U[1]), ylab = quote(U[2]))
> plot(U9, xlab = quote(U[1]), ylab = quote(U[2]))
```

Note that we have used the function setTheta() to change the parameter of the Frank copula.

As we can see from Fig. 2.4, by increasing the value of θ from 0 to 9, the components U_1 and U_2 of $\boldsymbol{U} \sim C_\theta^F$ become *positively dependent* in the sense that small (respectively, large) values of U_1 tend to be associated with small (respectively, large) values of U_2. On the contrary, for $\theta = -9$, U_1 and U_2 are *negatively dependent* in the sense that small (respectively, large) values of U_1 tend to be associated with large (respectively, small) values of U_2. □

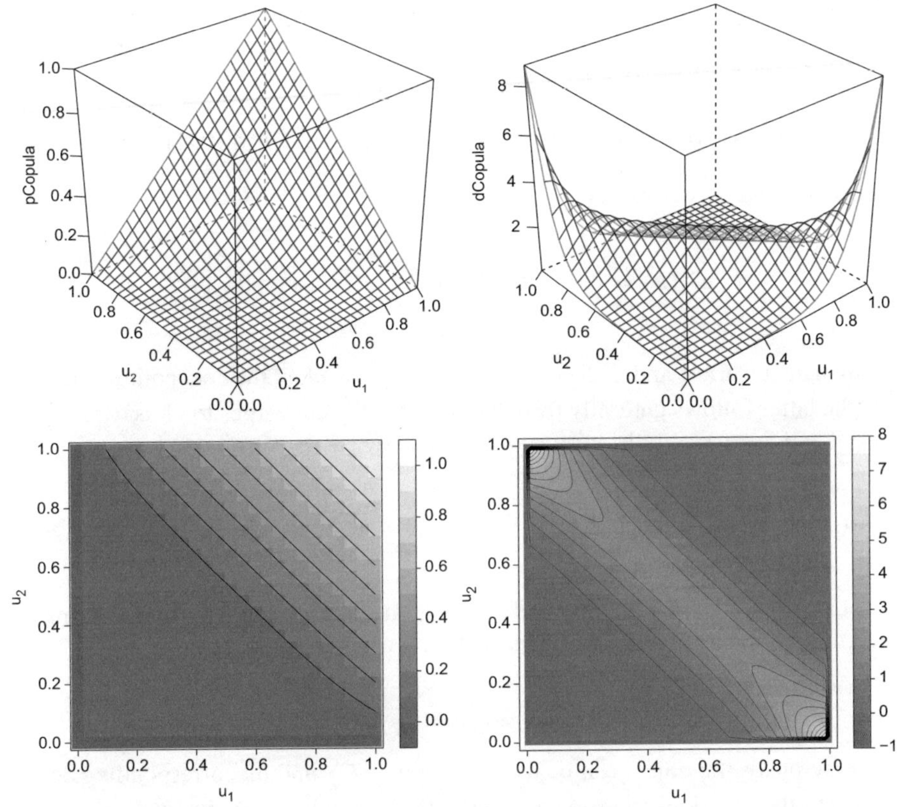

Fig. 2.3 Wireframe (top) and contour plots (bottom) of C_θ^{F} (left) and of the corresponding density c_θ^{F} (right) for $d = 2$ and $\theta = -9$

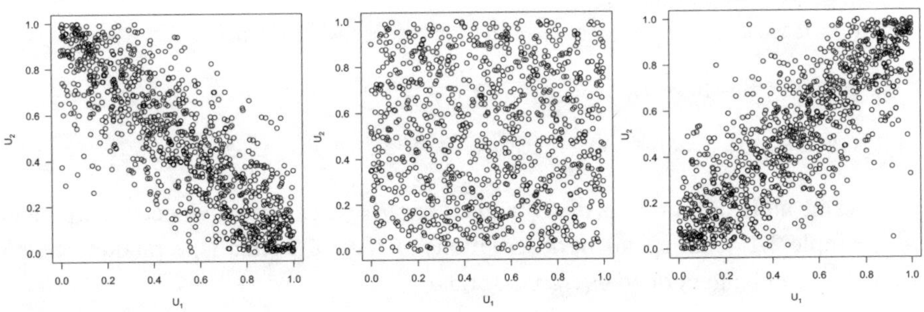

Fig. 2.4 $n = 1000$ independent observations of $(U_1, U_2) \sim C_\theta^{\mathrm{F}}$ for $\theta = -9$ (left), $\theta = 0$ (middle) and $\theta = 9$ (right)

Example 2.1.5 (Clayton Copula) As an another example of a parametric copula family, consider the *Clayton* family. For $d = 2$, its members, parametrized by $\theta \in [-1, \infty) \setminus \{0\}$, are defined by

$$C_\theta^{\mathrm{C}}(\boldsymbol{u}) = \max\left\{u_1^{-\theta} + u_2^{-\theta} - 1, 0\right\}^{-1/\theta}, \quad \boldsymbol{u} \in [0, 1]^2.$$

For $d \geq 3$, we only consider $\theta \in (0, \infty)$ and we have

$$C_\theta^{\mathrm{C}}(\boldsymbol{u}) = \left(1 - d + \sum_{j=1}^{d} u_j^{-\theta}\right)^{-1/\theta}, \quad \boldsymbol{u} \in [0, 1]^d.$$

Both in the bivariate and in the multivariate case, we adopt the convention that $C_0^{\mathrm{C}} = \Pi$ (the latter follows naturally from the fact that C_θ^{C} converges to Π as $\theta \to 0$). For $d \geq 2$ and $\theta \in (0, \infty)$, the corresponding density can be verified to be

$$c_\theta^{\mathrm{C}}(\boldsymbol{u}) = \left(\prod_{k=0}^{d-1}(\theta k + 1)\right)\left(\prod_{j=1}^{d} u_j\right)^{-(1+\theta)}\left(1 - d + \sum_{j=1}^{d} u_j^{-\theta}\right)^{-(d+1/\theta)}, \quad \boldsymbol{u} \in (0, 1)^d.$$

R objects allowing to manipulate Clayton copulas can be created using the function `claytonCopula()`. For instance:

```
> d <- 3
> cc <- claytonCopula(4, dim = d) # theta = 4
```

As previously, the object can be used to evaluate C_θ^{C} and its corresponding density c_θ^{C} using the functions `pCopula()` and `dCopula()`, respectively:

```
> set.seed(2013)
> n <- 5
> u <- matrix(runif(n * d), nrow = n) # random points in the unit hypercube
> pCopula(u, copula = cc) # copula values at u
```

```
[1] 0.004559638 0.247045204 0.634084005 0.702682364 0.250533621
```

```
> dCopula(u, copula = cc) # density values at u
```

```
[1] 1.947682e-15 1.251657e-03 5.108704e+00 5.341037e+00 1.931097e-02
```

Wireframe and contour plots of C_θ^{C} and c_θ^{C} for $d = 2$ can be produced as explained in Example 2.1.4. Independent observations of $\boldsymbol{U} \sim C_\theta^{\mathrm{C}}$ (that is, a random sample from C_θ^{C}) can be generated using `rCopula()`:

```
> set.seed(271)
> U <- rCopula(1000, copula = cc)
> splom2(U, cex = 0.3, col.mat = "black")
```

The resulting scatter-plot matrix (based on the function `splom2()` from the R package `copula`) is shown in Fig. 2.5.

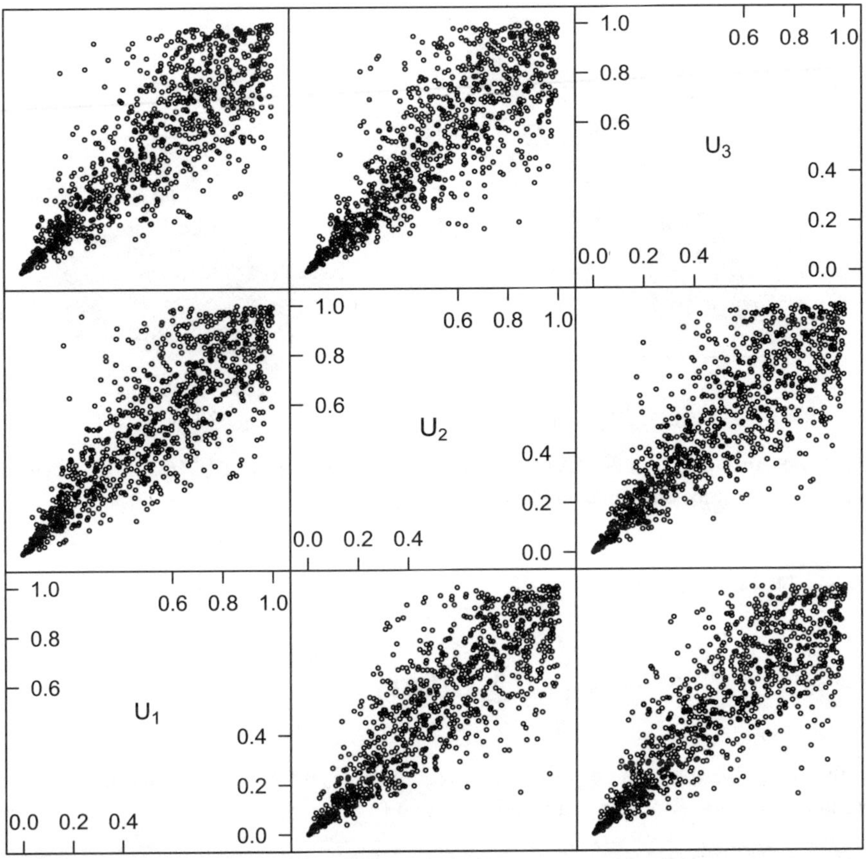

Fig. 2.5 Scatter-plot matrix of $n = 1000$ independent observations from a trivariate Clayton copula C_θ^{C} for $\theta = 4$

For $\theta \geq 0$, the larger the value of θ, the stronger the (positive) dependence between the components of $\boldsymbol{U} \sim C_\theta^{\mathrm{C}}$. For $d = 2$, it can additionally be verified (for instance, by adapting the previous code) that values of θ in $[-1, 0)$ lead to negative dependence between the components of $(U_1, U_2) \sim C_\theta^{\mathrm{C}}$. Later in this chapter (see Sect. 2.2), it will become clear why we can say that, in dimension two, the Clayton family "interpolates between perfect negative and perfect positive dependence," while in dimension three or higher, it "interpolates only between independence and perfect positive dependence." The same actually holds for the Frank family introduced in Example 2.1.4. □

Example 2.1.6 (Gumbel–Hougaard Copula) As a third example of a parametric copula family, consider the *Gumbel–Hougaard* family. Its members, parametrized

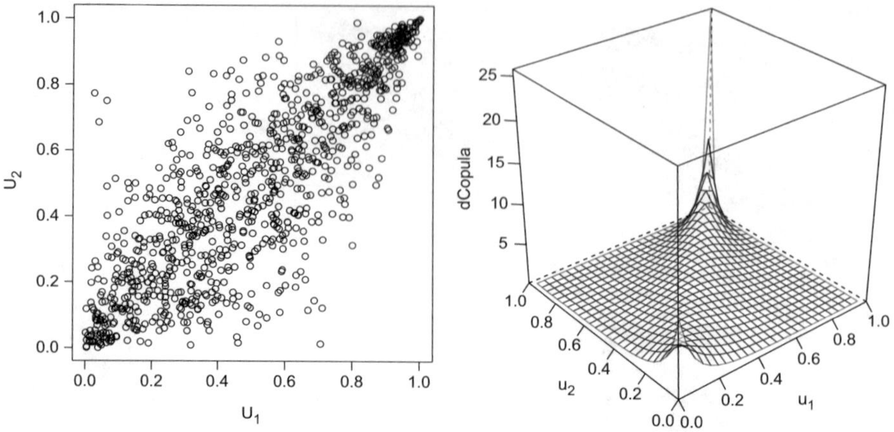

Fig. 2.6 Scatter plot of $n = 1000$ independent observations from C_θ^{GH} for $\theta = 3$ (left) and wireframe plot of the corresponding density c_θ^{GH} (right)

by $\theta \in [1, \infty)$, are defined by

$$C_\theta^{\mathrm{GH}}(\boldsymbol{u}) = \exp\left(-\Big(\sum_{j=1}^{d}(-\log u_j)^\theta\Big)^{1/\theta}\right), \quad \boldsymbol{u} \in [0,1]^d.$$

In the R package `copula`, objects representing Gumbel–Hougaard copulas can be created using the function `gumbelCopula()`:

```
> gc <- gumbelCopula(3) # theta = 3 (note the default dim = 2)
```

Evaluation of C_θ^{GH}, its density c_θ^{GH} or random number generation from C_θ^{GH} can be carried out as previously using the R functions `pCopula()`, `dCopula()` and `rCopula()`, respectively. For instance:

```
> set.seed(1993)
> U <- rCopula(1000, copula = gc)
> plot(U, xlab = quote(U[1]), ylab = quote(U[2]))
> wireframe2(gc, dCopula, delta = 0.025) # wireframe plot (density)
```

The resulting plots are shown in Fig. 2.6. □

2.2 The Fréchet–Hoeffding Bounds

The following theorem is one of the cornerstones of copula theory. It states that any copula C is pointwise bounded from below by the *lower Fréchet–Hoeffding bound* W and from above by the *upper Fréchet–Hoeffding bound* M, where W and M are defined by

$$W(\boldsymbol{u}) = \max\left\{\sum_{j=1}^{d} u_j - d + 1, 0\right\} \quad \text{and} \quad M(\boldsymbol{u}) = \min_{1\le j\le d}\{u_j\}, \quad \boldsymbol{u} \in [0,1]^d. \tag{2.6}$$

It is important to note that W is a copula only if $d = 2$ whereas M is a copula for all $d \geq 2$.

Theorem 2.2.1 (Fréchet–Hoeffding Bounds) *For any d-dimensional copula C,*

$$W(\boldsymbol{u}) \le C(\boldsymbol{u}) \le M(\boldsymbol{u}), \quad \boldsymbol{u} \in [0,1]^d. \qquad \square$$

Theorem 2.2.1 is attributed to Hoeffding (1940) and Fréchet (1951), although it is already partly appearing in Fréchet (1935).

Let $U \sim \mathrm{U}(0, 1)$. It is easy to verify that

$$(U, 1-U) \sim W \quad \text{and} \quad (U, \dots, U) \sim M. \tag{2.7}$$

Due to these *stochastic representations*, that is, representations in the distributional sense by simple random variables as building blocks, W (in dimension two only) is often called the *countermonotone copula* and M is often referred to as the *comonotone copula*. Furthermore, the dependence between the components of $(U, \dots, U)$ (modeled by M) is often termed *perfect positive dependence* (in the sense that if one component increases, all the other components increase as well *almost surely*, that is, with probability 1) whereas the dependence between the components of $(U, 1-U)$ (modeled by W) is often referred to as *perfect negative dependence* (in the sense that if one component increases, the other component almost surely decreases). This notion of perfect negative dependence cannot, however, be extended to the case $d \geq 3$: If two components of a random vector are perfectly negatively dependent, they cannot both be perfectly negatively dependent with a third component. In a related way, W is not a copula for $d \geq 3$: One can show that for $\boldsymbol{a} = (1/2, \dots, 1/2)$ and $\boldsymbol{b} = (1, \dots, 1)$, (2.2) implies that $\Delta_{(\boldsymbol{a},\boldsymbol{b}]}W < 0$.

Example 2.2.2 (Fréchet–Hoeffding Bounds) Due to the stochastic representations (2.7), it is easy to generate a random sample from M (in any dimension) and W (in dimension two):

```
> set.seed(1980)
> U <- runif(100)
> plot(cbind(U, 1-U), xlab = quote(U[1]), ylab = quote(U[2]))
> plot(cbind(U, U),   xlab = quote(U[1]), ylab = quote(U[2]))
```

The resulting plots are shown in Fig. 2.7.

The bivariate Fréchet–Hoeffding bound copulas W and M are among the rare copulas which can be identified from their wireframe and contour plots. They can

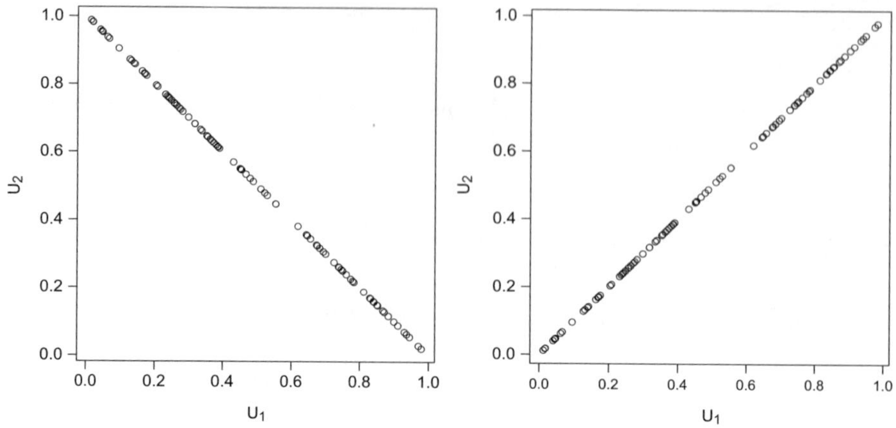

Fig. 2.7 Scatter plot of $n = 100$ independent observations from W (left) and M (right)

be produced as follows:

```
> u <- seq(0, 1, length.out = 40) # subdivision points in each dimension
> u12 <- expand.grid("u[1]" = u, "u[2]" = u) # build a grid
> W <- pmax(u12[,1] + u12[,2] - 1, 0) # values of W on grid
> M <- pmin(u12[,1], u12[,2]) # values of M on grid
> val.W <- cbind(u12, "W(u[1],u[2])" = W) # append grid
> val.M <- cbind(u12, "M(u[1],u[2])" = M) # append grid
> wireframe2(val.W)
> wireframe2(val.M)
> contourplot2(val.W, xlim = 0:1, ylim = 0:1)
> contourplot2(val.M, xlim = 0:1, ylim = 0:1)
```

The resulting plots are shown in Fig. 2.8. Note that for M, the contour lines should be perpendicular at the diagonal, but this is not perfectly captured numerically by the underlying `contourplot()` function of the R package `lattice` in this case. □

Starting from the code given in Example 2.1.4, one can verify that the larger (respectively, smaller) the parameter θ of a bivariate Frank copula, the closer a scatter plot of independent realizations from C_θ^{F} will be to the right (respectively, left) plot of Fig. 2.7. This illustrates the fact that the Frank copula family interpolates between the countermonotone copula W and the comonotone copula M.

From Fig. 2.7, it is clear that neither W nor M is absolutely continuous. Copulas such as W and M which put all probability mass on a set of (Lebesgue) measure 0 (for W and M these sets are the secondary and the primary diagonal, respectively) are called *singular*. There are also copulas C which are not singular but have a *singular component*, that is, copulas which put some probability mass in $(0, 1)$ on a set of (Lebesgue) measure 0. One such example is given by Marshall–Olkin copulas.

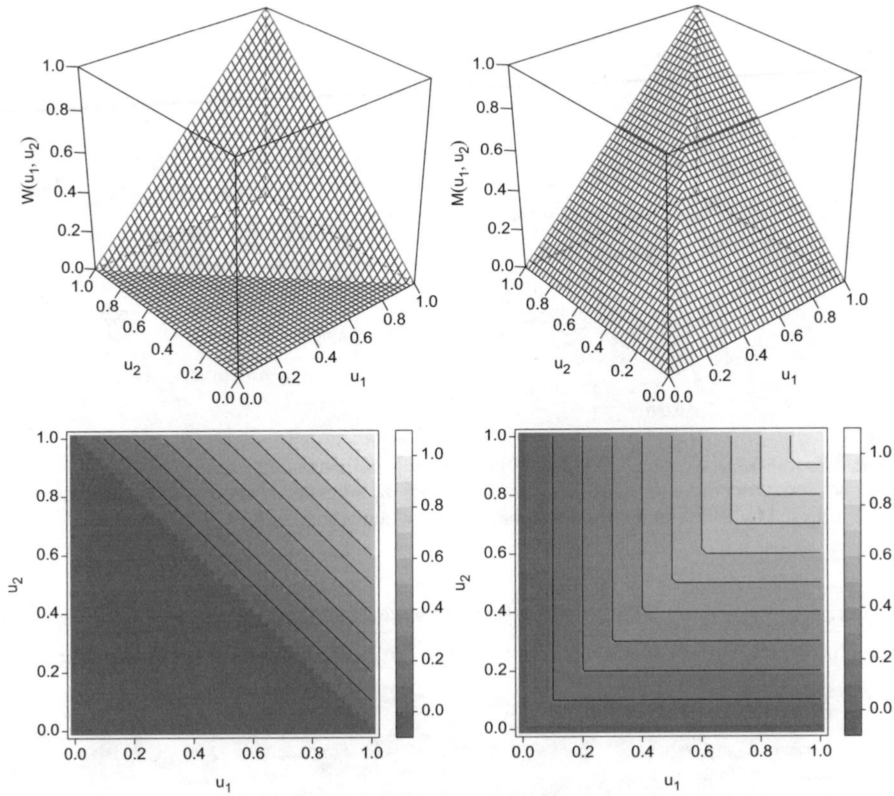

Fig. 2.8 Wireframe (top) and contour plots (bottom) of W (left) and M (right)

Example 2.2.3 (Marshall–Olkin Copulas) The bivariate *Marshall–Olkin* family of copulas is given by

$$C(u_1, u_2) = \min\{u_1 u_2^{1-\alpha_2}, u_1^{1-\alpha_1} u_2\}, \quad u_1, u_2 \in [0, 1], \tag{2.8}$$

with parameters $\alpha_1, \alpha_2 \in [0, 1]$; see Marshall and Olkin (1967). Notice that if $\alpha_1 = 0$ or $\alpha_2 = 0$, then $C = \Pi$. Furthermore, it is easy to verify that a random vector $(U_1, U_2) \sim C$ admits the stochastic representation

$$(U_1, U_2) = \left(\max\left\{V_1^{\frac{1}{1-\alpha_1}}, V_{12}^{\frac{1}{\alpha_1}}\right\}, \ \max\left\{V_2^{\frac{1}{1-\alpha_2}}, V_{12}^{\frac{1}{\alpha_2}}\right\} \right),$$

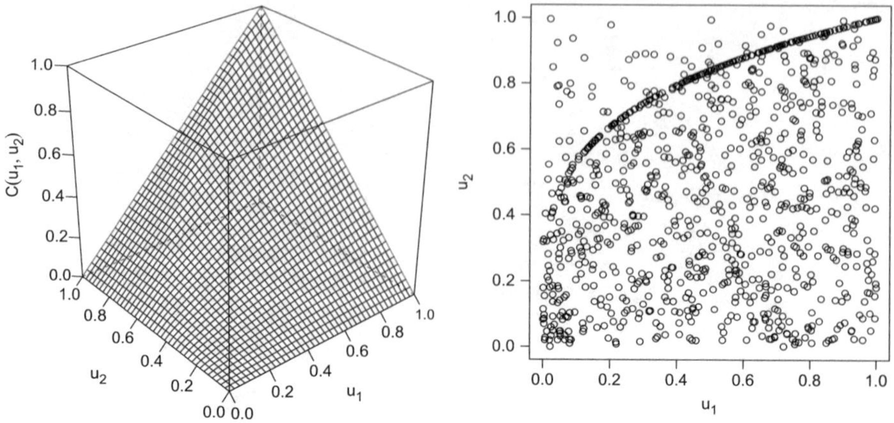

Fig. 2.9 Plot of a Marshall–Olkin copula (left) and corresponding scatter plot of $n = 1000$ independent observations (right). One can clearly see the singular component in the latter which is also reflected by a kink in the former (most easily spotted around $[0.4, 0.7] \times [0.8, 0.9]$)

where $V_1, V_2, V_{12} \overset{\text{ind.}}{\sim} \mathrm{U}(0, 1)$ and with the convention that, for any $v \in (0, 1)$, $v^{1/0} = 0$. This stochastic representation is used for sampling from C in the following code:

```
> ## A Marshall-Olkin copula
> C <- function(u, alpha)
      pmin(u[,1] * u[,2]^(1 - alpha[2]), u[,1]^(1 - alpha[1]) * u[,2])
> alpha <- c(0.2, 0.8)
> val <- cbind(u12, "C(u[1],u[2])" = C(u12, alpha = alpha)) # append C values
> ## Generate data
> set.seed(712)
> V <- matrix(runif(1000 * 3), ncol = 3)
> U <- cbind(pmax(V[,1]^(1/(1 - alpha[1])), V[,3]^(1/alpha[1])),
             pmax(V[,2]^(1/(1 - alpha[2])), V[,3]^(1/alpha[2])))
> ## Plots
> wireframe2(val)
> plot(U, xlab = quote(U[1]), ylab = quote(U[2]))
```

Figure 2.9 shows the copula C (left) and a sample from it (right). Intuitively, if U_1 takes on a certain value u_1, then, with a certain probability in $(0, 1)$, U_2 takes on the value $u_2 = u_1^{\alpha_1/\alpha_2}$ on the singular component, and with the remaining probability, U_2 takes on a value on the strip defined by $U_1 = u_1$ off the singular component. □

2.3 Sklar's Theorem

The following theorem, known as Sklar's Theorem, is the central theorem of copula theory. It explains why copulas determine the dependence between the components of a random vector. Sklar's Theorem is attributed to Sklar (1959). An analytical proof can be found in Sklar (1996), a probabilistic one in Rüschendorf (2009). As we continue, given a univariate df F, ran $F = \{F(x) : x \in \mathbb{R}\}$ denotes the *range* of F and $F^{\leftarrow}$ denotes the quantile function associated with F defined in (1.1). Recall that the latter is merely the ordinary inverse F^{-1} if F is continuous and strictly increasing.

Theorem 2.3.1 (Sklar's Theorem)

1) *For any d-dimensional df H with univariate margins $F_1, \ldots, F_d$, there exists a d-dimensional copula C such that*

$$H(\boldsymbol{x}) = C\big(F_1(x_1), \ldots, F_d(x_d)\big), \quad \boldsymbol{x} \in \mathbb{R}^d. \tag{2.9}$$

The copula C is uniquely defined on $\prod_{j=1}^{d}$ ran F_j and there given by

$$C(\boldsymbol{u}) = H(F_1^{\leftarrow}(u_1), \ldots, F_d^{\leftarrow}(u_d)), \quad \boldsymbol{u} \in \prod_{j=1}^{d} \operatorname{ran} F_j. \tag{2.10}$$

2) *Conversely, given a d-dimensional copula C and univariate dfs $F_1, \ldots, F_d$, H defined by* (2.9) *is a d-dimensional df with margins $F_1, \ldots, F_d$.*

□

In the rest of the book, if (2.9) holds for a df H with univariate margins $F_1, \ldots, F_d$ and if $\boldsymbol{X} \sim H$, we shall say that $\boldsymbol{X}$ (or H) *has copula C*.

Remark 2.3.2

1) By Sklar's Theorem, copulas are those functions which combine the univariate marginal dfs $F_1, \ldots, F_d$ to form the d-dimensional df H. In other words, they link (or "couple," hence the name) multivariate dfs to their univariate margins. They are thus precisely the functions to study if one is interested in the dependence between the components of a random vector.
2) If the univariate margins $F_1, \ldots, F_d$ in Part 1) of Sklar's Theorem are continuous, then ran F_j contains (0, 1) for all $j \in \{1, \ldots, d\}$, which can be used to show that C in (2.10) is uniquely defined on $[0, 1]^d$. Consequently, (2.9) holds for only one d-dimensional copula. This explains why statistical applications of copulas mostly concern the modeling of *continuous random vectors*, that is, random vectors with continuous marginal dfs.
3) Sklar's Theorem can be used to verify that a random vector has a continuous df H if and only if it has continuous univariate marginal dfs $F_1, \ldots, F_d$. The df H being continuous if $F_1, \ldots, F_d$ are continuous follows from the previous point and the fact that any d-dimensional copula C is uniformly continuous on $[0, 1]^d$. The margins $F_1, \ldots, F_d$ being continuous if H is continuous is a consequence of the fact that, for any $j \in \{1, \ldots, d\}$, $F_j(x_j) = H(\infty, \ldots, \infty, x_j, \infty, \ldots, \infty)$, $x_j \in \mathbb{R}$.

4) From (2.9), it also follows that H is absolutely continuous if and only if C and $F_1, \dots, F_d$ are absolutely continuous. In that case, the density h of H satisfies

$$h(\boldsymbol{x}) = c(F_1(x_1), \dots, F_d(x_d)) \prod_{j=1}^{d} f_j(x_j), \quad \boldsymbol{x} \in \prod_{j=1}^{d} \operatorname{ran} X_j,$$

where, for any $j \in \{1, \dots, d\}$, $\operatorname{ran} X_j = \{x \in \mathbb{R} : \mathbb{P}(X_j \in (x-h, x]) > 0 \text{ for all } h > 0\}$ is the *range* of the random variable X_j, f_j denotes the density of F_j and c denotes the density of C. Notice that absolute continuity of F_j implies that $\operatorname{ran} X_j = \{x \in \mathbb{R} : f_j(x) > 0\}$, $j \in \{1, \dots, d\}$. It also follows that c can be recovered from h via

$$c(\boldsymbol{u}) = \frac{h(F_1^{\leftarrow}(u_1), \dots, F_d^{\leftarrow}(u_d))}{f_1(F_1^{\leftarrow}(u_1)) \cdot \ldots \cdot f_d(F_d^{\leftarrow}(u_d))}, \quad \boldsymbol{u} \in (0,1)^d. \tag{2.11}$$

These expressions play a key role in the likelihood-based copula estimation methods presented in Chap. 4.

5) The decomposition of any d-dimensional df H into its univariate margins $F_1, \dots, F_d$ and a copula C given by (2.9) suggests to consider two classes of multivariate dfs: (i) the class of all d-dimensional dfs with given margins $F_1, \dots, F_d$ known as a *Fréchet class*; and (ii) the class of all dfs obtained from a given d-dimensional copula C known as *meta-C models*. For example, a d-dimensional meta-Π model consists of all multivariate df H such that

$$H(\boldsymbol{x}) = \prod_{j=1}^{d} F_j(x_j), \quad \boldsymbol{x} \in \mathbb{R}^d, \tag{2.12}$$

which recovers the well-known result that, under independence, a multivariate df H factors into its univariate margins $F_1, \dots, F_d$. As another example, a *meta-M model* consists of all dfs H such that

$$H(\boldsymbol{x}) = \min\{F_1(x_1), \dots, F_d(x_d)\}, \quad \boldsymbol{x} \in \mathbb{R}^d. \tag{2.13}$$

For fixed univariate dfs $F_1, \dots, F_d$, the multivariate dfs given by (2.12) and (2.13) are two members of the Fréchet class obtained from $F_1, \dots, F_d$. □

Sklar's Theorem is at the heart of the increasing use of copulas in areas such as statistics, risk management, finance, insurance, or engineering. Statistical applications exploit mostly Part 1) of Sklar's Theorem. For example, for a given continuous multivariate df, this part of Sklar's Theorem implies that the underlying unknown copula is unique, which justifies its estimation from available data. Once the margins and the copula are estimated, they are typically coupled as in (2.9), thereby providing the estimated multivariate df. The underlying statistical aspects are the subject of Chaps. 4, 5, and 6. In the context of risk management, finance, or insurance, Part 2) of Sklar's Theorem is often of more interest. For example, it allows one to construct flexible multivariate dfs with given univariate margins which then enter more complicated models, such as time series or pricing models. Another example of interest is *stress testing* which is applied in settings in which one does not necessarily have enough multivariate observations at hand to estimate C (consider, for instance, the situation where various types of losses are observed but not all simultaneously). Copulas are then classically used to formulate dependence scenarios and to propagate their effects on risk measures of interest, for example, by means of simulation.

Example 2.3.3 (First Part of Sklar's Theorem: Decomposition) Sklar's Theorem can be used to create copula families from existing families of multivariate dfs using (2.10). We illustrate this by using as H the df of the bivariate standard normal distribution $\mathrm{N}_2(\mathbf{0}, P)$ for which the off-diagonal elements of the correlation matrix P are $\rho = 0.7$. It is well known that the univariate margins of H are the df of the standard normal distribution $\mathrm{N}(0, 1)$. The corresponding copula of the form (2.10) can thus be evaluated, for instance, with the function `pmvnorm()` of the R package `mvtnorm` of Genz et al. (2017). For example:

```
> library(mvtnorm)
> d <- 2 # dimension
> rho <- 0.7 # off-diagonal entry of the correlation matrix P
> P <- matrix(rho, nrow = d, ncol = d) # build the correlation matrix P
> diag(P) <- 1
> set.seed(64)
> u <- runif(d) # generate a random evaluation point
> x <- qnorm(u)
> pmvnorm(upper = x, corr = P) # evaluate the copula C at u
```

```
[1] 0.04379748
attr(,"error")
[1] 1e-15
attr(,"msg")
[1] "Normal Completion"
```

This construction leads to the so-called *normal* (or *Gauss*) *copula* family. Its members are denoted by C^{n}_P in general and by C^{n}_ρ in the *homogeneous* case, that is, when the correlation matrix P is such that all of its off-diagonal elements are equal to ρ (this is always true in dimension two). In the R package `copula`, objects representing C^{n}_P can be created using the `normalCopula()` function:

```
> nc <- normalCopula(rho) # normal copula (note the default dim = 2)
> pCopula(u, copula = nc) # value of the copula at u
```

```
[1] 0.04379748
```

Similarly, starting from a multivariate t distribution leads to the *(Student) t copula* family parametrized by a correlation matrix P and a degrees of freedom parameter $\nu > 0$. Its members are denoted by $C^t_{P,\nu}$ in general and by $C^t_{\rho,\nu}$ in the homogeneous case. A t copula can be evaluated, for instance, with the function pmvt() from the R package mvtnorm (or, as before, with pCopula(); see below):

```
> nu <- 3 # degrees of freedom
> x. <- qt(u, df = nu)
> pmvt(upper = x., corr = P, df = nu) # evaluate the t copula at u
```

```
[1] 0.04239823
attr(,"error")
[1] 1e-15
attr(,"msg")
[1] "Normal Completion"
```

Notice that, in R, df stands for "degrees of freedom" and should not be confused with our abbreviation df for "distribution function."

For a step-by-step introduction to sampling from the multivariate t distribution in R and its many fallacies, see Hofert (2013). Also note that numerically evaluating the multivariate t df for non-integer degrees of freedom in R is currently still under development:

```
> try(pmvt(upper = x., corr = P, df = 3.5))
```

```
Error in pmvt(upper = x., corr = P, df = 3.5) : 'df' is not an integer
```

Objects representing t copulas can be created using the function tCopula():

```
> tc <- tCopula(rho, dim = d, df = nu)
> pCopula(u, copula = tc) # value of the copula at u
```

```
[1] 0.04239823
```

A more formal and complete introduction to the normal and t copula families will be given in Chap. 3. Note that the copula package also provides mechanisms for constructing and working with correlation matrices structured in specific ways and thus parametrized by less than $d(d-1)/2$ elements. □

Example 2.3.4 (Second Part of Sklar's Theorem: Composition) As already mentioned, Part 2) of Sklar's Theorem can be used to generate a whole variety of multivariate dfs. The key function in the R package copula to create dfs of the form (2.9) is the function mvdc(). The following code shows how to create an R object representing a bivariate df with Clayton copula C_1^{C}, and $\mathrm{N}(1, 4)$ and $\mathrm{Exp}(3)$ margins:

```
> H.obj <- mvdc(claytonCopula(1), margins = c("norm", "exp"),
                paramMargins = list(list(mean = 1, sd = 2), list(rate = 3)))
```

The resulting object H.obj can then be used to evaluate the df and the corresponding density using the functions pMvdc() and dMvdc(), respectively. For instance:

```
> set.seed(1979)
> z <- cbind(rnorm(5, mean = 1, sd = 2), rexp(5, rate = 3)) # evaluation points
> pMvdc(z, mvdc = H.obj) # values of the df at z
```

```
[1] 0.620039169 0.245159086 0.007856849 0.512835075 0.117870122
```

```
> dMvdc(z, mvdc = H.obj) # values of the corresponding density at z
```

```
[1] 0.10997504 0.03128093 0.19565908 0.17081495 0.40321721
```

A random sample from the created bivariate df H can be generated using the function `rMvdc()`:

```
> set.seed(1975)
> X <- rMvdc(1000, mvdc = H.obj)
```

The corresponding scatter plot is given in Fig. 2.10 along with a contourplot of the density:

```
> plot(X, cex = 0.5, xlab = quote(X[1]), ylab = quote(X[2]))
> contourplot2(H.obj, FUN = dMvdc, xlim = range(X[,1]), ylim = range(X[,2]),
               n.grid = 257)
```

As can be deduced from the call of `mvdc()` at the beginning of this example, its argument `margins` is a character vector specifying the R base names of the marginal distributions (`norm` and `exp` in this case) whose parameters can be set through the argument `paramMargins`. Any univariate distribution can be used, provided functions for evaluating the density, df, and quantile function are available. These functions must follow the R convention of being named `d<dist>()`, `p<dist>()`, and `q<dist>()`, respectively, where `<dist>` denotes the R base name of the distribution under consideration.

To illustrate this feature, we consider mixtures of normals for the margins as implemented in the R package `nor1mix` of Mächler (2017). Using the func-

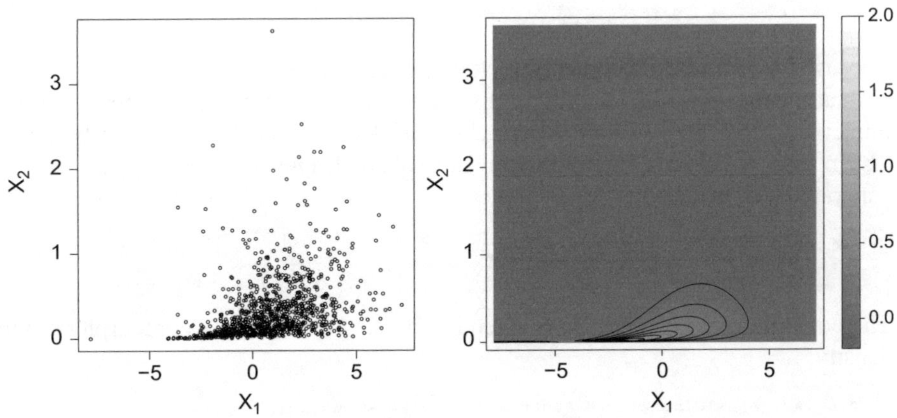

Fig. 2.10 Scatter plot of $n = 1000$ independent observations from a bivariate distribution with copula C_1^C and margins $N(1, 4)$ and $Exp(3)$ (left) and contour plot of the corresponding density (right)

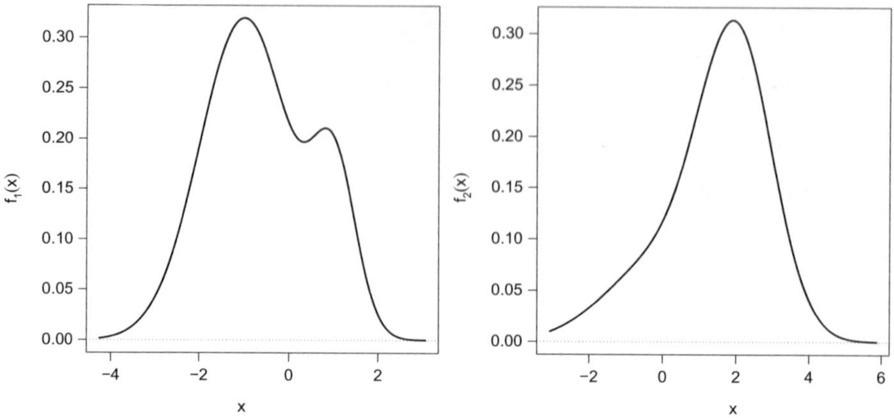

Fig. 2.11 Densities of the normal mixtures $0.2\,\mathrm{N}(1, 0.5^2) + 0.8\,\mathrm{N}(-1, 1)$ (left) and $0.3\,\mathrm{N}(0, 1.5^2) + 0.7\,\mathrm{N}(2, 1)$ (right) used as margins of the meta-Clayton model

tion `norMix()`, we first define the mixtures $0.2\,\mathrm{N}(1, 0.5^2) + 0.8\,\mathrm{N}(-1, 1)$ and $0.3\,\mathrm{N}(0, 1.5^2) + 0.7\,\mathrm{N}(2, 1)$, and plot their densities:

```
> library(nor1mix)
> ## Define and visualize two mixtures of normals
> plot(nm1 <- norMix(c(1, -1), sigma = c( .5, 1), w = c(.2, .8)))
> plot(nm2 <- norMix(c(0,  2), sigma = c(1.5, 1), w = c(.3, .7)))
```

The resulting graphs are shown in Fig. 2.11.

Next, we create an R object representing a bivariate df with these margins and the previously used Clayton copula C_1^{C}:

```
> H.obj.m <- mvdc(claytonCopula(1), margins = c("norMix", "norMix"),
                  paramMargins = list(nm1, nm2))
```

Because the required functions `dnorMix()`, `pnorMix()`, and `qnorMix()` (for evaluating the marginal densities, dfs, and quantile functions, respectively) are implemented in the package `nor1mix`, the density and the df of the bivariate distribution under consideration can be evaluated, and random samples can be generated. For instance:

```
> set.seed(271)
> X <- rMvdc(1000, mvdc = H.obj.m)
```

The corresponding scatter plot is given in Fig. 2.12 along with a contourplot of the density:

```
> plot(X, cex = 0.5, xlab = quote(X[1]), ylab = quote(X[2]))
> contourplot2(H.obj.m, FUN = dMvdc, xlim = range(X[,1]), ylim = range(X[,2]),
               n.grid = 129)
```

□

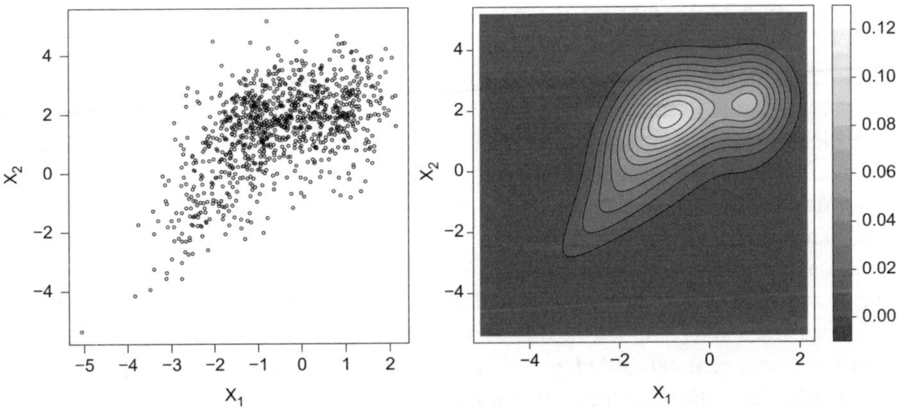

Fig. 2.12 Scatter plot of $n = 1000$ independent observations from a bivariate distribution with copula C_1^{C} and normal mixtures as margins (left) and contour plot of the corresponding density (right)

The following example illustrates an application of Part 2) of Sklar's Theorem to risk aggregation in quantitative risk management.

Example 2.3.5 (Risk Aggregation) Assume we have a continuous random vector $\boldsymbol{L} = (L_1, \dots, L_d)$, where L_j, $j \in \{1, \dots, d\}$, represents the nonnegative loss in the jth business line of a financial firm over a predetermined time period from now, say, 1 year ahead. In risk aggregation and, more generally, quantitative risk management, one is interested in the *aggregate loss*

$$L^+ = L_1 + \cdots + L_d$$

whose df F_{L^+} is known as *loss distribution*. A risk manager does not necessarily need to know F_{L^+} as a function, but at least certain quantities related to F_{L^+} that are to be interpreted as the amount of capital to put aside now in order to account for the losses incurred over the next year. More formally, a risk manager needs to be able to compute a *risk measure* $\varrho(L^+)$ based on F_{L^+}. One such measure is the so-called *value-at-risk* at *confidence level* $\alpha \in (0, 1)$, defined by

$$\mathrm{VaR}_\alpha(L^+) = F_{L^+}^{\leftarrow}(\alpha),$$

that is, the α-quantile of F_{L^+}. The risk measure $\mathrm{VaR}_\alpha(L^+)$ represents the smallest loss level which is exceeded with probability at most $1-\alpha$; see McNeil et al. (2015, Chap. 2) for advantages and drawbacks of this risk measure. In applications, α is typically a given number close to 1; for example, according to the Basel II guidelines (see Basel Committee on Banking Supervision (2009) and other documents), $\alpha = 0.99$ for market risk (over a 10 day time horizon) and $\alpha = 0.999$ for credit and operational risk (over a 1 year time horizon).

Estimating $\mathrm{VaR}_\alpha(L^+)$ is, in general, not an easy task. Based on historical losses $(L_{i1}, \dots, L_{id})$, $i \in \{1, \dots, n\}$, seen as realizations of $\boldsymbol{L} = (L_1, \dots, L_d)$, one could consider the empirical α-quantile of the sample $L_i^+ = L_{i1} + \dots + L_{id}$, $i \in \{1, \dots, n\}$, as an estimate of $\mathrm{VaR}_\alpha(L^+)$. This approach emerges from the historical simulation method; see McNeil et al. (2015, Sect. 9.2.3). However, the amount of available loss data is typically not sufficient to reliably estimate high quantiles such as $\mathrm{VaR}_\alpha(L^+)$. This partly follows from the way the available n realizations $(L_{i1}, \dots, L_{id})$, $i \in \{1, \dots, n\}$, of $\boldsymbol{L}$ are usually formed. Initially, for each business line $j \in \{1, \dots, d\}$, a longer series of losses is available. However, the d available series are typically of different lengths and losses from different business lines may happen at different time points. To form the aforementioned (approximate) realizations $(L_{i1}, \dots, L_{id})$, $i \in \{1, \dots, n\}$, of $\boldsymbol{L}$, the initial losses in each business line are then often aggregated according to some time period, say a week. Furthermore, as mentioned before, in stress testing, it is often of interest to estimate $\mathrm{VaR}_\alpha(L^+)$ under specific copulas C of $\boldsymbol{L}$. In order to do so for a target copula C, a first solution would be to simulate realizations of $\boldsymbol{L}$ under C and the (parametrically or nonparametrically) estimated marginal dfs, build the corresponding simulated realizations of L^+ and utilize, as before, the corresponding empirical α-quantile as an estimate of $\mathrm{VaR}_\alpha(L^+)$. This approach emerges from the Monte Carlo simulation method; see McNeil et al. (2015, Sect. 9.2.5). Once $\mathrm{VaR}_\alpha(L^+)$ is estimated for all the selected copulas (dependence scenarios), one option is to take as risk measure $\varrho(L^+)$ the maximal $\mathrm{VaR}_\alpha(L^+)$, related to the worst case dependence scenario.

Although a simulation-based estimation of $\mathrm{VaR}_\alpha(L^+)$ can easily be implemented with the R package `copula` based on `rCopula()`, we present hereafter a slightly different approach which does not require the estimation of the marginal loss distributions. We work with the matrix of losses $X = (L_{ij})$, $i \in \{1, \dots, n\}$, $j \in \{1, \dots, d\}$, and ask how we can create a matrix Y which contains, in each column (so for each marginal loss), the same numbers as X, but permuted within each column so that the rows of Y could be viewed as approximately being dependent according to some target copula C. Since we only permute the numbers within each column of X to get Y, the marginal loss dfs are not changed. Only their dependence is adjusted to resemble that of C. To this end, we assume n to be sufficiently large. A similar idea dates back to Iman and Conover (1982), where coordinate-wise independent samples are transformed into coordinate-wise dependent ones with a given Spearman rank correlation matrix based on its Cholesky decomposition (see Remark 3.1.2 Part 1) for the latter).

To demonstrate the idea of transforming X into Y, we consider a three-dimensional example based on simulated loss data from Pareto, lognormal, and loggamma marginal dfs. Let us start by generating X under this setup:

```
> ## Define parameters of the three margins
> th <- 2.5 # Pareto parameter
> m <- 10 # mean of the lognormal
> v <- 20 # variance of the lognormal
> s <- 4 # shape of the gamma underlying the loggamma
> r <- 5 # rate of the gamma underlying the loggamma
> ## Define list of marginal dfs
> qF <- list(qPar = function(p) (1 - p)^(-1/th) - 1,
             qLN  = function(p) qlnorm(p, meanlog = log(m)-log(1+v/m^2)/2,
                                          sdlog = sqrt(log(1+v/m^2))),
             qLG  = function(p) exp(qgamma(p, shape = s, rate = r)))
> ## Generate the data
> set.seed(271) # for reproducibility
> X <- sapply(qF, function(mqf) mqf(runif(2500))) # (2500, 3)-matrix
```

As dependence structures coupling the independently simulated marginal losses, we consider homogeneous t copulas $C^t_{\rho,\nu}$ with $\nu = 3.5$ degrees of freedom and correlation parameter $\rho \in \{0, 0.1, \dots, 1\}$. As confidence level α for estimating $\mathrm{VaR}_\alpha(L^+)$, we consider $\alpha \in \{0.001, 0.01, 0.05, 0.1, 0.2, \dots, 0.9, 0.95, 0.99, 0.999\}$. Note that confidence levels smaller than 90% are rather unrealistic in applications but are included here to study the behavior of $\mathrm{VaR}_\alpha(L^+)$ as a function of α. We also study $\mathrm{VaR}_\alpha(L^+)$ as a function of ρ for the aforementioned ρ values. For convenience, we implement an auxiliary function which computes, for a given matrix X, (a vector of) confidence level(s) α, correlation parameter ρ, and degrees of freedom $\nu = 3.5$ of a t copula, the empirical quantile estimator based on the row sums of Y, that is, a nonparametric estimator of $\mathrm{VaR}_\alpha(L^+)$ under the dependence scenario $C^t_{\rho,3.5}$. Since this is a longer function with several arguments, we properly document it with the Roxygen documentation system. Note that the function is vectorized in α, thus allowing us to estimate $\mathrm{VaR}_\alpha(L^+)$ for all α based on the same t copula sample. We also set the seed inside the auxiliary function; this is fine here as we only call this function once per value of ρ. Both of these design decisions aim at reducing the variance across different ρ.

```
> ##' @title Nonparametric VaR estimate under a t copula
> ##' @param X loss matrix
> ##' @param alpha confidence level(s)
> ##' @param rho correlation parameter of the t copula
> ##' @param df degrees of freedom parameter of the t copula
> ##' @return Nonparametric VaR estimate under the t copula (numeric)
> VaR <- function(X, alpha, rho, df = 3.5)
  {
      stopifnot(is.matrix(X), 0 <= rho, rho <= 1, length(rho) == 1,
                0 < alpha, alpha < 1, length(alpha) >= 1)
      n <- nrow(X) # sample size
      d <- ncol(X) # dimension
      ## Simulate from a t copula with d.o.f. parameter 3.5 and exchangeable
      ## correlation matrix with off-diagonal entry rho. Also compute the
      ## componentwise ranks.
      ## Note: We can set the seed here as we can estimate VaR for all
      ##       confidence levels based on the same copula sample. We
      ##       even *should* set the seed here to minimize the variance
      ##       of the estimator  and make the results more comparable.
      set.seed(271)
      U <- rCopula(n, copula = tCopula(rho, dim = d, df = df))
      rk <- apply(U, 2, rank)
      ## Componentwise reorder the data according to these ranks to
      ## mimic the corresponding t copula dependence among the losses
      Y <- sapply(1:d, function(j) sort(X[,j])[rk[,j]])
      ## Build row sums to mimic a sample from the distribution of the
      ## sum under the corresponding t copula.
      S <- rowSums(Y)
      ## Nonparametrically estimate VaR for all confidence levels alpha
      ## Note: We use the mathematical definition ('type = 1') of a
      ##       quantile function here
      quantile(S, probs = alpha, type = 1, names = FALSE)
  }
```

We can now build a grid of the aforementioned α and ρ values and compute $\mathrm{VaR}_\alpha(L^+)$ for all combinations of these two inputs:

```
> alpha <- c(0.001, 0.01, 0.05, 0.1, 0.2, 0.3, 0.4, 0.5,
             0.6, 0.7, 0.8, 0.9, 0.95, 0.99, 0.999) # confidence levels
> rho <- seq(0, 1, by = 0.1) # parameter of the homogeneous t copula
> grid <- expand.grid("alpha" = alpha, "rho" = rho)[,2:1] # build a grid
> VaR.fit <- sapply(rho, function(r)
      VaR(X, alpha = alpha, rho = r)) # (alpha, rho)
> res <- cbind(grid, "VaR[alpha](L^'+')" = as.vector(VaR.fit))
```

The left-hand side of Fig. 2.13 shows the nonparametrically estimated $\mathrm{VaR}_\alpha(L^+)$ for the given marginal dfs and $C^t_{\rho,3.5}$ copula as a function of ρ and α. For larger α, the estimation using the empirical quantile function is based on less and less data and thus affected by a larger variance; this is also why we chose the rather large sample size of $n = 2500$ here. Especially for large α, other techniques are available; see, for example, McNeil et al. (2015, Sect. 5.2.3). Note furthermore that for large (small) α, $\mathrm{VaR}_\alpha(L^+)$ seems to be increasing (decreasing) in ρ.

The right-hand side of Fig. 2.13 shows, for each considered α, the maximal $\mathrm{VaR}_\alpha(L^+)$ over all considered ρ (and thus the worst $\mathrm{VaR}_\alpha(L^+)$ over all the

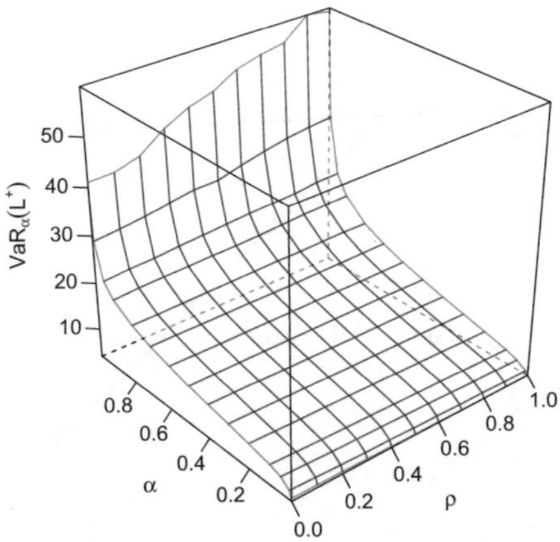

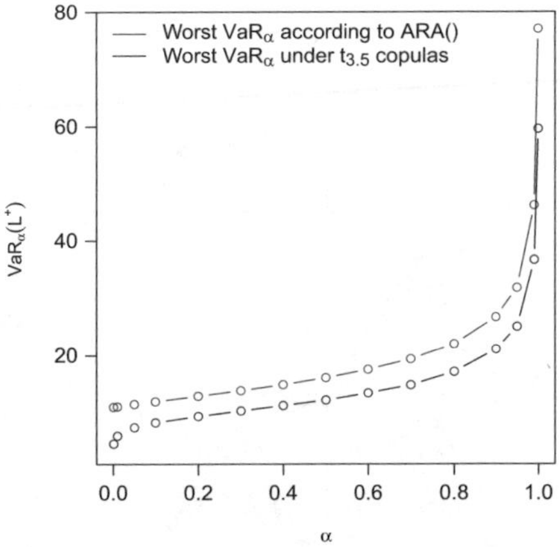

Fig. 2.13 Nonparametrically estimated $\mathrm{VaR}_\alpha(L^+)$ as a function of ρ and α based on rearranging the given Pareto, lognormal, and loggamma marginal losses according to the component ranks of a sample from a homogeneous $C^t_{\rho,3.5}$ copula (left). Worst $\mathrm{VaR}_\alpha(L^+)$ over all considered t copulas $C^t_{\rho,3.5}$ and approximate worst $\mathrm{VaR}_\alpha(L^+)$ over all possible copulas according to the ARA (right)

considered $C^t_{\rho,3.5}$ copulas for the given marginal dfs). Included is another curve, obtained from the adaptive rearrangement algorithm (ARA) of the R package `qrmtools` of Hofert and Hornik (2017); see Embrechts et al. (2013) and Hofert et al. (2017) for the underlying ideas. This algorithm determines an approximation of the so-called *worst value-at-risk*, that is, the largest $\mathrm{VaR}_\alpha(L^+)$ for our given marginal dfs under all possible copulas (not just t copulas).

```
> wireframe2(res)
> library(qrmtools)
> worst.VaR <- sapply(alpha, function(a) mean(ARA(a, qF = qF)$bounds))
> plot(alpha, worst.VaR, type = "b", col = 2,
       xlab = quote(alpha), ylab = quote(VaR[alpha](L^'+')),
       ylim = range(VaR.fit, worst.VaR)) # computed with the ARA
> lines(alpha, apply(VaR.fit, 1, max), type = "b", col = 1) # simulated
> legend("topleft", bty = "n", lty = rep(1, 2), col = 2:1,
         legend = c(expression("Worst"~VaR[alpha]~"according to ARA()"),
                    expression("Worst"~VaR[alpha]~"under"~t[3.5]~"copulas")))
```

As we can see from the right-hand side of Fig. 2.13, for no α does the worst $\mathrm{VaR}_\alpha(L^+)$ over all considered $C^t_{\rho,3.5}$ copulas attain the worst $\mathrm{VaR}_\alpha(L^+)$ computed over all copulas, not even for $\rho = 1$ (which results in $C^t_{\rho,3.5}$ being the comonotone copula M). In particular, the strongest possible positive dependence ($\rho = 1$) does not lead to the largest $\mathrm{VaR}_\alpha(L^+)$. It is a common misbelief that maximal correlation leads to maximal $\mathrm{VaR}_\alpha(L^+)$. The reason why this argument fails is that although value-at-risk is comonotone additive (see McNeil et al. 2015, Sect. 7.2.1), it is well known to be superadditive under certain scenarios; see McNeil et al. (2015, Sect. 2.3.5) for more details.

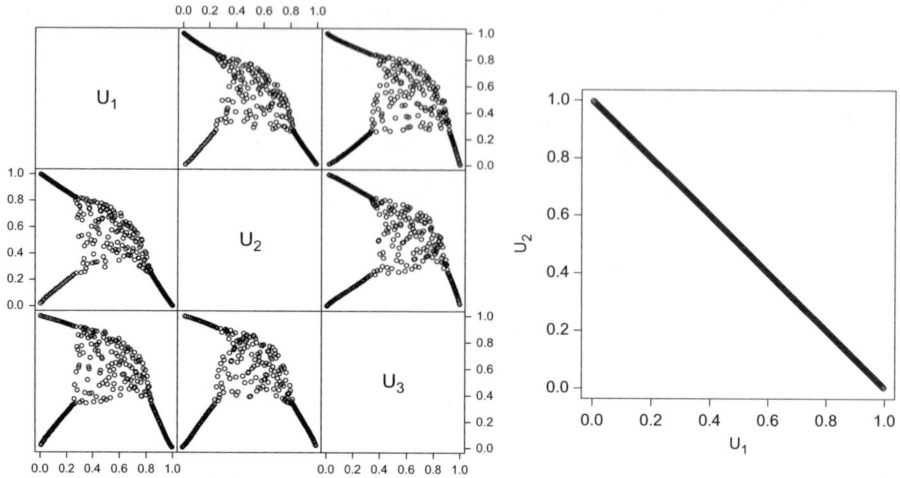

Fig. 2.14 Approximate sample from the copula (of $\boldsymbol{L} = (L_1, \ldots, L_d)$ conditional on L_j being greater than or equal to its α-quantile for all j) which leads to the worst $\mathrm{VaR}_\alpha(L^+)$ for the given Pareto, lognormal, and loggamma marginal dfs (left) and the same for the bivariate restriction of the problem for the given Pareto and lognormal marginal dfs (right)

One might ask what the worst $\mathrm{VaR}_\alpha(L^+)$ dependence structure may look like for our given marginal distributions. The left-hand side of Fig. 2.14 shows a scatter-plot matrix of an approximate sample from this worst $\mathrm{VaR}_\alpha(L^+)$ copula for $\alpha = 0.99$ (such a pseudo-sample can be obtained by forming componentwise scaled ranks with the function `pobs()` from the R package `copula`; this sample will be referred to as pseudo-observations in Sect. 4.1.2). The right-hand side of Fig. 2.14 shows a scatter plot of a pseudo-sample from the worst $\mathrm{VaR}_\alpha(L^+)$ copula in the bivariate case (corresponding to the first two of the three considered marginal dfs); as we can see, it corresponds to the countermonotone copula W (see Embrechts et al. (2013) for the corresponding analytical result). These two worst $\mathrm{VaR}_\alpha(L^+)$ copula pseudo-samples were obtained after applying the ARA as implemented in the R function `ARA()` of `qrmtools`. This algorithm starts by setting up a matrix X containing, in the jth column, a parametric estimate of the quantile function of L_j evaluated on an equidistant grid of probabilities in $[\alpha, 1]$. Similar to the rearrangement of losses according to a $C^t_{\rho,3.5}$ copula before, the (iterative) ARA procedure aims at mimicking the dependence which leads to the worst (that is, the largest) $\mathrm{VaR}_\alpha(L^+)$; see Embrechts et al. (2013) and Hofert et al. (2017) for more details. Note that, since our marginal dfs live on the positive real line, the ARA only needs to consider the estimated marginal dfs above their α-quantiles; see Puccetti and Rüschendorf (2013, Theorem 2.1). The two worst $\mathrm{VaR}_\alpha(L^+)$ dependence structures shown in Fig. 2.14 are thus actually the copulas of $\boldsymbol{L} = (L_1, \ldots, L_d)$ conditional on L_j being greater than or equal to its α-quantile for all j. The copula of $\boldsymbol{L}$ conditional on $\boldsymbol{L}$ being in the remaining region is simply irrelevant for this dependence problem (and could equally well be taken as the independence copula).

```
> ## Computing worst VaR in the three-dimensional case
> wVaR <- ARA(0.99, qF = qF) # compute worst VaR (bounds)
> X <- wVaR[["X.rearranged"]]$up # extract rearranged matrix (upper bound)
> U <- pobs(X) # compute pseudo-observations
> pairs2(U) # approx. sample of a copula leading to worst VaR for our marg. dfs
> ## Computing worst VaR in the bivariate case
> wVaR. <- ARA(0.99, qF = qF[1:2]) # compute worst VaR (bounds)
> X. <- wVaR.[["X.rearranged"]]$up # extract rearranged matrix (upper bound)
> U. <- pobs(X.) # compute pseudo-observations
> plot(U., xlab = quote(U[1]), ylab = quote(U[2]))
```

□

2.4 The Invariance Principle

The power of Sklar's Theorem also manifests itself when moving from multivariate dfs to the corresponding random vectors. The latter can be seen as adopting a "stochastic" perspective instead of an "analytical" one. The first result of this section can be regarded as a stochastic analog of Sklar's Theorem for continuous random vectors. It is used to show the invariance principle stated later in this section.

Lemma 2.4.1 (Stochastic Analog of Sklar's Theorem) *Let $\boldsymbol{X}$ be a d-dimensional random vector with continuous univariate marginal dfs $F_1, \dots, F_d$. Then $\boldsymbol{X}$ has copula C if and only if $(F_1(X_1), \dots, F_d(X_d)) \sim C$.* □

On the one hand, given $\boldsymbol{X} \sim H$ with continuous univariate margins $F_1, \dots, F_d$, the necessity in Lemma 2.4.1 allows us to construct a random vector $\boldsymbol{U}$ distributed according to the underlying copula C, where "underlying" refers to the (unique) copula C which couples $F_1, \dots, F_d$ to H by Sklar's Theorem. Since we can recover C from H in this sense, the necessity in Lemma 2.4.1 provides a stochastic analog of Part 1) of Sklar's Theorem. On the other hand, given a random vector $\boldsymbol{U} \sim C$ and univariate dfs $F_1, \dots, F_d$, the sufficiency in Lemma 2.4.1 allows us to construct a random vector $\boldsymbol{X} \sim H$ with copula C, namely via

$$(F_1^{\leftarrow}(U_1), \dots, F_d^{\leftarrow}(U_d)) \sim H.$$

In this sense, the sufficiency in Lemma 2.4.1 is a stochastic analog of Part 2) of Sklar's Theorem. As such Lemma 2.4.1 is an important and useful result to keep in mind. The following two examples address both of its implications.

Example 2.4.2 (Sampling from a Normal or t Copula) An application of the necessity in Lemma 2.4.1 leads to a sampling algorithm for the normal or the t copula family. We shall illustrate this for the t family in the homogeneous case, again using the `mvtnorm` package. One first needs to generate independent observations from a multivariate t df and then apply, to each component sample, the corresponding marginal df in order to obtain a sample from $C^t_{\rho,\nu}$:

```
> n <- 1000 # sample size
> d <- 2 # dimension
> rho <- 0.7 # off-diagonal entry in the correlation matrix P
> P <- matrix(rho, nrow = d, ncol = d) # build the correlation matrix P
> diag(P) <- 1
> nu <- 3.5 # degrees of freedom
> set.seed(271)
> X <- rmvt(n, sigma = P, df = nu) # n ind. multivariate t observations
> U <- pt(X, df = nu) # n ind. realizations from the corresponding copula
```

The function `rCopula()` for the t copula family is actually just a wrapper of the previous code, as verified below:

```
> set.seed(271)
> U. <- rCopula(n, tCopula(rho, dim = d, df = nu))
> stopifnot(all.equal(U, U.)) # test of (numerical) equality
```

The previous numerical equality is also confirmed by the scatter plots shown in Fig. 2.15:

```
> plot(U., xlab = quote(U[1]), ylab = quote(U[2]))
> plot(U,  xlab = quote(U[1]), ylab = quote(U[2]))
```

□

Example 2.4.3 (From a Multivariate t Distribution to a t Copula to a Meta-t Model) We have already seen in Example 2.4.2 how the necessity in Lemma 2.4.1 can be used to obtain a sample from a bivariate t copula $C^t_{\rho,\nu}$ with correlation parameter $\rho = 0.7$ and degrees of freedom $\nu = 3.5$. Once we have a sample from $C^t_{\rho,\nu}$, we can use the sufficiency in Lemma 2.4.1 and apply, to each component sample,

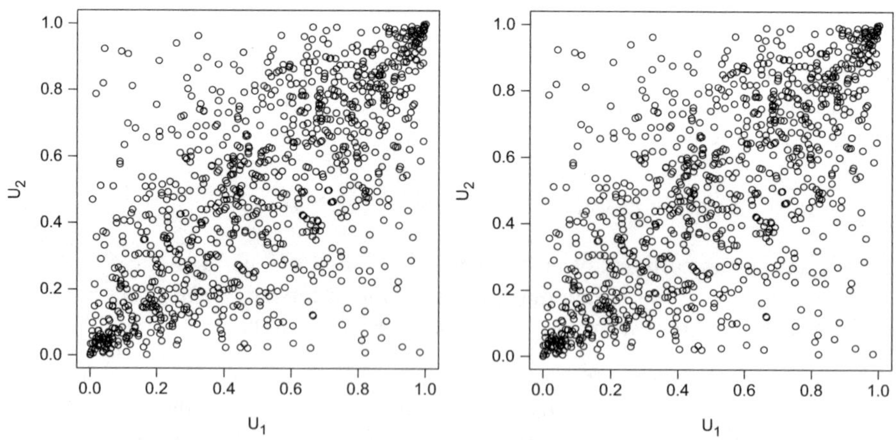

Fig. 2.15 Scatter plots of $n = 1000$ independent observations generated from $C^t_{\rho,\nu}$ for $\rho = 0.7$ and $\nu = 3.5$ manually (left) and via `rCopula()` (right)

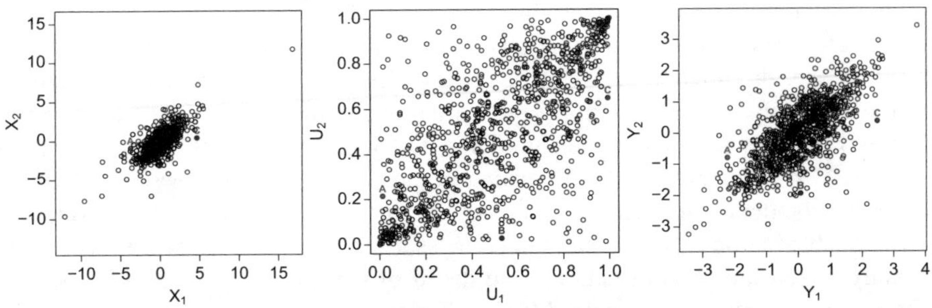

Fig. 2.16 Scatter plots of $n = 1000$ independent observations from a multivariate t distribution (left), the corresponding (probability transformed) sample from the t copula (middle) and the corresponding marginally normally distributed (quantile transformed) meta-t sample (right)

a quantile function of interest to obtain multivariate observations from a meta-t model; note that it is precisely the aforementioned implication that tells us that the underlying dependence structure, here given by the copula $C^t_{\rho,\nu}$, remains unchanged when doing so.

The three plots in Fig. 2.16 illustrate this. On the left-hand side, we see a sample from a multivariate t distribution, in the middle is the corresponding sample from the t copula $C^t_{\rho,\nu}$ (after applying the marginal dfs) and the right-hand side shows the sample obtained after applying the quantile function of the N(0, 1) distribution to all component series of the previous sample, giving multivariate observations from a meta-t model with N(0, 1) margins. Note that if we had used t quantile functions with degrees of freedom different than ν, we would have obtained a sample from a bivariate distribution that, as the bivariate t distribution, has a t copula and univariate t margins, but which is more flexible in the sense of allowing for different degrees of freedom for each margin and the t copula.

Here is the code for producing the plots based on the auxiliary function `plotABC()`:

```
> ## Plot function highlighting three points
> plotABC <- function(x, ind3, col = adjustcolor("black", 1/2), pch = 19, ...)
  {
      cols <- adjustcolor(c("red","blue","magenta"), offset = -c(1,1,1,1.5)/4)
      par(pty = "s")
      plot(x, col = col, asp = 1,...)
      xy <- x[ind3, , drop = FALSE]
      points(xy, pch = pch, col = cols)
      text(xy, label = names(ind3), adj = c(0.5, -0.6), col = cols, font = 2)
  }
> ind3 <- c(A = 725, B = 351, C = 734) # found via 'plot(X); identify(X)'
> ## Scatter plot of observations from the multivariate t distribution
> plotABC(X, ind3 = ind3, xlab = quote(X[1]), ylab = quote(X[2]))
> ## Scatter plot of observations from the corresponding t copula
> plotABC(U, ind3 = ind3, xlab = quote(U[1]), ylab = quote(U[2]))
> ## Scatter plot of observations from the meta-t distribution
> Y <- qnorm(U) # transform U (t copula) to normal margins
> plotABC(Y, ind3 = ind3, xlab = quote(Y[1]), ylab = quote(Y[2]))
```

Notice that we also colorized three points A, B, and C in Fig. 2.16 to see how the strictly increasing transformations under consideration act on the sample. As one can see, the relative locations of the three points A, B, and C are unchanged in the sense that in the three plots, their order according to their first (respectively, second) coordinate remains the same: The order according to the x-coordinate is always A, B, C, and the order according to the y-coordinate is always B, A, C. In other words, the componentwise ranks of the three points were not affected by the two transformations. This should not come as a surprise since ranks are invariant to strictly increasing transformations, and the df of the t and the quantile function of the normal distribution are indeed strictly increasing. □

Because of their importance in practice, we now quickly provide the two sampling algorithms which follow from Lemma 2.4.1. Algorithm 2.4.4 can be used, for instance, to sample *implicit copulas* (that is, copulas defined by (2.10) in Sklar's Theorem) such as the normal and t copulas (as done in Example 2.4.2). Algorithm 2.4.5 can be used to sample meta-C models (as done in Example 2.4.3). The latter algorithm is actually implemented in the function `rMvdc()` and was used to generate the sample shown on the left-hand side of Fig. 2.10.

Algorithm 2.4.4 (Simulation of Implicit Copulas)

1) Sample $\boldsymbol{X} \sim H$, where H is a d-dimensional df with continuous margins $F_1, \ldots, F_d$.
2) Return $\boldsymbol{U} = (F_1(X_1), \ldots, F_d(X_d))$. □

Algorithm 2.4.5 (Simulation of Meta-C Models)

1) Sample $\boldsymbol{U} \sim C$.
2) Return $\boldsymbol{X} = (F_1^{\leftarrow}(U_1), \ldots, F_d^{\leftarrow}(U_d))$. □

Remark 2.4.6 (Stochastic Interpretations of W and M) Lemma 2.4.1 can also be used to show the following stochastic interpretations of the Fréchet–Hoeffding bounds:

1) Let (X_1, X_2) be a continuous random vector. The copula of (X_1, X_2) is the lower Fréchet–Hoeffding bound W if and only if X_2 is (almost surely) a strictly decreasing function of X_1.
2) Let $d \geq 2$ and $\boldsymbol{X} = (X_1, \ldots, X_d)$ be a continuous random vector. The copula of $\boldsymbol{X}$ is the upper Fréchet–Hoeffding bound M if and only if $X_2, \ldots, X_d$ are (almost surely) strictly increasing functions of X_1.

Note that these results apply, in particular, to the random vectors $(U, 1 - U)$ and $(U, \ldots, U)$, with $U \sim \mathrm{U}(0, 1)$ as we have already seen in (2.7). □

The random variables X_1 and X_2 in Remark 2.4.6 Part 1) are said to be *perfectly negatively dependent* or *countermonotone* (as their copula is the countermonotone copula W; see also Theorem 2.2.1 and thereafter); the ones in Part 2) are said to be *perfectly positively dependent* or *comonotone* (as their copula is the comonotone copula M; see also Theorem 2.2.1 and thereafter).

We now present the invariance principle, which states that copulas are invariant under strictly increasing transformations on the ranges of the underlying random variables; see Schweizer and Wolff (1981).

Theorem 2.4.7 (Invariance Principle) *Let* $\boldsymbol{X} \sim H$ *with continuous univariate margins* $F_1, \dots, F_d$ *and copula* C. *If, for any* $j \in \{1, \dots, d\}$, T_j *is a strictly increasing transformation on* $\operatorname{ran} X_j$, *then* $(T_1(X_1), \dots, T_d(X_d))$ *(also) has copula* C. □

For any d-dimensional continuous random vector whose marginal dfs $F_1, \dots, F_d$ are strictly increasing, the invariance principle immediately applies with $T_j = F_j$, $j \in \{1, \dots, d\}$. Similarly, for any d-dimensional random vector with standard uniform margins, it immediately applies with quantile functions which are strictly increasing on $(0, 1)$. Figure 2.16 can be seen as an illustration of these two settings.

Example 2.4.8 (Verifying the Invariance Principle) To illustrate the invariance principle (stochastically and analytically), let us consider a bivariate random vector (X_1, X_2) with df Φ_P, the df of the bivariate standard normal distribution whose correlation matrix P has off-diagonal elements ρ. Since both marginal dfs of Φ_P are the df Φ of the univariate standard normal distribution, Part 1) of Sklar's Theorem implies that the copula C of (X_1, X_2) is the normal copula C_ρ^{n}. Consider the transformations $T_1(x) = \exp(x)$ and $T_2(x) = 1/(1 + \exp(-x))$, $x \in \mathbb{R}$, which are both strictly increasing on the whole real line, and thus on the ranges of X_1 and X_2.

We first verify the invariance principle analytically. The df $\tilde{H}$ of $(T_1(X_1), T_2(X_2))$ is

$$\begin{aligned}\tilde{H}(x_1, x_2) &= \mathbb{P}(T_1(X_1) \le x_1, T_2(X_2) \le x_2) = \mathbb{P}(X_1 \le T_1^{-1}(x_1), X_2 \le T_2^{-1}(x_2)) \\ &= \Phi_P\Bigl(\log x_1, -\log\Bigl(\frac{1 - x_2}{x_2}\Bigr)\Bigr), \quad (x_1, x_2) \in (0, \infty) \times (0, 1).\end{aligned}$$

The corresponding marginals $\tilde{F}_1(x_1) = \Phi(\log x_1)$, $x_1 > 0$ (obtained for $x_2 \uparrow 1$) and $\tilde{F}_2(x_2) = \Phi(-\log((1 - x_2)/x_2))$, $x_2 \in (0, 1)$ (obtained for $x_1 \uparrow \infty$) have quantile functions $\tilde{F}_1^{-1}(u) = \exp(\Phi^{-1}(u))$ and $\tilde{F}_2^{-1}(u) = 1/\bigl(1 + \exp(-\Phi^{-1}(u))\bigr)$, $u \in (0, 1)$, respectively. By Sklar's Theorem, the copula of $(T_1(X_1), T_2(X_2))$ is thus

$$\begin{aligned}\tilde{C}(u_1, u_2) &= \tilde{H}(\tilde{F}_1^{-1}(u_1), \tilde{F}_2^{-1}(u_2)) \\ &= \Phi_P\Biggl(\log\bigl(\exp(\Phi^{-1}(u_1))\bigr), -\log\Biggl(\frac{1 - 1/\bigl(1 + \exp(-\Phi^{-1}(u_2))\bigr)}{1/\bigl(1 + \exp(-\Phi^{-1}(u_2))\bigr)}\Biggr)\Biggr) \\ &= \Phi_P\Bigl(\Phi^{-1}(u_1), -\log\bigl(\exp(-\Phi^{-1}(u_2))\bigr)\Bigr) = \Phi_P(\Phi^{-1}(u_1), \Phi^{-1}(u_2)) \\ &= C(u_1, u_2), \quad (u_1, u_2) \in (0, 1)^2,\end{aligned}$$

that is, the (same) copula C as the one of (X_1, X_2). This verifies the invariance principle analytically (the proof of which roughly follows along the same lines). Let us now verify the equality of $\tilde{C}$ and C at selected points in R:

```
> rho <- 0.6
> P <- matrix(c(1, rho, rho, 1), ncol = 2) # the correlation matrix
> C <- function(u) pCopula(u, copula = normalCopula(rho)) # normal copula
> Htilde <- function(x)
      apply(cbind(log(x[,1]), -log((1-x[,2])/x[,2])), 1, function(x.)
            pmvnorm(upper = x., corr = P))
> qF1tilde <- function(u) exp(qnorm(u))
> qF2tilde <- function(u) 1/(1+exp(-qnorm(u)))
> Ctilde <- function(u) Htilde(cbind(qF1tilde(u[,1]), qF2tilde(u[,2])))
> set.seed(31)
> u <- matrix(runif(5 * 2), ncol = 2) # 5 random evaluation points
> stopifnot(all.equal(Ctilde(u), C(u)))
```

Let us now verify the invariance principle stochastically. To form independent observations from the copula of $(T_1(X_1), T_2(X_2))$, we use the necessity in Lemma 2.4.1. Applying the dfs $\tilde{F}_j$ to $T_j(X_j)$, $j \in \{1, 2\}$, should give us a sample from the copula $\tilde{C}$ of $(T_1(X_1), T_2(X_2))$, which we can then compare (in distribution) to a sample of C. We can check this here with exact (numerical) equality (not just in distribution):

```
> set.seed(721)
> X <- rmvnorm(1000, mean = c(0,0), sigma = P) # sample from N(0, P)
> ## 'Sample' the copula of X directly
> U <- pnorm(X)
> ## Transform the sample X componentwise
> TX <- cbind(exp(X[,1]), plogis(X[,2])) # note: plogis(x) = 1/(1+exp(-x))
> ## Apply the marginal dfs to get a sample from the copula of TX
> ## Note: qlogis(p) == logit(p) == log(p/(1-p))
> V <- cbind(pnorm(log(TX[,1])), pnorm(qlogis(TX[,2])))
> stopifnot(all.equal(V, U)) # => the samples of the two copulas are the same
```

□

2.5 Survival Copulas and Copula Symmetries

The *survival function* $\bar{F}$ corresponding to a univariate df $F(x) = \mathbb{P}(X \le x)$, $x \in \mathbb{R}$, is defined by $\bar{F}(x) = \mathbb{P}(X > x)$, $x \in \mathbb{R}$, and thus satisfies $\bar{F}(x) = 1 - F(x)$, $x \in \mathbb{R}$. Similarly, the *multivariate survival function* $\bar{H}$ corresponding to a d-dimensional df $H(\boldsymbol{x}) = \mathbb{P}(\boldsymbol{X} \le \boldsymbol{x})$ is defined by $\bar{H}(\boldsymbol{x}) = \mathbb{P}(\boldsymbol{X} > \boldsymbol{x})$, $\boldsymbol{x} \in \mathbb{R}^d$; note that if $d > 1$, $\bar{H}(\boldsymbol{x}) \ne 1 - H(\boldsymbol{x})$ in general. The survival function $\bar{F}_j$ of the jth component of the d-dimensional random vector $\boldsymbol{X} \sim H$ can either be obtained from the corresponding marginal df F_j of H as explained above or from $\bar{H}$ via $\bar{F}_j(x_j) = \bar{H}(-\infty, \dots, -\infty, x_j, -\infty, \dots, -\infty)$, $x_j \in \mathbb{R}$. The latter explains why $\bar{F}_1, \dots, \bar{F}_d$ are also called the *univariate margins* of $\bar{H}$ or the *marginal survival functions* of $\boldsymbol{X}$.

Having the notion of survival function available, it should not come as a surprise that Sklar's Theorem can equally well be formulated in terms of survival functions.

The resulting version of the theorem is also used to construct copula classes and families; in that case, the marginal survival functions are often concentrated on the nonnegative real line.

Theorem 2.5.1 (Sklar's Theorem for Survival Functions)

1) *For any d-dimensional survival function $\bar{H}$ with margins $\bar{F}_1, \dots, \bar{F}_d$, there exists a copula $\bar{C}$ such that*

$$\bar{H}(\boldsymbol{x}) = \bar{C}(\bar{F}_1(x_1), \dots, \bar{F}_d(x_d)), \quad \boldsymbol{x} \in \mathbb{R}^d. \tag{2.14}$$

The copula $\bar{C}$ is uniquely defined on $\prod_{j=1}^d \operatorname{ran} \bar{F}_j$ and there given by

$$\bar{C}(\boldsymbol{u}) = \bar{H}(F_1^{\leftarrow}(1-u_1), \dots, F_d^{\leftarrow}(1-u_d)), \quad \boldsymbol{u} \in \prod_{j=1}^d \operatorname{ran} \bar{F}_j. \tag{2.15}$$

2) *Conversely, given any d-dimensional copula $\bar{C}$ and univariate survival functions $\bar{F}_1, \dots, \bar{F}_d$, $\bar{H}$ defined by* (2.14) *is a survival function with margins $\bar{F}_1, \dots, \bar{F}_d$.*

□

Let $\boldsymbol{X} \sim H$ with copula C and marginal dfs $F_1, \dots, F_d$. Since it couples the univariate marginal survival functions $\bar{F}_1, \dots, \bar{F}_d$ to their joint survival function $\bar{H}$, the copula $\bar{C}$ in (2.14) is called the *survival copula* of $\boldsymbol{X}$ (or H). Note that $\bar{C}$ is a copula (and thus a df). However, neither $\bar{H}$ nor $\bar{F}_1, \dots, \bar{F}_d$ are dfs.

As mentioned before, many copula families were actually derived through (2.15). For example, bivariate Marshall–Olkin copulas as in (2.8) were originally constructed as survival copulas of lifetimes of the form $X_1 = \min\{Z_1, Z_{12}\}$ and $X_2 = \min\{Z_2, Z_{12}\}$, where Z_1, Z_2, Z_{12} are independent exponentials with parameters $\lambda_1, \lambda_2, \lambda_{12}$, respectively, representing arrival times of two individual and one joint fatal shock to a system. By using the parametrization $\alpha_j = \lambda_j/(\lambda_j+\lambda_{12})$, $j \in \{1, 2\}$, and including the boundary cases $\alpha_j \in \{0, 1\}$, $j \in \{1, 2\}$, Marshall–Olkin copulas arise as survival copulas of (X_1, X_2). Other prominent examples of copula families constructed as survival copulas are the Clayton, Frank, and Gumbel–Hougaard families, and, more generally, the class of Archimedean copulas to which they belong; see Sect. 3.2 for more details.

One might wonder whether there is a stochastic representation of a random vector distributed according to $\bar{C}$ given a random vector distributed according to C. Also, one might be interested in the connection between $\bar{C}$ and C, and between their corresponding densities if they exist. The following proposition provides answers. Its first part is a consequence of Sklar's Theorem for survival functions, its second part follows from an application of the inclusion–exclusion principle (see, for example, Durrett 2010, Exercise 1.6.9), and its third part is a direct consequence of the second part.

Proposition 2.5.2 (Properties of Survival Copulas) *Let C be a copula and let $\boldsymbol{U} \sim C$. Then:*

1) *$\mathbf{1} - \boldsymbol{U} \sim \bar{C}$, that is, $\mathbf{1} - \boldsymbol{U} = (1 - U_1, \ldots, 1 - U_d)$ is a random vector whose df is the survival copula $\bar{C}$ corresponding to C.*
2) *The survival copula $\bar{C}$ itself can be computed from C via*

$$\bar{C}(\boldsymbol{u}) = \sum_{J \subseteq \{1,\ldots,d\}} (-1)^{|J|} C((1-u_1)^{1(1\in J)}, \ldots, (1-u_d)^{1(d\in J)}), \quad \boldsymbol{u} \in [0,1]^d, \tag{2.16}$$

where the sum extends over all 2^d subsets J of $\{1, \ldots, d\}$, $|J|$ denotes the number of elements of J and $1(j \in J)$ is the indicator of $j \in J$. For $d = 2$, $\bar{C}(\boldsymbol{u}) = 1 - (1 - u_1) - (1 - u_2) + C(1 - u_1, 1 - u_2) = -1 + u_1 + u_2 + C(1 - u_1, 1 - u_2)$.
3) *If C admits a density c, so does $\bar{C}$ and the density $\bar{c}$ of $\bar{C}$ is given by*

$$\bar{c}(\boldsymbol{u}) = c(1 - u_1, \ldots, 1 - u_d), \quad \boldsymbol{u} \in (0,1)^d. \qquad \square$$

Example 2.5.3 (Survival Copulas) It follows from Proposition 2.5.2 that sampling from $\bar{C}$ and evaluating its density $\bar{c}$ is not more complicated than it is for C and c, respectively. Figure 2.17 provides an illustration in the bivariate case for the Clayton copula C_2^C and exploits the fact that, in arbitrary dimensions, observations of $\boldsymbol{V} \sim \bar{C}$ can be immediately obtained from observations of $\boldsymbol{U} \sim C$ using the relationship $\boldsymbol{V} = \mathbf{1} - \boldsymbol{U}$:

```
> cc <- claytonCopula(2)
> set.seed(271)
> U <- rCopula(1000, copula = cc) # sample from the Clayton copula
> V <- 1 - U # sample from the survival Clayton copula
> plot(U, xlab = quote(U[1]), ylab = quote(U[2])) # scatter plot
> plot(V, xlab = quote(V[1]), ylab = quote(V[2])) # for the survival copula
```

Equivalently, from a geometrical perspective, for any $d \geq 2$, a sample from $\bar{C}$ can be obtained from a sample from C by point reflection with respect to $\mathbf{1/2} = (1/2, \ldots, 1/2)$.

In the R package `copula`, objects representing survival copulas can be created using the function `rotCopula()`. More details will be given in Sect. 3.4.1. The following code exploits this feature to produce a wireframe plot of $\bar{c}_2^C$ (see Fig. 2.17):

```
> wireframe2(cc,              FUN = dCopula, delta = 0.025)
> wireframe2(rotCopula(cc), FUN = dCopula, delta = 0.025)
```

Computing $\bar{C}$ from C becomes prohibitive as d increases. Additionally, summing up the terms in (2.16) (even when sorted according to their signs) can be affected by round-off errors and may result in a number outside $[0, 1]$. This holds even more so for copulas that cannot be evaluated explicitly, such as normal or t copulas (in which case $\bar{C}$ is also affected by Monte Carlo errors in dimensions as small as three). $\square$

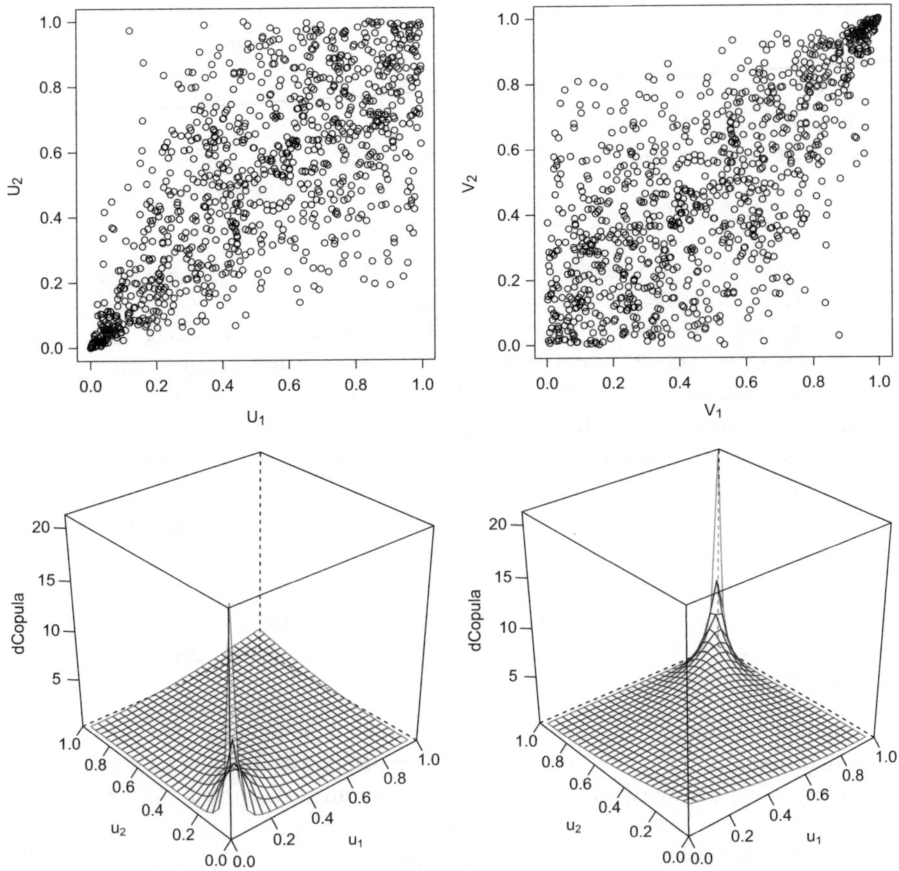

Fig. 2.17 Scatter plots (top) and wireframe plots of the densities (bottom) of a Clayton copula C_θ^{C} (left) and survival Clayton copula $\bar{C}_\theta^{\mathrm{C}}$ (right) with parameter $\theta = 2$

We now introduce two kinds of symmetries that can appear in multivariate distributions. They are linked to elliptical copulas and Archimedean copulas; see Chap. 3.

Definition 2.5.4 (Radial Symmetry, Exchangeability)

1) A random vector $\boldsymbol{X}$ is called *radially symmetric* (for $d = 1$ simply *symmetric*) about $\boldsymbol{a} \in \mathbb{R}^d$ if $\boldsymbol{X} - \boldsymbol{a} \overset{\mathrm{d}}{=} \boldsymbol{a} - \boldsymbol{X}$, that is, if $\boldsymbol{X} - \boldsymbol{a}$ and $\boldsymbol{a} - \boldsymbol{X}$ are equal in distribution.
2) The random vector $\boldsymbol{X}$ is called *exchangeable* if $(X_{j_1}, \dots, X_{j_d}) \overset{\mathrm{d}}{=} (X_1, \dots, X_d)$ for all permutations $(j_1, \dots, j_d)$ of $\{1, \dots, d\}$. □

In terms of copulas, one can derive the following results around radial symmetry and exchangeability; see Nelsen (2006, Sect. 2.7) for the bivariate case (the arguments readily extend to the case $d > 2$).

Proposition 2.5.5 (Characterization of Radial Symmetry and Exchangeability) *Let $\boldsymbol{X} \sim H$ with continuous univariate margins $F_1, \dots, F_d$ and copula C.*

1) *If X_j is symmetric about a_j, $j \in \{1, \dots, d\}$, then $\boldsymbol{X}$ is radially symmetric about $\boldsymbol{a}$ if and only if $C = \bar{C}$.*
2) *The random vector $\boldsymbol{X}$ is exchangeable if and only if $F_1 = \cdots = F_d$ and $C(u_{j_1}, \dots, u_{j_d}) = C(u_1, \dots, u_d)$ for all permutations $(j_1, \dots, j_d)$ of $\{1, \dots, d\}$ and all $u_1, \dots, u_d \in [0, 1]$.* □

If $\bar{C} = C$, we call C *radially symmetric*. If $C(u_{j_1}, \dots, u_{j_d}) = C(u_1, \dots, u_d)$ for all $u_1, \dots, u_d \in [0, 1]$ and all permutations $(j_1, \dots, j_d)$ of $\{1, \dots, d\}$, we call C *exchangeable*. Examples of both radially symmetric and exchangeable copulas are W (in dimension two only), Π, and M. So are the normal and t copulas which are homogeneous (that is, which have correlation matrices P with equal off-diagonal entries). Gumbel–Hougaard copulas, for example, are easily seen to be exchangeable, however, they are not radially symmetric; this will become clear from the following example. Hence, exchangeability does not imply radial symmetry. Note also that radial symmetry does not imply exchangeability (consider, for instance, (2.20) in Example 2.6.3 below with $h_1(u) = \sin(\pi u)/\pi$ and $h_2(u) = u(1-u)$, $u \in [0, 1]$) and that there are copulas which are neither radially symmetric nor exchangeable (consider, for example, Marshall–Olkin copulas with distinct parameters $\alpha_1, \alpha_2 \in (0, 1)$; see Example 2.2.3 and, in particular, the right panel of Fig. 2.9).

Example 2.5.6 (Visually Assessing Radial Symmetry and Exchangeability) Radial symmetry and exchangeability are the two most prominent types of symmetries involved in copula modeling. Understanding visually what these notions imply is crucial when carrying out statistical inference on copulas.

Figure 2.18 shows contour plots of the densities $c^t_{\rho,\nu}$ and c^{GH}_θ of a t copula $C^t_{\rho,\nu}$ with correlation parameter $\rho = 0.7$ and degrees of freedom $\nu = 3.5$ (left) and a

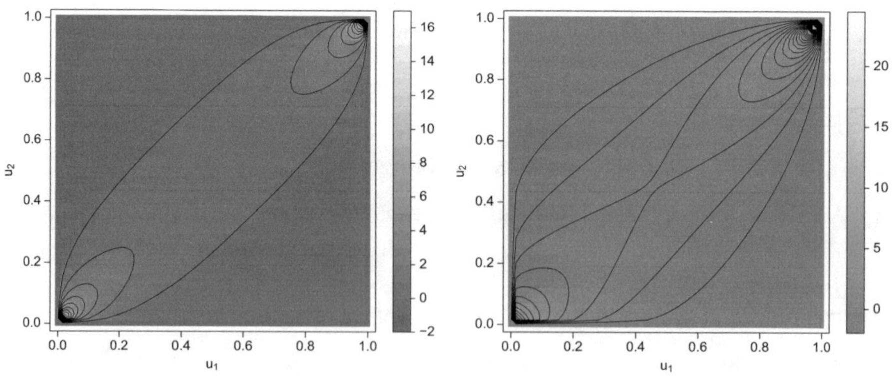

Fig. 2.18 Contour plots of the densities $c^t_{\rho,\nu}$ for $\rho = 0.7$, $\nu = 3.5$ (left) and c^{GH}_θ for $\theta = 2$ (right)

Gumbel–Hougaard copula C_θ^{GH} with parameter $\theta = 2$ (right):

```
> contourplot2(tCopula(0.7, df = 3.5), FUN = dCopula, n.grid = 64, lwd = 1/2)
> contourplot2(gumbelCopula(2),         FUN = dCopula, n.grid = 64, lwd = 1/4,
               pretty = FALSE, cuts = 42,
               col.regions = gray(seq(0.5, 1, length.out = 128)))
```

Geometrically, if its density exists, radial symmetry of a copula means symmetry of the density with respect to the point $\mathbf{1/2} = (1/2, \ldots, 1/2)$; see the left-hand side of Fig. 2.18. Exchangeability can be inferred from the symmetry of the density (or of the copula) with respect to the main diagonal; see the right-hand side of Fig. 2.18. It can be shown that bivariate t copulas are both radially symmetric and exchangeable and that Gumbel–Hougaard copulas C_θ^{GH} are exchangeable but not radially symmetric unless $\theta = 1$. Given bivariate data from a copula, radial symmetry and exchangeability can be assessed from a scatter plot provided a sufficiently large sample is available. Related graphical diagnostics and formal tests for observations not necessarily arising from marginally standard uniform random vectors will be discussed in Chap. 5. □

2.6 Measures of Association

With applications in mind, it is often desirable to summarize the dependence between components of a random vector by a real number. Such numerical summaries of (aspects of) dependence are known as *measures of association* and are mostly studied in the bivariate case; for higher-dimensional extensions, see, for example, Jaworski et al. (2010, Chap. 10).

One widely known measure of association is *Pearson's* (or the *linear*) *correlation coefficient* defined, for a random vector (X_1, X_2) whose components have finite variances (or, equivalently, such that $\mathbb{E}(X_1^2) < \infty$ and $\mathbb{E}(X_2^2) < \infty$), by

$$\mathrm{Cor}(X_1, X_2) = \frac{\mathrm{Cov}(X_1, X_2)}{\sqrt{\mathrm{Var}(X_1)}\sqrt{\mathrm{Var}(X_2)}} = \frac{\mathbb{E}\big((X_1 - \mathbb{E}(X_1))(X_2 - \mathbb{E}(X_2))\big)}{\sqrt{\mathbb{E}\big((X_1 - \mathbb{E}(X_1))^2\big)}\sqrt{\mathbb{E}\big((X_2 - \mathbb{E}(X_2))^2\big)}}. \tag{2.17}$$

Some well-known properties of the linear correlation coefficient are summarized in the following proposition.

Proposition 2.6.1 (Properties of Linear Correlation) *Let (X_1, X_2) be a random vector whose components have finite variances. Then,*

1) $-1 \le \mathrm{Cor}(X_1, X_2) \le 1$*;*
2) $|\mathrm{Cor}(X_1, X_2)| = 1$ *if and only if there exist* $a, b \in \mathbb{R}$, $a \neq 0$, *such that* $X_2 = aX_1 + b$ *almost surely, with* $a < 0$ *if and only if* $\mathrm{Cor}(X_1, X_2) = -1$ *and* $a > 0$ *if*

and only if $\text{Cor}(X_1, X_2) = 1$. *In any of these two cases,* X_1 *and* X_2 *are* perfectly linearly dependent.

3) *If* X_1 *and* X_2 *are independent, then* $\text{Cor}(X_1, X_2) = 0$.
4) *For any* $a_1 > 0$, $a_2 > 0$, *or any* $a_1 < 0$, $a_2 < 0$, *and for any* $b_1, b_2 \in \mathbb{R}$,

$$\text{Cor}(a_1 X_1 + b_1,\ a_2 X_2 + b_2) = \text{Cor}(X_1, X_2).$$

In particular, Pearson's correlation coefficient is invariant under strictly increasing linear transformations. □

As becomes clear from Proposition 2.6.1, Pearson's correlation coefficient is merely a measure of linear dependence and its usefulness for quantifying dependence is truly meaningful only in the context of the so-called elliptical distributions (see Sect. 3.1.1) such as the multivariate normal or t distributions. Indeed, correlation matrices have a natural parametric role in these distributions, which they do not necessarily have in more general multivariate models; see the fallacies in Sect. 2.6.1.

The other kinds of measures of association which we present (in the bivariate case) are two rank correlation coefficients and the so-called tail-dependence coefficients. We focus on continuous random vectors, which implies that these measures of association only depend on the underlying unique copula and, hence, are of much broader applicability than the linear correlation coefficient.

2.6.1 Fallacies Related to the Correlation Coefficient

Before presenting more adequate measures of association for bivariate continuous random vectors, we briefly point out fallacies related to the use of the linear correlation coefficient as a measure of dependence; a must-read on this subject is Embrechts et al. (2002).

The first fallacy can be avoided by carefully examining (2.17).

Fallacy 1 (Existence)
$\text{Cor}(X_1, X_2)$ exists for every random vector (X_1, X_2). □

From (2.17), we see that $\text{Cor}(X_1, X_2)$ is a normalized covariance. A well-known consequence of the Cauchy–Schwarz inequality is that $|\text{Cov}(X_1, X_2)| \le \sqrt{\text{Var}(X_1)\,\text{Var}(X_2)}$. Hence, both the numerator and the denominator in (2.17) exist if the second moments of X_1 and X_2 are finite. If one of these moments does not exist, $\text{Cor}(X_1, X_2)$ is not defined. As a counterexample to Fallacy 1, consider $X_1, X_2 \overset{\text{ind.}}{\sim} F(x) = 1 - x^{-3}$, $x \in [1, \infty)$. By independence (see Proposition 2.6.1), $\text{Cor}(X_1, X_2) = 0$ but $\text{Cor}(X_1^3, X_2)$ does not exist since neither $\mathbb{E}((X_1^3)^2)$ nor $\mathbb{E}(X_1^3)$ are finite. This counterexample can also serve as a counterexample to the following fallacy.

Fallacy 2 (Invariance)
$\mathrm{Cor}(X_1, X_2)$ is invariant under strictly increasing transformations on $\operatorname{ran} X_1$ or $\operatorname{ran} X_2$. □

The following proposition provides a useful representation of the correlation coefficient. It originally appeared in a slightly different form in Hoeffding (1940); see McNeil et al. (2015, Lemma 7.27).

Proposition 2.6.2 (Hoeffding's Formula) *Let (X_1, X_2) be a random vector with df H with margins F_1, F_2 and whose components have finite variances. Then*

$$\mathrm{Cor}(X_1, X_2) = \frac{\int_{-\infty}^{\infty}\int_{-\infty}^{\infty}\big(H(x_1, x_2) - F_1(x_1)F_2(x_2)\big)\mathrm{d}x_1\mathrm{d}x_2}{\sqrt{\mathrm{Var}(X_1)}\sqrt{\mathrm{Var}(X_2)}}. \tag{2.18}$$

□

By Sklar's Theorem, (2.18) can be rewritten as

$$\mathrm{Cor}(X_1, X_2) = \frac{\int_{-\infty}^{\infty}\int_{-\infty}^{\infty}\big(C(F_1(x_1), F_2(x_2)) - F_1(x_1)F_2(x_2)\big)\mathrm{d}x_1\mathrm{d}x_2}{\sqrt{\mathrm{Var}(X_1)}\sqrt{\mathrm{Var}(X_2)}}, \tag{2.19}$$

where C is a copula satisfying (2.9). This representation leads to a counterexample to the following two fallacies and can also be used to refute Fallacy 5 below.

Fallacy 3 (Uniqueness)
The marginal distributions and the correlation coefficient uniquely determine the joint distribution. □

Fallacy 4 (Uncorrelatedness Implies Independence)
$\mathrm{Cor}(X_1, X_2) = 0$ implies that X_1 and X_2 are independent. □

Example 2.6.3 (Counterexample to Fallacies 3 and 4) By checking groundedness and uniform margins, as well as by computing the density, it can be verified that

$$C(u_1, u_2) = u_1u_2 + h_1(u_1)h_2(u_2), \quad u_1, u_2 \in [0, 1], \tag{2.20}$$

is a copula for all continuously differentiable $h_1, h_2 : [0, 1] \to \mathbb{R}$ such that $h_1(0) = h_2(0) = h_1(1) = h_2(1) = 0$ and $1 + h_1'(u_1)h_2'(u_2) \geq 0$ for all $u_1, u_2 \in (0, 1)$.

Let $h_1(u) = 2u(u - 1/2)(u - 1)$ and $h_2(u) = \theta u(1 - u)$, $u \in [0, 1]$, $\theta \in [-1, 1]$, and let $(U_1, U_2) \sim C$. Then, Hoeffding's formula implies that

$$\begin{aligned}\mathrm{Cor}(U_1, U_2) &= 12\int_0^1\int_0^1 (C(u_1, u_2) - u_1u_2)\mathrm{d}u_1\mathrm{d}u_2\\ &= 12\int_0^1 h_1(u_1)\mathrm{d}u_1\int_0^1 h_2(u_2)\mathrm{d}u_2.\end{aligned}$$

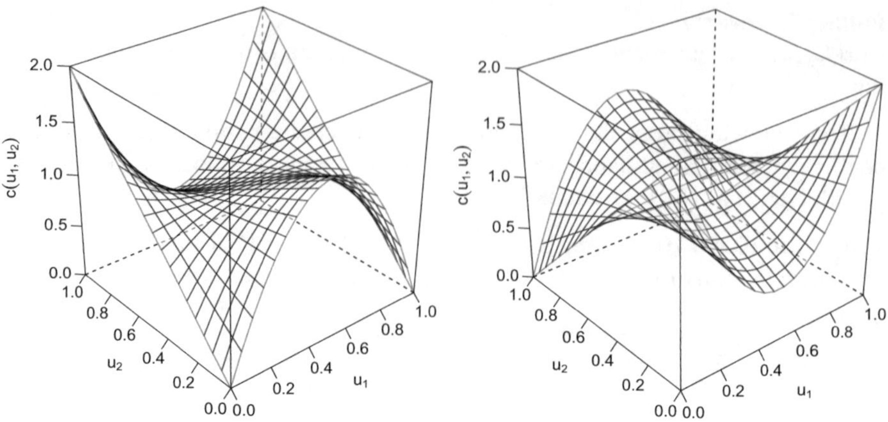

Fig. 2.19 Density c of C in (2.20) for $\theta = -1$ (left) and $\theta = 1$ (right)

It can be checked that $\mathrm{Cor}(U_1, U_2) = 0$ for all $\theta \in [-1, 1]$. In other words, the margins and the correlation coefficient do not change for $\theta \in [-1, 1]$, while the bivariate distribution C does change with θ as can, for instance, be verified from Fig. 2.19 produced by the following code:

```
> ## Evaluate the density of C for h_1(u) = 2*u*(u-1/2)*(u-1),
> ## h_2(u) = theta*u*(1-u) and two different thetas
> u <- seq(0, 1, length.out = 20) # subdivision points in each dimension
> u12 <- expand.grid("u[1]" = u, "u[2]" = u) # build a grid
> dC <- function(u, th) 1 + th * (6 * u[,1] * (u[,1]-1) + 1) * (1 - 2*u[,2])
> wireframe2(cbind(u12, "c(u[1],u[2])" = dC(u12, th = -1)))
> wireframe2(cbind(u12, "c(u[1],u[2])" = dC(u12, th =  1)))
```

This example therefore constitutes a counterexample to Fallacy 3. Since $C \neq \Pi$ for all $\theta \in [-1, 1] \setminus \{0\}$, this example also serves as a counterexample to Fallacy 4. Note that Fallacy 4 alone can more easily be falsified by taking $X_1 \sim \mathrm{N}(0, 1)$ and $X_2 = X_1^2$; similarly, taking $X_1 \sim \mathrm{U}(0, 1)$ and $X_2 = |X_1 - 1/2|$ also provides a valid counterexample. □

Example 2.6.4 (Uncorrelatedness Versus Independence) Another counterexample to Fallacies 3 and 4 can be constructed by a mixture of the two Fréchet–Hoeffding bounds W and M. To this end, let $Z \sim \mathrm{N}(0, 1)$ and define $X_1 = Z$ and $X_2 = VZ$, where V takes its values in $\{-1, 1\}$, each with probability $1/2$. Conditional on $V = -1$ (respectively, $V = 1$), the copula C of (X_1, X_2) is thus W (respectively, M). In other words, $C(u_1, u_2) = \lambda M(u_1, u_2) + (1 - \lambda) W(u_1, u_2)$ with $\lambda = 1/2$. Since $\mathrm{Cor}(X_1, X_2) = \mathrm{Cov}(X_1, X_2) = \mathbb{E}(X_1 X_2) = \mathbb{E}(V)\mathbb{E}(Z^2) = \mathbb{E}(V) = 0$, X_1 and X_2 are uncorrelated. Note that X_1 and X_2 are not independent though, since C is not equal to the independence copula Π.

Let us verify this empirically in R. The plot on the left-hand side of Fig. 2.20 shows $n = 1000$ realizations of (X_1, X_2), from which we clearly see that uncorrelatedness does not necessarily imply independence. As a comparison, the

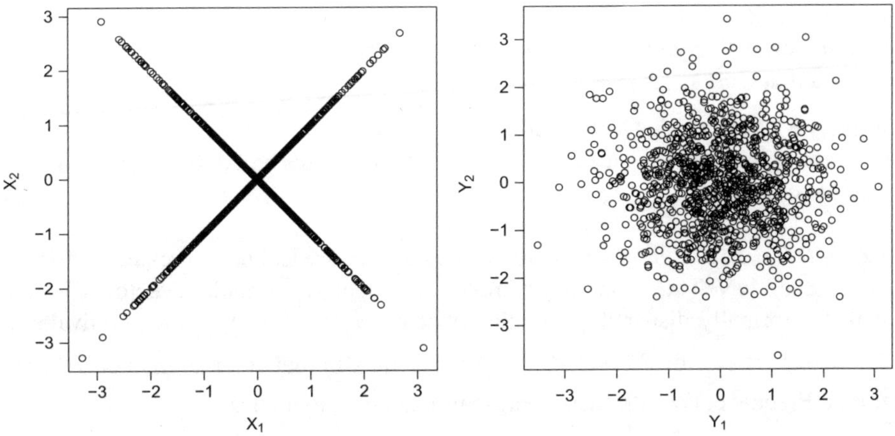

Fig. 2.20 $n = 1000$ independent realizations from (X_1, X_2) (left) and (Y_1, Y_2) (right), both having N(0, 1) margins and zero correlation; the copula of (X_1, X_2) (respectively, (Y_1, Y_2)) is a mixture between the Fréchet–Hoeffding bounds W and M (respectively, is the independence copula)

plot on the right-hand side of Fig. 2.20 displays $n = 1000$ realizations of (Y_1, Y_2) for independent $Y_1, Y_2 \sim \mathrm{N}(0, 1)$. The random vectors (X_1, X_2) and (Y_1, Y_2) have thus the same margins and have both zero correlation. In addition to producing the aforementioned plots, the following code performs a test of uncorrelatedness on the generated independent observations of (X_1, X_2) using the function `cor.test()`:

```
> n <- 1000
> set.seed(314)
> Z <- rnorm(n)
> U <- runif(n)
> V <- rep(1, n)
> V[U < 1/2] <- -1 # => V in {-1,1}, each with probability 1/2
> X <- cbind(Z, Z*V) # (X_1,X_2)
> stopifnot(cor.test(X[,1], X[,2])$p.value >= 0.05) # H0:'cor=0' not rejected
> Y <- matrix(rnorm(n * 2), ncol = 2) # independent N(0,1)
> ## Plots
> plot(X, xlab = quote(X[1]), ylab = quote(X[2]))
> plot(Y, xlab = quote(Y[1]), ylab = quote(Y[2]))
```

□

Let (X_1, X_2) and (X'_1, X'_2) be continuous random vectors whose components have finite variances and the same marginal dfs F_1, F_2 but different copulas C and C', respectively, such that, for any $\boldsymbol{u} \in [0, 1]^d$, $C(\boldsymbol{u}) \leq C'(\boldsymbol{u})$. Hoeffding's formula as given in (2.19) then implies that $\mathrm{Cor}(X_1, X_2) \leq \mathrm{Cor}(X'_1, X'_2)$. It follows from Theorem 2.2.1 on the Fréchet–Hoeffding bounds that, for any continuous random vector (X_1, X_2) with marginal dfs F_1, F_2 and whose components have finite variances,

$$\underline{\mathrm{Cor}}(X_1, X_2) \leq \mathrm{Cor}(X_1, X_2) \leq \overline{\mathrm{Cor}}(X_1, X_2),$$

where $\underline{\mathrm{Cor}}(X_1, X_2)$ and $\overline{\mathrm{Cor}}(X_1, X_2)$ are obtained by replacing C in (2.19) by W and M, respectively. It is a fallacy to believe that, for any margins, $\underline{\mathrm{Cor}}(X_1, X_2) = -1$ and that $\overline{\mathrm{Cor}}(X_1, X_2) = 1$.

Fallacy 5 (Attainable Correlations)
Given margins F_1, F_2, all $\mathrm{Cor}(X_1, X_2) \in [-1, 1]$ can be attained by choosing a suitable copula for (X_1, X_2). □

Example 2.6.5 (Counterexample to Fallacy 5) A classical counterexample for Fallacy 5, see McNeil et al. (2015, Example 7.29), consists of random vectors (X_1, X_2) with lognormally distributed margins, that is, $X_j \sim \mathrm{LN}(0, \sigma_j^2)$ or, equivalently, $X_j \stackrel{\mathrm{d}}{=} \exp(\sigma_j Z_j)$ for $Z_j \sim \mathrm{N}(0, 1)$, $j \in \{1, 2\}$. Minimal and maximal correlations for the Fréchet class with these marginals can be computed as

$$\underline{\mathrm{Cor}}(X_1, X_2) = \rho_{\sigma_1, -\sigma_2} \quad \text{and} \quad \overline{\mathrm{Cor}}(X_1, X_2) = \rho_{\sigma_1, \sigma_2},$$

where

$$\rho_{x,y} = \frac{\exp((x + y)^2/2) - \exp((x^2 + y^2)/2)}{\sqrt{(\exp(x^2) - 1)\exp(x^2)}\sqrt{(\exp(y^2) - 1)\exp(y^2)}}.$$

Figure 2.21, produced as follows based on the auxiliary function `corBoundLN()`, shows the correlation bounds $\underline{\mathrm{Cor}}(X_1, X_2)$ and $\overline{\mathrm{Cor}}(X_1, X_2)$ as functions of σ_1

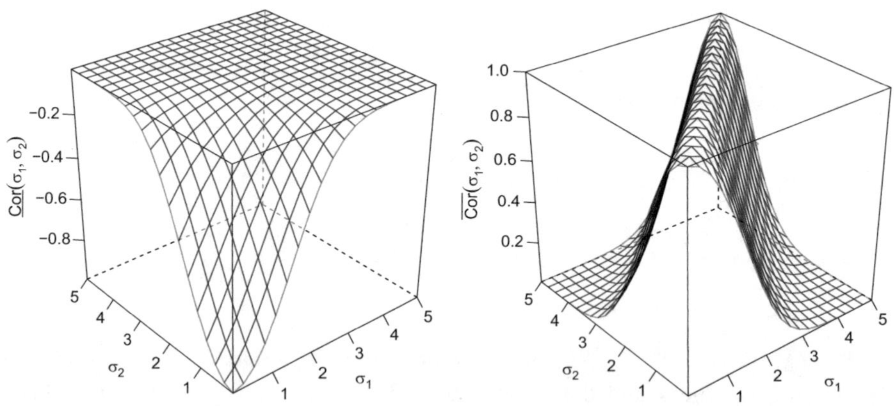

Fig. 2.21 Minimal (left) and maximal (right) attainable correlations under $\mathrm{LN}(0, \sigma_1^2)$ and $\mathrm{LN}(0, \sigma_2^2)$ margins

and σ_2:

```
> ## Function to compute the correlation bounds for LN(0, sigma_.^2) margins
> corBoundLN <- function(s, bound = c("max", "min"))
  {
      ## s = (sigma_1, sigma_2)
      if(!is.matrix(s)) s <- rbind(s)
      bound <- match.arg(bound)
      if(bound == "min") s[,2] <- -s[,2]
      (exp((s[,1]+s[,2])^2/2)-exp((s[,1]^2+s[,2]^2)/2)) /
          sqrt(expm1(s[,1]^2)*exp(s[,1]^2)*expm1(s[,2]^2)*exp(s[,2]^2))
  }
> ## Evaluate correlation bounds on a grid
> s <- seq(0.01, 5, length.out = 20) # subdivision points in each dimension
> s12 <- expand.grid("sigma[1]" = s, "sigma[2]" = s) # build a grid
> ## Plots
> wireframe2(cbind(s12, `underline(Cor)(sigma[1],sigma[2])` =
                                  corBoundLN(s12, bound = "min")))
> wireframe2(cbind(s12, `bar(Cor)(sigma[1],sigma[2])` = corBoundLN(s12)))
```

Clearly, σ_1 and σ_2 can be chosen so that not all correlations in $[-1, 1]$ are attainable, for instance, by taking $\sigma_1 = 1$ and $\sigma_2 = 4$. □

To conclude, the main limitations of the linear correlation coefficient $\mathrm{Cor}(X_1, X_2)$ as a measure of association are:

1) $\mathrm{Cor}(X_1, X_2)$ does not exist for all random vectors (X_1, X_2) (only for those with finite second moments);
2) $\mathrm{Cor}(X_1, X_2)$ depends on the marginal dfs of (X_1, X_2) even when the latter are continuous (it can thus not be expressed in terms of the unique underlying copula alone);
3) $\mathrm{Cor}(X_1, X_2)$ is invariant only under strictly increasing linear transformations (not under strictly increasing transformations in general).

2.6.2 *Rank Correlation Measures*

By only depending on the underlying copula C in the case of continuous random vectors, *rank correlation coefficients* overcome several of the aforementioned limitations of the linear correlation coefficient. In this section, we focus on the two best-known examples of rank correlation measures: Spearman's rho and Kendall's tau.

Definition 2.6.6 (Spearman's Rho, Kendall's Tau) Let (X_1, X_2) be a bivariate random vector with continuous marginal dfs F_1 and F_2.

1) The *(population version of) Spearman's rho* is defined by

$$\rho_s = \rho_s(X_1, X_2) = \mathrm{Cor}(F_1(X_1), F_2(X_2)).$$

2) Let (X_1', X_2') be an independent copy of (X_1, X_2). The *(population version of) Kendall's tau* is defined by

$$\tau = \tau(X_1, X_2) = \mathbb{E}\big(\operatorname{sign}((X_1 - X_1')(X_2 - X_2'))\big),$$

where $\operatorname{sign}(x)$ denotes the *sign* of x, that is,

$$\operatorname{sign}(x) = \begin{cases} -1, & \text{if } x < 0, \\ 0, & \text{if } x = 0, \\ 1, & \text{if } x > 0. \end{cases}$$ □

Spearman's rho is simply the linear correlation coefficient of the random vector $(F_1(X_1), F_2(X_2))$ obtained from (X_1, X_2) by marginally applying the probability transformation. In comparison to the linear correlation coefficient, Spearman's rho thus always exists. Since, by Lemma 2.4.1, the df of $(F_1(X_1), F_2(X_2))$ is the copula C, it is also immediate that Spearman's rho only depends on the underlying copula C, and thus not on the marginal dfs F_1 and F_2. Hence, ρ_s measures dependence independently of the margins (as opposed to the linear correlation coefficient); in particular, Spearman's rho is invariant under strictly increasing transformations of X_1 and X_2. These immediate results already show that Spearman's rho is not affected by Fallacies 1 and 2. One can also show that Spearman's rho and Kendall's tau are *measures of concordance*; see Scarsini (1984) for an axiomatic definition. Several properties of these two measures of association can be derived from this fact.

To define the notion of *concordance* (and also) to better understand the definition of Kendall's tau, consider two points in $\mathbb{R}^2$ denoted by (x_1, x_2) and (x_1', x_2'). These points are said to be *concordant* if $(x_1 - x_1')(x_2 - x_2') > 0$ (so if the slope of the line through the two points is positive) and to be *discordant* if $(x_1 - x_1')(x_2 - x_2') < 0$. By the definition of the sign function, Kendall's tau can be written as

$$\tau = \mathbb{P}\big((X_1 - X_1')(X_2 - X_2') > 0\big) - \mathbb{P}\big((X_1 - X_1')(X_2 - X_2') < 0\big),$$

that is, Kendall's tau is the probability of concordance minus the probability of discordance of the random pairs (X_1, X_2) and (X_1', X_2'). If X_2 tends to increase (respectively, decrease) with X_1, then the probability of concordance (relative to the probability of discordance) is high (respectively, low). This is the intuition behind the definition of Kendall's tau. By noting that under the assumption of continuous margins, $(X_1 - X_1')(X_2 - X_2') \neq 0$ with probability one, one immediately obtains that

$$\tau = 2\mathbb{P}\big((X_1 - X_1')(X_2 - X_2') > 0\big) - 1. \tag{2.21}$$

Like Spearman's rho, Kendall's tau can be expressed in terms of the linear correlation coefficient. This can be seen from (2.21) via

$$\begin{aligned}\tau = 4\mathbb{P}\big(X_1 \le X_1',\ X_2 \le X_2'\big) - 1 &= \frac{\mathbb{E}\big(1(X_1 \le X_1',\ X_2 \le X_2')\big) - \frac{1}{2}\cdot\frac{1}{2}}{\sqrt{\frac{1}{2}\big(1-\frac{1}{2}\big)\frac{1}{2}\big(1-\frac{1}{2}\big)}}\\ &= \frac{\mathbb{E}\big(1(X_1 \le X_1')1(X_2 \le X_2')\big) - \mathbb{E}\big(1(X_1 \le X_1')\big)\mathbb{E}\big(1(X_2 \le X_2')\big)}{\sqrt{\operatorname{Var}\big(1(X_1 \le X_1')\big)\operatorname{Var}\big(1(X_2 \le X_2')\big)}}\\ &= \operatorname{Cor}\big(1(X_1 \le X_1'), 1(X_2 \le X_2')\big).\end{aligned}$$

Kendall's tau thus equals the correlation coefficient of the indicators $1(X_1 \le X_1')$ and $1(X_2 \le X_2')$.

For an appropriate definition of Kendall's tau for discontinuous random variables, see, for example, Genest and Nešlehová (2007). For a similar discussion on Spearman's rho, see also Genest et al. (2013).

The following proposition provides representations of Spearman's rho and Kendall's tau in terms of the underlying copula C; see Nelsen (2006, Theorem 5.1.3 and 5.1.6).

Proposition 2.6.7 (Representation of Spearman's Rho and Kendall's Tau) *Let (X_1, X_2) be a bivariate random vector with continuous marginal dfs and copula C. Then,*

$$\rho_s = \rho_s(C) = 12\int_{[0,1]^2} C(\boldsymbol{u})\,\mathrm{d}\boldsymbol{u} - 3 = 12\int_{[0,1]^2} u_1u_2\,\mathrm{d}C(\boldsymbol{u}) - 3 \tag{2.22}$$

and

$$\tau = \tau(C) = 4\int_{[0,1]^2} C(\boldsymbol{u})\,\mathrm{d}C(\boldsymbol{u}) - 1. \tag{2.23}$$

□

As becomes clear from Proposition 2.6.7, Spearman's rho and Kendall's tau only depend on the underlying copula; they can be viewed as moments of the copula. This proposition can also be used to show that both measures of association satisfy the defining axioms of a measure of concordance. As such they share the property that $\rho_s(W) = \tau(W) = -1$, $\rho_s(M) = \tau(M) = 1$, that is, the minimal and maximal values of ρ_s and τ are attained for the lower and upper Fréchet–Hoeffding bounds, so for counter- and comonotone random variables, respectively. This shows that an analog of Fallacy 5 does not apply to rank correlation measures. For ρ_s and τ in particular, one can also show the converse: If one of these measures is -1 or 1, then the copula C of (X_1, X_2) must be W or M, respectively. It also follows from the axioms of a measure of concordance that if $C = \Pi$, then the measure is 0. Similar to Fallacy 4, however, the converse is not true in general which can be shown, for instance, by considering the uniform distribution on the unit circle; see Embrechts et al. (2002, Proposition 3).

Consider next a one-parameter family of copulas $\{C_\theta : \theta \in \Theta\}$, where the parameter space Θ is a subset of $\mathbb{R}$. For many such copula families, the functions

$$g_{\rho_s}(\theta) = \rho_s(C_\theta) \quad \text{and} \quad g_\tau(\theta) = \tau(C_\theta), \quad \theta \in \Theta,$$

are one-to-one. For example:

- Clayton family: $g_\tau(\theta) = \theta/(\theta + 2), \theta \in (0, \infty)$;
- Gumbel–Hougaard family: $g_\tau(\theta) = 1 - 1/\theta, \theta \in [1, \infty)$;
- Normal family: $g_{\rho_s}(\theta) = (6/\pi) \arcsin(\theta/2)$ and $g_\tau(\theta) = (2/\pi) \arcsin \theta, \theta \in [-1, 1]$.

By considering the corresponding inverses $g_{\rho_s}^{-1}$ and g_τ^{-1}, one can thus obtain the unique value of θ corresponding to an admissible value of ρ_s or τ. This idea lies at the core of the well-known method-of-moments estimators of copula parameters; see Sect. 4.1.2 for more details.

Example 2.6.8 (`rho()`, `iRho()`, `tau()` and `iTau()`) In the R package `copula`, the functions ρ_s and τ are called `rho()` and `tau()`, respectively. They can be used to evaluate g_{ρ_s} and g_τ:

```
> theta <- -0.7
> stopifnot(all.equal(rho(normalCopula(theta)), 6 / pi * asin(theta / 2)))
> stopifnot(all.equal(tau(normalCopula(theta)), 2 / pi * asin(theta)))
> theta <- 2
> stopifnot(all.equal(tau(claytonCopula(theta)), theta / (theta + 2)))
> stopifnot(all.equal(tau(gumbelCopula(theta)), 1 - 1 / theta))
```

The inverse functions $g_{\rho_s}^{-1}$ and g_τ^{-1} are called `iRho()` and `iTau()`, respectively, and are "vectorized":

```
> theta <- (0:8)/16
> stopifnot(all.equal(iRho(normalCopula(), rho = 6/pi * asin(theta/2)), theta))
> stopifnot(all.equal(iTau(normalCopula(), tau = 2/pi * asin(theta)),   theta))
> theta <- 1:20
> stopifnot(all.equal(iTau(claytonCopula(), theta / (theta + 2)), theta))
> stopifnot(all.equal(iTau(gumbelCopula(),  1 - 1 / theta),       theta))
```

Although, for certain parametric copula families, the function g_{ρ_s} and g_τ are one-to-one, they do not necessarily admit an analytical expression. In that case, a numerical approximation is used (see Kojadinovic and Yan 2010, Appendix A):

```
> theta <- 3
> iRho(claytonCopula(), rho = rho(claytonCopula(theta)))
```

```
[1] 3.000007
```

□

A natural next question concerns estimation: Given a random sample $(X_{11}, X_{12}), \dots, (X_{n1}, X_{n2})$ from (X_1, X_2), how could Spearman's rho or Kendall's tau be estimated? We start with Spearman's rho. Being defined as the linear correlation coefficient of $(F_1(X_1), F_2(X_2))$, ρ_s can be estimated by first estimating the unknown marginal dfs F_1 and F_2, then estimating the unobservable random sample $(F_1(X_{11}), F_2(X_{12})), \dots, (F_1(X_{n1}), F_2(X_{n2}))$ and, finally, computing the linear correlation coefficient from the resulting "pseudo-observations" of $(F_1(X_1), F_2(X_2))$. Estimating F_1 and F_2 by the empirical dfs of the corresponding component samples amounts to working with the sample of bivariate ranks $(R_{11}, R_{12}), \dots, (R_{n1}, R_{n2})$, where R_{ij} is the rank of X_{ij} among $X_{1j}, \dots, X_{nj}$, $j \in \{1, 2\}$. A natural estimator of ρ_s (referred to as the *sample version of Spearman's rho*) is then simply

$$\rho_{s,n} = \frac{\sum_{i=1}^{n}(R_{i1} - \bar{R}_1)(R_{i2} - \bar{R}_2)}{\sqrt{\sum_{i=1}^{n}(R_{i1} - \bar{R}_1)^2}\sqrt{\sum_{i=1}^{n}(R_{i2} - \bar{R}_2)^2}}, \tag{2.24}$$

where $\bar{R}_1 = \bar{R}_2 = (n+1)/2$ is the mean rank of the two component series. In other words, $\rho_{s,n}$ is merely the sample linear correlation coefficient computed from the sample of bivariate ranks. Interestingly, this estimator can also be seen as resulting from another "plug-in principle": if one starts from (2.22) and replaces C by a nonparametric estimator, see Sect. 4.2.1, one recovers (2.24).

Let us now turn to the estimation of Kendall's tau. Starting from a bivariate random sample $(X_{11}, X_{12}), \dots, (X_{n1}, X_{n2})$ from (X_1, X_2) and replacing the expectation operator in Definition 2.6.6 by a sample mean, a natural estimator of Kendall's tau (referred to as the *sample version of Kendall's tau*) is given by

$$\tau_n = \frac{1}{\binom{n}{2}} \sum_{1 \le i < j \le n} \operatorname{sign}\big((X_{i1} - X_{j1})(X_{i2} - X_{j2})\big).$$

The latter can be rewritten as a sample version of (2.21):

$$\tau_n = 2\frac{p_n}{\binom{n}{2}} - 1 = \frac{4p_n}{n(n-1)} - 1, \tag{2.25}$$

where p_n is the number of concordant pairs in the sample. Similar to Spearman's rho, τ_n can be seen as resulting from plugging-in a nonparametric estimator of C (see Sect. 4.2.1) into (2.23).

Example 2.6.9 (Estimating Spearman's Rho and Kendall's Tau) In R, Spearman's rho can be estimated with `cor(, method = "spearman")`:

```
> theta <- iRho(claytonCopula(), rho = 0.6) # true Spearman's rho = 0.6
> set.seed(974)
> U <- rCopula(1000, copula = claytonCopula(theta))
> rho.def <- cor(apply(U, 2, rank))[1,2]       # Spearman's rho manually
> rho.R   <- cor(U, method = "spearman")[1,2] # Spearman's rho from R
> stopifnot(all.equal(rho.def, rho.R)) # the same
> rho.R  # indeed close to 0.6
```

```
[1] 0.5915302
```

Kendall's tau can be estimated with `cor(, method = "kendall")`:

```
> theta <- iTau(normalCopula(), tau = -0.5) # true Kendall's tau = -0.5
> set.seed(974)
> U <- rCopula(1000, copula = normalCopula(theta))
> p.n <- 0
> for(i in 1:(n-1)) # number of concordant pairs (obviously inefficient)
      for(j in (i+1):n)
          if(prod(apply(U[c(i,j),], 2, diff)) > 0) p.n <- p.n + 1
> tau.def <- 4 * p.n / (n * (n - 1)) - 1   # Kendall's tau manually
> tau.R <- cor(U, method = "kendall")[1,2] # Kendall's tau from R
> stopifnot(all.equal(tau.def, tau.R)) # the same
> tau.R # close to -0.5
```

```
[1] -0.5308388
```

□

Example 2.6.10 (Spearman's Rho and Kendall's Tau Under Counter- and Comonotonicity) We can also verify in R that $\rho_{s,n}$ and τ_n are indeed -1 under countermonotonicity and 1 under comonotonicity as mentioned above for the population versions of Spearman's rho and Kendall's tau.

```
> set.seed(75)
> X <- rnorm(100)
> Y <- -X^3 # perfect negative dependence
> rho.counter <- cor(X, Y, method = "spearman")
> tau.counter <- cor(X, Y, method = "kendall")
> stopifnot(rho.counter == -1, tau.counter == -1)
> Z <- exp(X) # perfect positive dependence
> rho.co <- cor(X, Z, method = "spearman")
> tau.co <- cor(X, Z, method = "kendall")
> stopifnot(rho.co == 1, tau.co == 1)
```

□

To conclude, because they only depend on the underlying copula, Spearman's rho and Kendall's tau

1) always exist (and are not limited to continuous random vectors with finite second moments; compare with Fallacy 1);
2) are invariant under strictly increasing transformations (compare with Fallacy 2); and
3) can reach any value in $[-1, 1]$ (compare with Fallacy 5).

Analogs of Fallacies 3 and 4 still apply to rank correlation coefficients, though. This cannot be expected to be avoided since we know by Sklar's Theorem that a function on $[0, 1]^d$ completely determines the dependence between the components of a continuous random vector, not a single number. Summarizing a copula this way causes, as with any summary, a loss of information. Nevertheless, rank correlation measures are useful. For instance, for bivariate data sets that exhibit monotone trends, sample versions of Spearman's rho and Kendall's tau provide an idea of the "average strength of association."

Concerning which rank correlation measure to use, various comparisons between Spearman's rho and Kendall's tau and discussions concerning their efficiency, robustness, application, and popularity in various fields of statistics can be found in the literature. From a copula modeling point of view, one advantage of Kendall's tau (over Spearman's rho, for example) is that there are (semi-)analytical formulas known for two popular classes of copulas, namely elliptical and Archimedean copulas; see Sects. 3.1.2 and 3.2, respectively. For a comparison of these two rank correlation measures, including their differences in specific examples as well as two general bounds involving the two, see Nelsen (2006, Sect. 5.1.3).

The following example provides Spearman's rho and Kendall's tau in terms of the correlation parameter ρ of a normal copula C^{n}_{ρ}. It shows that ρ is rather well approximated by Spearman's rho (for normal copulas), which has sometimes been used for estimating the parameter correlation matrix of normal copulas in the spirit of method-of-moments estimators; see Sect. 4.1.2.

Example 2.6.11 (Spearman's Rho and Kendall's Tau for Normal Copulas) Figure 2.22, generated with the following code, shows Spearman's rho and Kendall's tau as functions of the correlation parameter ρ of a normal copula C^{n}_{ρ}. As we can see on the left-hand side, ρ can be approximated quite well by Spearman's rho in this case. The right-hand side shows the corresponding error including the absolute error bound of 0.0181 as given in McNeil et al. (2015, Sect. 7.3.2).

```
> rho <- seq(-1, 1, by = 0.01) # correlation parameters of normal copulas
> rho.s <- (6/pi) * asin(rho/2) # corresponding Spearman's rho
> tau <- (2/pi) * asin(rho) # corresponding Kendall's tau
> plot(rho, rho.s, type = "l", col = 2, lwd = 2,
       xlab = expression("Correlation parameter"~rho~"of"~C[rho]^n),
       ylab = expression("Corresponding"~rho[s]~"and"~tau))
> abline(a = 0, b = 1, col = 1, lty = 2, lwd = 2)
> lines(rho, tau, col = 3, lwd = 2)
> legend("bottomright", bty = "n", col = 1:3, lty = c(2, 1, 1), lwd = 2,
         legend = c("Diagonal", expression(rho[s]), expression(tau)))
> plot(rho, rho.s - rho, type = "l", yaxt = "n", lwd = 2,
       xlab = expression(rho), ylab = expression(rho[s]-rho))
> mdiff <- max(rho.s - rho)
> abline(h = c(-1, 1) * mdiff, lty = 2, lwd = 2)
> rmdiff <- round(mdiff, 4)
> axis(2, at = c(-mdiff, -0.01, 0, 0.01, mdiff),
       labels = as.character(c(-rmdiff, -0.01, 0, 0.01, rmdiff)))
```

As we will learn in Sect. 3.1.2, all elliptical copulas share the same (namely the above) formula of Kendall's tau and thus the graph of τ in terms of ρ as shown in the left-hand side of Fig. 2.22 is the same for all elliptical copulas. □

2.6.3 Tail Dependence Coefficients

Coefficients of tail dependence aim at summarizing the dependence in the (joint) tails of bivariate distributions. Financial risk managers, for example, are well aware of the fact that the tails of a bivariate t distribution with low degrees of freedom and correlation $|\rho| < 1$ are "heavier" (having more probability mass) than the tails

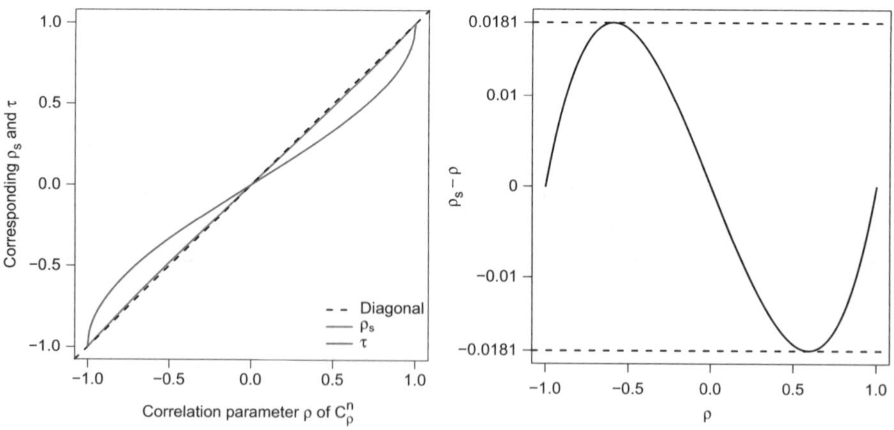

Fig. 2.22 Spearman's rho ρ_s and Kendall's tau τ as functions of the correlation parameter ρ of a normal copula C^{n}_{ρ} (left) and $\rho_s - \rho$ as a function of ρ (right)

of a bivariate normal distribution with the same correlation ρ. This and many other tail-related applications of copula theory can be put on solid ground with the notion of tail dependence discussed in this section.

Example 2.6.12 (Four Distributions with N(0, 1) *Margins and a Kendall's Tau of 0.7)* As an introductory example, let us consider the Fréchet class of bivariate distributions with standard normal margins, and let us investigate how the choice of the copula affects the dependence in the tails. Following McNeil et al. (2015, Chap. 7), we consider four one-parameter copulas with a Kendall's tau of 0.7 and generate random samples of size $n = 10{,}000$ from the corresponding bivariate distributions:

```
> ## Kendall's tau and corresponding copula parameters
> tau <- 0.7
> th.n <- iTau(normalCopula(),  tau = tau)
> th.t <- iTau(tCopula(df = 3), tau = tau)
> th.c <- iTau(claytonCopula(), tau = tau)
> th.g <- iTau(gumbelCopula(),  tau = tau)
> ## Samples from the corresponding 'mvdc' objects
> set.seed(271)
> n <- 10000
> N01m <- list(list(mean = 0, sd = 1), list(mean = 0, sd = 1)) # margins
> X.n <- rMvdc(n, mvdc = mvdc(normalCopula(th.n),    c("norm", "norm"), N01m))
> X.t <- rMvdc(n, mvdc = mvdc(tCopula(th.t, df = 3), c("norm", "norm"), N01m))
> X.c <- rMvdc(n, mvdc = mvdc(claytonCopula(th.c),   c("norm", "norm"), N01m))
> X.g <- rMvdc(n, mvdc = mvdc(gumbelCopula(th.g),    c("norm", "norm"), N01m))
> ##' @title Function for producing one scatter plot
> ##' @param X data
> ##' @param qu (lower and upper) quantiles to consider
> ##' @param lim (x- and y-axis) limits
> ##' @param ... additional arguments passed to the underlying plot functions
> ##' @return invisible()
> plotCorners <- function(X, qu, lim, smooth = FALSE, ...)
  {
      plot(X, xlim = lim, ylim = lim, xlab = quote(X[1]), ylab = quote(X[2]),
           col = adjustcolor("black", 0.5), ...) # or pch = 16
      abline(h = qu, v = qu, lty = 2, col = adjustcolor("black", 0.6))
      ll <- sum(apply(X <= qu[1], 1, all)) * 100 / n
      ur <- sum(apply(X >= qu[2], 1, all)) * 100 / n
      mtext(sprintf("Lower left: %.2f%%, upper right: %.2f%%", ll, ur),
            cex = 0.9, side = 1, line = -1.5)
      invisible()
  }
> ## Plots
> a. <- 0.005
> q <- qnorm(c(a., 1 - a.)) # a- and (1-a)-quantiles of N(0,1)
> lim <- range(q, X.n, X.t, X.c, X.g)
> lim <- c(floor(lim[1]), ceiling(lim[2]))
> plotCorners(X.n, qu = q, lim = lim, cex = 0.4)
> plotCorners(X.t, qu = q, lim = lim, cex = 0.4)
> plotCorners(X.c, qu = q, lim = lim, cex = 0.4)
> plotCorners(X.g, qu = q, lim = lim, cex = 0.4)
```

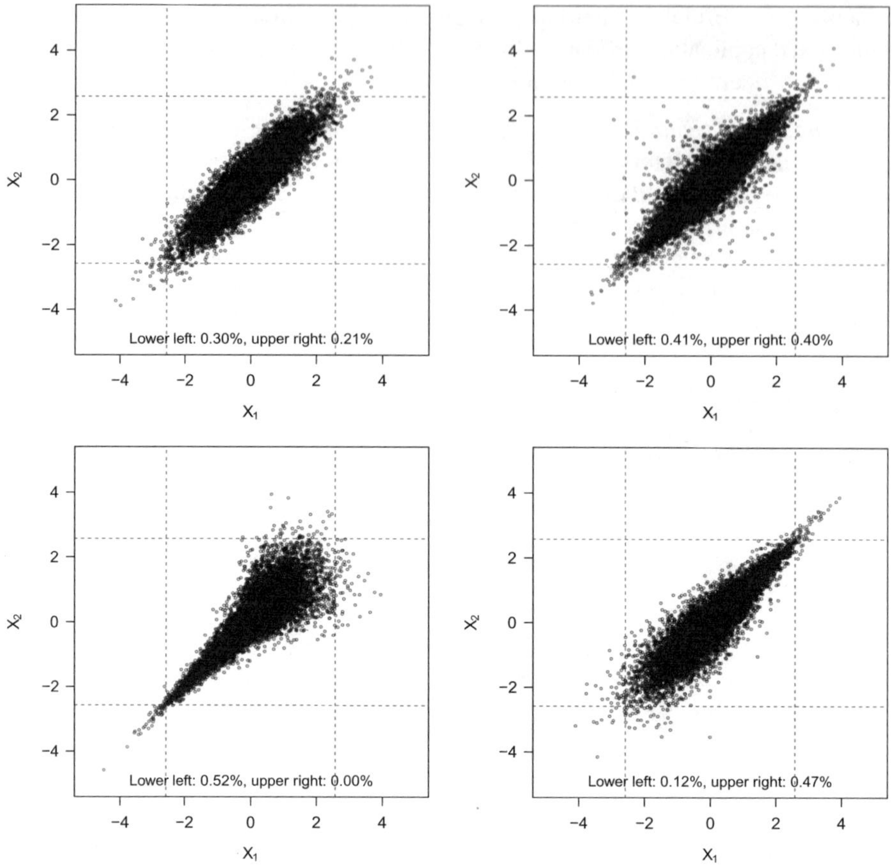

Fig. 2.23 Scatter plots of four samples of size $n = 10{,}000$ from bivariate distributions with N(0, 1) margins and a Kendall's tau of 0.7 but different copulas. The top left plot corresponds to the normal copula, the top right to the t_3 copula, the bottom left to the Clayton copula, and the bottom right to the Gumbel–Hougaard copula. The dashed lines indicate the N(0, 1) 0.005- and 0.995-quantiles. The percentages correspond to the proportions of generated bivariate observations which fall in the lower left and upper right corners

The scatter plots produced with the auxiliary function `plotCorners()` are shown in Fig. 2.23. As one can notice from the lower left and upper right corners of the four scatter plots, the four constructed bivariate distributions have a significantly different behavior in their bivariate tails although they all share the same margins and even the same Kendall's tau. The differences in these bivariate tails are solely a consequence of the underlying copulas. Informally, in the case of the normal copula (top left), there seems to be no strong dependence in the lower left and upper right corners, while the t copula with three degrees of freedom (top right) appears to have more mass and more structure in the lower and upper tail (thus hinting at the notion of "tail dependence"). In the case of the Clayton copula (bottom left), small values of one variable frequently seem to be accompanied by small values of the other variable (thus hinting at "lower tail dependence"), whereas large values of one variable do not necessarily seem to imply large values of the other variable (thus hinting at "upper tail independence"). For the Gumbel–Hougaard copula (bottom right), this behavior seems to be reversed. □

The coefficients of tail dependence, see Nelsen (2006, Sect. 5.4), are limits of conditional probabilities of quantile exceedances. In the case of upper tail dependence, for instance, one considers the limit for q tending to 1 of the probability that X_2 exceeds its q-quantile given that X_1 exceeds its q-quantile.

Definition 2.6.13 (Coefficients of Tail Dependence) Let (X_1, X_2) be a random vector with marginal dfs F_1, F_2. Provided that the limits exist, the *coefficient of lower* and *upper tail dependence* of X_1 and X_2 are defined by

$$\lambda_l = \lambda_l(X_1, X_2) = \lim_{q \downarrow 0} \mathbb{P}\big(X_2 \le F_2^{\leftarrow}(q) \mid X_1 \le F_1^{\leftarrow}(q)\big),$$

$$\lambda_u = \lambda_u(X_1, X_2) = \lim_{q \uparrow 1} \mathbb{P}\big(X_2 > F_2^{\leftarrow}(q) \mid X_1 > F_1^{\leftarrow}(q)\big),$$

respectively. If $\lambda_l(X_1, X_2) \in (0, 1]$ (respectively, $\lambda_u(X_1, X_2) \in (0, 1]$), then X_1 and X_2 are said to be *lower* (respectively, *upper*) *tail dependent*. If $\lambda_l(X_1, X_2) = 0$ (respectively, $\lambda_u(X_1, X_2) = 0$), X_1 and X_2 are *lower* (respectively, *upper*) *tail independent*. □

Like rank correlations, when the marginal dfs F_1 and F_2 are continuous, coefficients of tail dependence are measures of association that only depend on the underlying copula.

Proposition 2.6.14 (Representation of the Coefficients of Tail Dependence) *Let (X_1, X_2) be a bivariate random vector with continuous marginal dfs and copula C. Then,*

$$\lambda_l = \lambda_l(C) = \lim_{q \downarrow 0} \frac{C(q, q)}{q} \quad \textit{and}$$

$$\lambda_u = \lambda_u(C) = \lim_{q \uparrow 1} \frac{1 - 2q + C(q, q)}{1 - q} = \lim_{q \downarrow 0} \frac{\bar{C}(q, q)}{q}. \tag{2.26}$$

□

If the copula has a simple closed form, calculation of the coefficients of tail dependence λ_l and λ_u is often straightforward, typically by using (2.26) and l'Hôpital's rule. For example, for Archimedean copulas with differentiable generator ψ

(see Sect. 3.2), one has the typically easy to evaluate formulas

$$\lambda_l = \lim_{t\to\infty} \frac{\psi(2t)}{\psi(t)} = 2 \lim_{t\to\infty} \frac{\psi'(2t)}{\psi'(t)} \quad \text{and}$$
$$\lambda_u = 2 - \lim_{t\downarrow 0} \frac{1-\psi(2t)}{1-\psi(t)} = 2 - 2\lim_{t\downarrow 0} \frac{\psi'(2t)}{\psi'(t)}. \tag{2.27}$$

For a Clayton copula with parameter $\theta \in (0, \infty)$, we obtain that

$$\lambda_l = 2^{-1/\theta} \quad \text{and} \quad \lambda_u = 0.$$

In other words, when used to model positive association, bivariate Clayton copulas are lower tail dependent but upper tail independent. The strength of lower tail dependence tends to 1 as $\theta \to \infty$ since Clayton copulas converge to the comonotone copula M as $\theta \to \infty$. For a Gumbel–Hougaard copula with parameter θ, one has

$$\lambda_l = 0 \quad \text{and} \quad \lambda_u = 2 - 2^{1/\theta}, \qquad \theta \in [1, \infty).$$

A Gumbel–Hougaard copula is thus upper tail dependent unless it is the independence copula, and is always lower tail independent. The strength of upper tail dependence tends to 1 as $\theta \to \infty$ since Gumbel–Hougaard copulas converge to the comonotone copula M as $\theta \to \infty$.

Equation (2.26) implies that the coefficients of lower and upper tail dependence for a radially symmetric copula are equal. Using known results about conditional normal and t distributions, one can obtain the coefficients of tail dependence for normal and t copulas. For the bivariate normal copula C^{n}_{ρ}, this coefficient $\lambda = \lambda_l = \lambda_u$ of (lower or upper) tail dependence is given by

$$\lambda = 1(\rho = 1).$$

Normal copulas are thus tail independent unless equal to the upper Fréchet–Hoeffding bound M. For the bivariate t copula $C^t_{\rho,\nu}$, this coefficient is given by

$$\lambda = 2t_{\nu+1}\left(-\sqrt{\frac{(\nu+1)(1-\rho)}{1+\rho}}\right),$$

where $t_{\nu+1}$ denotes the df of the Student t distribution with $\nu+1$ degrees of freedom.

Example 2.6.15 (Computing the Coefficients of Tail Dependence) The key function for computing the coefficients of tail dependence is `lambda()`:

```
> ## Clayton copula
> theta <- 3
> lam.c <- lambda(claytonCopula(theta))
> stopifnot(all.equal(lam.c[["lower"]], 2^(-1/theta)),
            all.equal(lam.c[["upper"]], 0))
> ## Gumbel--Hougaard copula
> lam.g <- lambda(gumbelCopula(theta))
> stopifnot(all.equal(lam.g[["lower"]], 0),
            all.equal(lam.g[["upper"]], 2-2^(1/theta)))
> ## Normal copula
> rho <- 0.7
> nu <- 3
> lam.n <- lambda(normalCopula(rho))
> stopifnot(all.equal(lam.n[["lower"]], 0),
            all.equal(lam.n[["lower"]], lam.n[["upper"]]))
> ## t copula
> lam.t <- lambda(tCopula(rho, df = nu))
> stopifnot(all.equal(lam.t[["lower"]],
                      2*pt(-sqrt((nu+1)*(1-rho)/(1+rho)), df = nu + 1)),
            all.equal(lam.t[["lower"]], lam.t[["upper"]]))
```

□

Example 2.6.16 (Tail Dependence of t Copulas) For the t copula $C^t_{\rho,\nu}$ with correlation parameter ρ and degrees of freedom ν, Fig. 2.24, produced with the following code, displays graphs of the coefficient of tail dependence λ as a function of ρ (left) and ν (right):

```
> ## Coefficient of tail dependence as a function of rho
> rho <- seq(-1, 1, by = 0.01)
> nu <- c(3, 4, 8, Inf)
> n.nu <- length(nu)
> lam.rho <- sapply(nu, function(nu.) # (rho, nu) matrix
      sapply(rho, function(rho.) lambda(tCopula(rho., df = nu.))[["lower"]]))
> expr.rho <- as.expression(lapply(1:n.nu, function(j)
      bquote(nu == .(if(nu[j] == Inf) quote(infinity) else nu[j]))))
> matplot(rho, lam.rho, type = "l", lty = 1, lwd = 2, col = 1:n.nu,
          xlab = quote(rho), ylab = quote(lambda))
> legend("topleft", legend = expr.rho, bty = "n", lwd = 2, col = 1:n.nu)
> ## Coefficient of tail dependence as a function of nu
> nu. <- c(seq(3, 12, by = 0.2), Inf)
> rho. <- c(-1, -0.5, 0, 0.5, 1)
> n.rho <- length(rho.)
> lam.nu <- sapply(rho., function(rh) # (nu, rho) matrix
      sapply(nu., function(nu) lambda(tCopula(rh, df = nu))[["lower"]]))
> expr <- as.expression(lapply(1:n.rho, function(j) bquote(rho == .(rho.[j]))))
> matplot(nu., lam.nu, type = "l", lty = 1, lwd = 2, col = 1:n.rho,
          xlab = quote(nu), ylab = quote(lambda))
> legend("right", expr, bty = "n", lwd = 2, col = 1:n.rho)
```

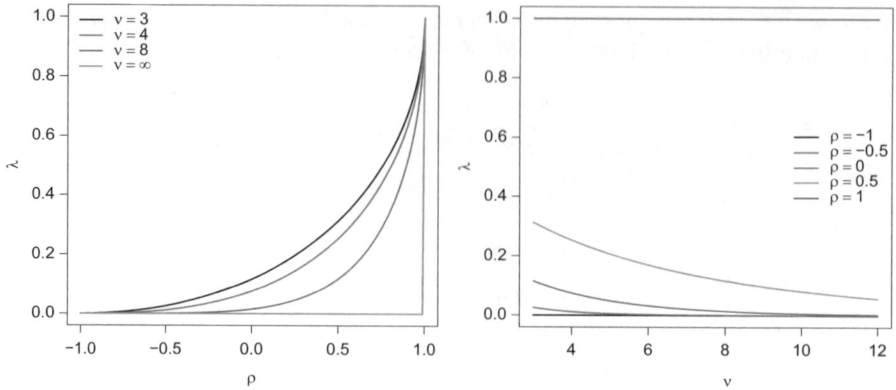

Fig. 2.24 Coefficient of tail dependence $\lambda(= \lambda_l = \lambda_u)$ of $C^t_{\rho,\nu}$ as a function of the correlation parameter ρ (left) and the degrees of freedom ν (right)

For fixed, finite ν, tail dependence increases as ρ increases; see the left-hand side of Fig. 2.24. For fixed $|\rho| < 1$, tail dependence increases as ν decreases; see the right-hand side of Fig. 2.24. □

The following two examples address tail probabilities and highlight tail differences between t copulas and normal copulas.

Example 2.6.17 (Effect of ρ and ν on $u \mapsto \mathbb{P}(U_1 > u, U_2 > u)$ for t Copulas) We now illustrate how the correlation parameter ρ and the degrees of freedom ν affect the (joint) tail probability function $u \mapsto \mathbb{P}(U_1 > u, U_2 > u)$ of the t copula $C^t_{\rho,\nu}$. Note that, by radial symmetry, $\mathbb{P}(U_1 > u, U_2 > u) = \mathbb{P}(1 - U_1 > u, 1 - U_2 > u) = \mathbb{P}(U_1 \le 1 - u, U_2 \le 1 - u) = C^t_{\rho,\nu}(1 - u, 1 - u)$. The left-hand side of Fig. 2.25 shows that the larger ρ for fixed ν or the smaller ν for fixed ρ, the larger $u \mapsto \mathbb{P}(U_1 > u, U_2 > u)$. The right-hand side shows this tail probability function divided by the one of the normal copula (that is, graphs of $u \mapsto C^t_{\rho,\nu}(1 - u, 1 - u)/C^{\mathrm{n}}_\rho(1 - u, 1 - u)$) and implies that the difference between the tails of the normal and the t copula becomes more pronounced the smaller ρ or the smaller ν or the larger u, all other factors remaining unchanged.

Figure 2.25 is produced with the following code:

```
> ## Note: All calculations here are deterministic
> u <- seq(0.95, to = 0.9999, length.out = 128) # levels u of P(U_1> u, U_2> u)
> rho <- c(0.75, 0.5) # correlation parameter rho
> nu <- c(3, 4, 8, Inf) # degrees of freedom
> len <- length(rho) * length(nu)
> tail.prob <- matrix(u, nrow = length(u), ncol = 1 + len) # tail probabilities
> expr <- vector("expression", length = len) # vector of expressions
```

```
> ltys <- cols <- numeric(len) # line types and colors
> for(i in seq_along(rho)) { # rho
      for(j in seq_along(nu)) { # degrees of freedom
          k <- length(nu) * (i - 1) + j
          ## Create the copula
          cop <- ellipCopula("t", param = rho[i], df = nu[j])
          ## Evaluate P(U_1 > u, U_2 > u) = P(U_1 <= 1 - u, U_2 <= 1 - u)
          tail.prob[,k+1] <- pCopula(cbind(1 - u, 1 - u), copula = cop)
          ## Create plot information
          expr[k] <- as.expression(
              substitute(group("(",list(rho, nu), ")") ==
                         group("(", list(RR, NN), ")"),
                         list(RR = rho[i],
                              NN = if(is.infinite(nu[j]))
                                       quote(infinity) else nu[j])))
          ltys[k] <- length(rho) - i + 1
          cols[k] <- j
      }
  }
> ## Standardize w.r.t. Gauss case
> tail.prob.fact <- tail.prob # for comparison to Gauss case
> tail.prob.fact[,2:5] <- tail.prob[,2:5] / tail.prob[,5]
> tail.prob.fact[,6:9] <- tail.prob[,6:9] / tail.prob[,9]
> ## Plot tail probabilities
> matplot(tail.prob[,1], tail.prob[,-1], type = "l", lwd = 2, lty = ltys,
          col = cols, xlab = quote(P(U[1]>u, U[2]>u)~~"as a function of u"),
          ylab = "")
> legend("topright", expr, bty = "n", lwd = 2, lty = ltys, col = cols)
> ## Plot standardized tail probabilities
> matplot(tail.prob.fact[,1], tail.prob.fact[,-1], log = "y", type = "l",
          lty = ltys, col = cols, lwd = (wd <- 2*c(1,1,1,1.6,1,1,1,1)),
          xlab = quote(P(U[1]>u, U[2]>u)~~
                       "as a function of u standardized by Gauss case"),
          ylab = "")
> legend("topleft", expr, bty = "n", lwd = wd, lty = ltys, col = cols)
```

□

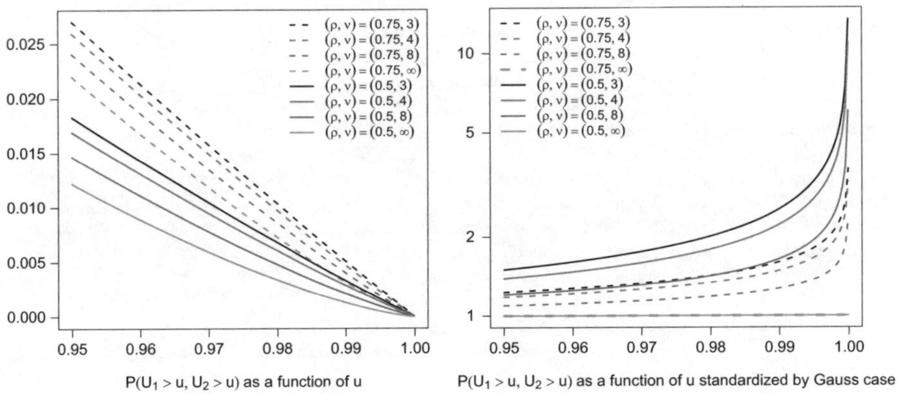

Fig. 2.25 Tail probability $\mathbb{P}(U_1 > u, U_2 > u)$ as a function of u for the t copula $C^t_{\rho,\nu}$ with correlation parameter ρ and degrees of freedom ν (left) and divided by the corresponding tail probability of the normal copula (right)

Example 2.6.18 (Effect of ρ and ν on $d \mapsto \mathbb{P}(U_1 > 0.99, \ldots, U_d > 0.99)$ for t Copulas) This example illustrates how the correlation parameter ρ and the degrees of freedom ν affect the (joint) tail probability function $d \mapsto \mathbb{P}(U_1 > 0.99, \ldots, U_d > 0.99)$ for $(U_1, \ldots, U_d) \sim C^t_{\rho,\nu}$, where $C^t_{\rho,\nu}$ denotes the homogeneous t copula. Clearly, by radial symmetry, $\mathbb{P}(U_1 > 0.99, \ldots, U_d > 0.99)$ is equal to $C^t_{\rho,\nu}(0.01, \ldots, 0.01)$. The left-hand side of Fig. 2.26 shows that the larger ρ for fixed ν or the smaller ν for fixed ρ, the larger $d \mapsto \mathbb{P}(U_1 > 0.99, \ldots, U_d > 0.99)$. The right-hand side plot displays the corresponding graphs of the function $d \mapsto C^t_{\rho,\nu}(0.01, \ldots, 0.01)/C^{\mathrm{n}}_{\rho}(0.01, \ldots, 0.01)$ and implies that the difference between the tails of the normal and the t copula becomes more pronounced the smaller ρ or the smaller ν or the larger d, all other factors remaining unchanged. Note that, for $d \geq 3$, t and normal copulas are evaluated by simulation (with an integration method known as randomized quasi-Monte Carlo based on quasi-random numbers, see also Example 2.7.3 for the latter) which may affect both precision and run time.

```
> d <- 2:20 # dimensions
> u <- 0.99 # level u of P(U_1 > u, ..., U_d > u)
> tail.pr.d <- matrix(d, nrow = length(d), ncol = 1+len)# tail prob; P[,1] = d
> set.seed(271) # set seed due to MC randomness here
> for(i in seq_along(rho)) { # rho
      for(j in seq_along(nu)) { # degrees of freedom
          k <- length(nu) * (i-1) + j
          for(l in seq_along(d)) { # dimension
              ## Create the copula
              cop <- ellipCopula("t", param = rho[i], dim = d[l], df = nu[j])
              ## Evaluate P(U_1 > u,...,U_d > u) = P(U_1 <= 1-u,...,U_d <= 1-u)
              tail.pr.d[l, k+1] <- pCopula(rep(1-u, d[l]), copula = cop)
          }
      }
  }
> ## Standardize w.r.t. Gauss case
> tail.pr.d.fact <- tail.pr.d # for comparison to Gauss case
> tail.pr.d.fact[,2:5] <- tail.pr.d[,2:5] / tail.pr.d[,5]
> tail.pr.d.fact[,6:9] <- tail.pr.d[,6:9] / tail.pr.d[,9]
> ## Plot tail probabilities
> matplot(tail.pr.d[,1], tail.pr.d[,-1], type = "l", log = "y", yaxt = "n",
          lty = ltys, col = cols, lwd = 2, ylab = "",
          xlab = quote(P(U[1] > 0.99, ..., U[d] > 0.99)~~
                       "as a function of d"))
> sfsmisc::eaxis(2, cex.axis = 0.8)
> axis(1, at = 2)
> legend("topright",   expr[1:4], bty="n", lty=ltys[1:4], col=cols[1:4], lwd=2)
> legend("bottomleft", expr[5:8], bty="n", lty=ltys[5:8], col=cols[5:8], lwd=2)
> ## Plot standardized tail probabilities
> matplot(tail.pr.d.fact[,1], tail.pr.d.fact[,-1], log = "y", type = "l",
          las = 1, lty = ltys, col = cols,
          lwd = (wd <- 2*c(1,1,1,1.6,1,1,1,1)), ylab = "",
          xlab = quote(P(U[1] > 0.99,..,U[d] > 0.99)~~
                       "as a function of d standardized by Gauss case"))
> legend("topleft", expr, bty = "n", lty = ltys, lwd = wd, col = cols)
> axis(1, at = 2)
```

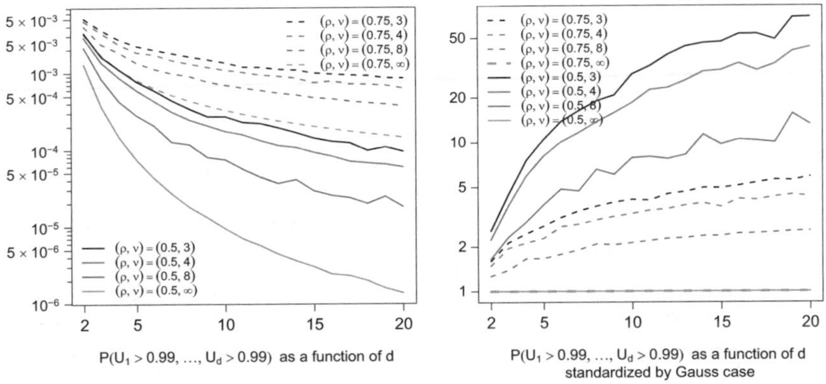

Fig. 2.26 Tail probability $\mathbb{P}(U_1 > 0.99, \ldots, U_d > 0.99)$ as a function of the dimension d for homogeneous t copulas with correlation parameter ρ and degrees of freedom ν (left) and divided by the corresponding tail probability for the normal copula (right)

As a possible application of the previous results, let us consider a setting similar to the one of McNeil et al. (2015, Example 7.40): Let us assume that we are given five daily negative log-returns $\boldsymbol{X} = (X_1, \ldots, X_5)$ with fixed continuous marginal dfs and fixed common pairwise Kendall's tau equal to 1/3. In addition, suppose that we are unsure whether a normal or a t copula should be used as underlying dependence structure C. If C is the homogeneous normal copula $C^{\mathrm{n}}_{0.5}$, the probability that, on any day, all five negative log-returns lie above their $u = 0.99$ quantiles is (up to the Monte Carlo error involved in evaluating $C^{\mathrm{n}}_{0.5}$ in three or more dimensions) $\mathbb{P}(X_1 > F_1^{\leftarrow}(u), \ldots, X_5 > F_5^{\leftarrow}(u)) = \mathbb{P}(F_1(X_1) > u, \ldots, F_5(X_5) > u) = C^{\mathrm{n}}_{0.5}(1-u, \ldots, 1-u)$, where the last equality follows by radial symmetry. Such an event happens about once every $1/C^{\mathrm{n}}_{0.5}(1-u, \ldots, 1-u)$ trading days on average, so, approximately, once every $1/(260\, C^{\mathrm{n}}_{0.5}(1-u, \ldots, 1-u))$ years assuming 260 trading days in a year. However, if the copula of $\boldsymbol{X}$ is $C^{t}_{0.5,3}$, the right-hand side of Fig. 2.26 implies that such an event will happen approximately 10 times more often, that is, approximately once every $1/(2600\, C^{\mathrm{n}}_{0.5}(1-u, \ldots, 1-u))$ years. Note that this factor gets larger the larger d.

The aforementioned calculations can be carried out using the following code:

```
> ## Joint exceedance probability under the normal copula
> d <- 5
> rho <- 0.5
> u <- 0.99
> set.seed(271)
> ex.prob.norm <- pCopula(rep(1 - u, d), copula = normalCopula(rho, dim = d))
> 1 / (260 * ex.prob.norm) # ~ 51.72 years
```

```
[1] 51.72134
```

```
> ## Joint exceedance probability under the t copula model with 3 df
> ## 1) Via scaling of the probability obtained from the normal copula
> ##    Note that the scaling factor was read off from the previous plot
> 1 / (2600 * ex.prob.norm) # ~ 5.17 years
```

```
[1] 5.172134
```

```
> ## 2) Directly using the t copula
> ex.prob.t3 <- pCopula(rep(1 - u, d), copula = tCopula(rho, dim = d, df = 3))
> 1 / (260 * ex.prob.t3) # ~ 5.91 years
```

```
[1] 5.911367
```

□

2.7 Rosenblatt Transform and Conditional Sampling

Similar to multivariate dfs, one can consider conditional dfs derived from copulas. To this end, let C be a d-dimensional copula, let $\boldsymbol{U} \sim C$, and, for $j \in \{2, \dots, d\}$, $u_1, \dots, u_{j-1} \in (0, 1)$ and $u_j \in [0, 1]$, let us introduce the notation

$$C_{j|1,\dots,j-1}(u_j \mid u_1, \dots, u_{j-1}) = \mathbb{P}(U_j \le u_j \mid U_1 = u_1, \dots, U_{j-1} = u_{j-1}). \tag{2.28}$$

Fix $u_1, \dots, u_{j-1} \in (0, 1)$. Provided the conditional probability on the right-hand side of the previous displayed equation is well defined, the function $C_{j|1,\dots,j-1}(\cdot \mid u_1, \dots, u_{j-1})$ is a df on $[0, 1]$. As it shall be illustrated in the forthcoming examples, conditional dfs of this form are of interest for sampling and goodness-of-fit testing.

To understand how (2.28) can be evaluated, we shall consider in the rest of this section that C has a density c such that $c(\boldsymbol{u}) > 0$ for all $\boldsymbol{u} \in (0, 1)^d$. For each $j \in \{2, \dots, d\}$, denote by $C^{(1,\dots,j)}$ the j-variate margin of C corresponding to the first j arguments, that is, $C^{(1,\dots,j)}(u_1, \dots, u_j) = C(u_1, \dots, u_j, 1, \dots, 1)$, $u_1, \dots, u_j \in [0, 1]$, and by $c^{(1,\dots,j)}$ the corresponding density. One can then show that, for any $u_1, \dots, u_{j-1} \in (0, 1)$ and $u_j \in [0, 1]$,

$$C_{j|1,\dots,j-1}(u_j \mid u_1, \dots, u_{j-1}) = \frac{\int_0^{u_j} c^{(1,\dots,j)}(u_1, \dots, u_{j-1}, u)\,\mathrm{d}u}{c^{(1,\dots,j-1)}(u_1, \dots, u_{j-1})}, \tag{2.29}$$

with the convention that $c^{(1)} = 1$. It follows from (2.29) that, for any $u_1, \dots, u_{j-1} \in (0, 1)$, $C_{j|1,\dots,j-1}(\cdot \mid u_1, \dots, u_{j-1})$ is a continuous, strictly increasing df on $[0, 1]$.

Some thought reveals that the display in (2.29) can be conveniently rewritten as

$$C_{j|1,\dots,j-1}(u_j \mid u_1, \dots, u_{j-1}) = \frac{\mathrm{D}_{j-1,\dots,1}\, C^{(1,\dots,j)}(u_1, \dots, u_j)}{\mathrm{D}_{j-1,\dots,1}\, C^{(1,\dots,j-1)}(u_1, \dots, u_{j-1})}, \tag{2.30}$$

where, in the numerator, we have used the notation

$$\mathrm{D}_{j-1,\dots,1}\, C^{(1,\dots,j)}(u_1,\dots,u_j) = \frac{\partial^{j-1}}{\partial u_{j-1}\dots\partial u_1} C^{(1,\dots,j)}(u_1,\dots,u_j),$$

and similarly for the denominator. Formula (2.30) may still hold if C does not admit a density; see, for instance, Schmitz (2003, Theorem 2.27) for more details. For $d = 2$, (2.30) reads $C_{2|1}(u_2 \mid u_1) = \mathrm{D}_1\, C(u_1, u_2)$, while for $d = 3$ and $j = 3$, (2.30) reads $C_{3|1,2}(u_3 \mid u_1, u_2) = \mathrm{D}_{2,1}\, C(u_1, u_2, u_3) / \mathrm{D}_{2,1}\, C^{(1,2)}(u_1, u_2)$.

Example 2.7.1 (Evaluation of and Sampling from $C_{j|1,\dots,j-1}(\cdot \mid u_1, \dots, u_{j-1})$) In R, we can evaluate $C_{j|1,\dots,j-1}(\cdot \mid u_1, \dots, u_{j-1})$ with the function `cCopula()`. Note that its argument `indices` is a vector whose components specify the indices j for which $C_{j|1,\dots,j-1}(\cdot \mid u_1, \dots, u_{j-1})$ is to be computed. As an example, consider $j = 2$ and a t copula $C^t_{\rho,\nu}$ with $\nu = 3.5$ degrees of freedom and ρ chosen such that Kendall's tau equals 0.5:

```
> ## Define the copula
> nu <- 3.5
> theta <- iTau(tCopula(df = nu), tau = 0.5)
> tc <- tCopula(theta, df = nu)
> ## Evaluate the df C(.|u_1) at u for several u_1
> u <- c(0.05, 0.3, 0.7, 0.95)
> u2 <- seq(0, 1, by = 0.01)
> ccop <- sapply(u, function(u.)
      cCopula(cbind(u., u2), copula = tc, indices = 2))
> ## Evaluate the function C(u_2|.) at u for several u_2
> u1 <- seq(0, 1, by = 0.01)
> ccop. <- sapply(u, function(u.)
      cCopula(cbind(u1, u.), copula = tc, indices = 2))
```

The left-hand side of Fig. 2.27 shows graphs of the df $C_{2|1}(\cdot \mid u_1)$ for several u_1:

```
> matplot(ccop, type = "l", lty = 1, lwd = 2,
          col = (cols <- seq_len(ncol(ccop))), ylab = "",
          xlab = substitute(C["2|1"](u[2]~"|"~u[1])~~"as a function of"~
                            u[2]~"for a"~{C^italic(t)}[list(rho,nu)]~"copula",
                            list(nu = nu)))
> legend("bottomright", bty = "n", lwd = 2, col = cols,
         legend = as.expression(lapply(seq_along(u), function(j)
             substitute(u[1] == u1, list(u1 = u[j])))))
```

The right-hand side of Fig. 2.27 shows graphs of the function $C_{2|1}(u_2 \mid \cdot)$ for several u_2:

```
> matplot(ccop., type = "l", lty = 1, lwd = 2,
          col = (cols <- seq_len(ncol(ccop.))), ylab = "",
          xlab = substitute(C["2|1"](u[2]~"|"~u[1])~~"as a function of"~
                            u[1]~"for a"~{C^italic(t)}[list(rho,nu)]~"copula",
                            list(nu = nu)))
> legend("center", bty = "n", lwd = 2, col = cols,
         legend = as.expression(lapply(seq_along(u), function(j)
             substitute(u[2] == u2, list(u2 = u[j])))))
```

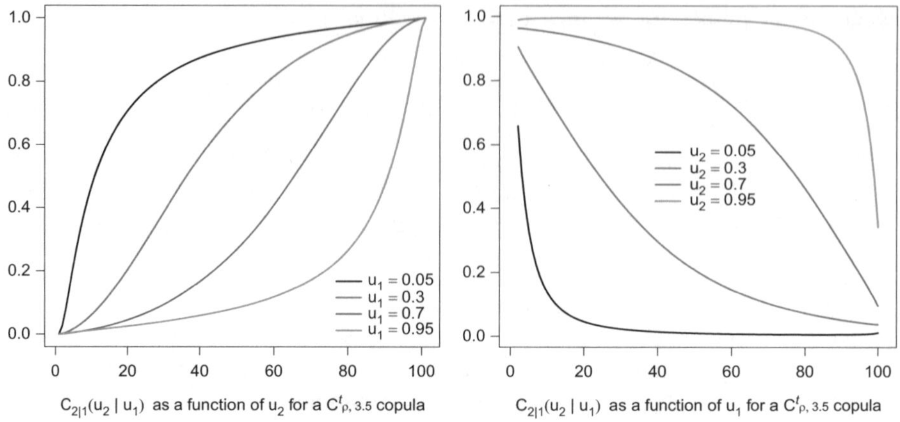

Fig. 2.27 Graphs of the df $C_{2|1}(\cdot \,|\, u_1)$ for several u_1 (left) and of the function $C_{2|1}(u_2 \,|\, \cdot)$ for several u_2 (right) for a t copula $C^t_{\rho,\nu}$ with $\nu = 3.5$ degrees of freedom and $\rho \approx 0.7071$

A sample from the df $C_{2|1}(\cdot \,|\, u_1)$ based on the t copula under consideration can be obtained using the inversion method. To evaluate the quantile function $C^{\leftarrow}_{2|1}(\cdot \,|\, u_1)$ corresponding to $C_{2|1}(\cdot \,|\, u_1)$, one can use the function `cCopula()` with argument `inverse = TRUE`. The values 0.05 and 0.95 are considered for u_1 hereafter:

```
> ## Sample from C_{2|1}(.|u_1)
> set.seed(271)
> u2 <- runif(1000)
> ## Small u_1
> u1 <- 0.05
> U2 <- cCopula(cbind(u1, u2), copula = tc, indices = 2, inverse = TRUE)
> ## Large u_1
> u1. <- 0.95
> U2. <- cCopula(cbind(u1., u2), copula = tc, indices = 2, inverse = TRUE)
> plot(U2, ylab = substitute(U[2]~"|"~U[1]==u, list(u = u1)))
> plot(U2., ylab = substitute(U[2]~"|"~U[1]==u, list(u = u1.)))
```

The generated sample for $u_1 = 0.05$ (respectively, $u_1 = 0.95$) is displayed on the left-hand side (respectively, right-hand side) of Fig. 2.28.

Note that the function `cCopula()` is implemented for the Archimedean and elliptical copula families available in the R package `copula`. These classes of copulas will be formally introduced in Chap. 3. □

Given a d-dimensional copula C, the functions $(u_1, \ldots, u_j) \mapsto C_{j|1,\ldots,j-1}(u_j \,|\, u_1, \ldots, u_{j-1})$, $j \in \{2, \ldots, d\}$, appear in a transformation due to Rosenblatt (1952). This transformation, denoted by R_C, transforms a vector $\boldsymbol{u} \in (0,1)^d$ into a vector $\boldsymbol{u}' \in (0,1)^d$ such that

$$u'_1 = u_1,$$

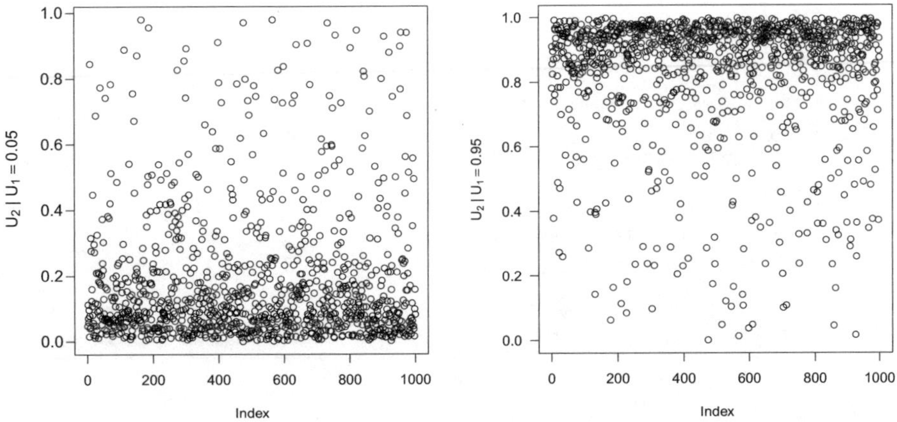

Fig. 2.28 Samples from $C_{2|1}(\cdot \mid u_1)$ for $u_1 = 0.05$ (left) and $u_1 = 0.95$ (right) for a t copula $C^t_{\rho,\nu}$ with $\nu = 3.5$ degrees of freedom and $\rho \approx 0.7071$

$$u'_2 = C_{2|1}(u_2 \mid u_1),$$
$$\vdots$$
$$u'_d = C_{d|1,\ldots,d-1}(u_d \mid u_1, \ldots, u_{d-1}).$$

The *Rosenblatt transform* of a d-dimensional random vector $\boldsymbol{U} \sim C$ is then the d-dimensional random vector $\boldsymbol{U}' = R_C(\boldsymbol{U})$. A property of the Rosenblatt transform is that $R_C(\boldsymbol{U}) \sim \Pi$. More precisely, given a d-dimensional copula C' and $\boldsymbol{U} \sim C$, $R_{C'}(\boldsymbol{U}) \sim \Pi$ if and only if $C' = C$. The latter result can be used for a graphical goodness-of-fit test as demonstrated in the following example.

Example 2.7.2 (Rosenblatt Transform) Let us illustrate how the transformation R_C can be used as a basic graphical goodness-of-fit test. To this end, we first sample from the bivariate Gumbel–Hougaard copula C_2^{GH}. We then apply the transformations $R_{C_2^{\mathrm{GH}}}$ and $R_{C_4^{\mathrm{GH}}}$ to each generated bivariate observation:

```
> ## Sample from a Gumbel-Hougaard copula
> gc <- gumbelCopula(2)
> set.seed(271)
> U <- rCopula(1000, copula = gc)
> ## Apply the transformation R_C with the correct copula
> U. <- cCopula(U, copula = gc)
> ## Apply the transformation R_C with a wrong copula
> gc. <- setTheta(gc, value = 4) # larger theta
> U.. <- cCopula(U, copula = gc.)
> plot(U., xlab = quote(U*"'"[1]), ylab = quote(U*"'"[2]))
> plot(U.., xlab = quote(U*"'"[1]), ylab = quote(U*"'"[2]))
```

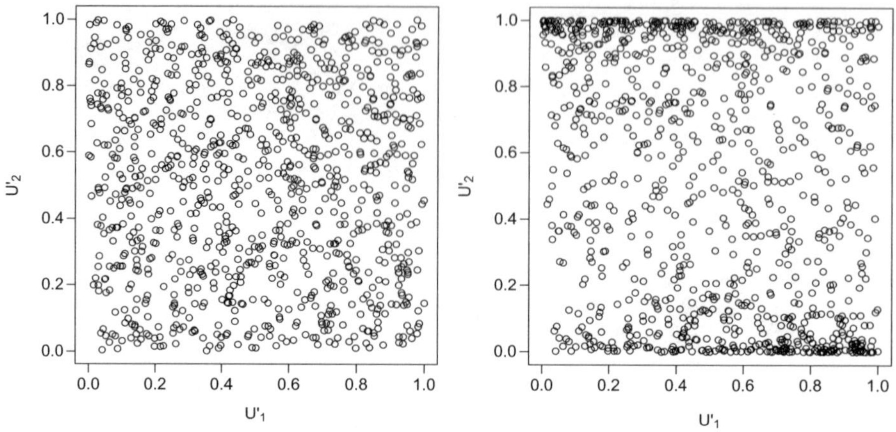

Fig. 2.29 Starting from 1000 independent observations of $\boldsymbol{U} \sim C_2^{\mathrm{GH}}$, scatter plots of the corresponding realizations of $R_{C_2^{\mathrm{GH}}}(\boldsymbol{U})$ (left) and of $R_{C_4^{\mathrm{GH}}}(\boldsymbol{U})$ (right; the departure from bivariate uniformity is clearly visible)

The scatter plot on the left-hand side (respectively, right-hand side) of Fig. 2.29 displays the sample transformed with the copula C_2^{GH} (respectively, C_4^{GH}). The departure from Π, that is, bivariate standard uniformity, is clearly visible in the right-hand side plot. For an application of this idea to higher dimensions, see Hofert and Mächler (2014). □

The Rosenblatt transform is closely related to the so-called *conditional distribution method* for sampling from a given d-dimensional copula C. As previously, given $u_1, \dots, u_{j-1} \in (0, 1)$, the quantile function of the df $C_{j|1,\dots,j-1}(\cdot \mid u_1, \dots, u_{j-1})$ will be denoted by $C^{\leftarrow}_{j|1,\dots,j-1}(\cdot \mid u_1, \dots, u_{j-1})$. Then, let $R^{\leftarrow}_C$ denote the mapping that transforms a vector $\boldsymbol{u}' \in (0, 1)^d$ into a vector $\boldsymbol{u} \in (0, 1)^d$ such that

$$
\begin{aligned}
u_1 &= u'_1, \\
u_2 &= C^{\leftarrow}_{2|1}(u'_2 \mid u_1), \\
&\ \ \vdots \\
u_d &= C^{\leftarrow}_{d|1,\dots,d-1}(u'_d \mid u_1, \dots, u_{d-1}).
\end{aligned}
$$

The notation $R^{\leftarrow}_C$ stems from the fact that if $\boldsymbol{U}' \sim \Pi$, then $\boldsymbol{U} = R^{\leftarrow}_C(\boldsymbol{U}') \sim C$; see, for instance, Embrechts et al. (2003) and Hofert (2010, Sect. 1.8). The latter property is at the heart of the conditional distribution method. More generally, it can be verified that, for $\boldsymbol{U}' \sim \Pi$ and any $\boldsymbol{U} \sim C$,

$$R^{\leftarrow}_C(R_C(\boldsymbol{U})) = \boldsymbol{U} \quad \text{and} \quad R_C(R^{\leftarrow}_C(\boldsymbol{U}')) = \boldsymbol{U}', \quad \text{almost surely.} \tag{2.31}$$

It is important to note that even if the conditional dfs $C_{j|1,...,j-1}(\cdot \mid u_1, \dots, u_{j-1})$ have explicit expressions, their corresponding quantile functions $C^{\leftarrow}_{j|1,...,j-1}(\cdot \mid u_1, \dots, u_{j-1})$ might not be analytically available, in which case their evaluation might require numerical root finding.

Although there are typically faster and numerically more robust methods available for sampling from specific copulas, the conditional distribution method has its advantages. For example, it applies to any copula and can also preserve low discrepancy as the following example visually demonstrates; see Cambou et al. (2016) for more details.

Example 2.7.3 (Conditional Distribution Method and Quasi-Random Copula Sampling) For a given copula C, the mapping $R^{\leftarrow}_C$ is implemented in the R function `cCopula(, inverse = TRUE)`. We illustrate it for sampling from a Clayton copula with parameter 2 in two different ways: First, based on a pseudo-random sample from Π and second, based on a quasi-random sample from Π generated with the function `ghalton()` of the R package `qrng` of Hofert and Lemieux (2016). Informally, *quasi-random numbers*, also known as low-discrepancy sequences, aim at more evenly distributing the points in $[0, 1]^d$ (less gaps, less clusters) rather than producing perfectly random samples from Π. They can help to reduce the variance of Monte Carlo estimators.

```
> ## Define the Clayton copula to be sampled
> cc <- claytonCopula(2)
> ## Pseudo-random sample from the Clayton copula via CDM
> set.seed(271)
> U.pseudo <- rCopula(1000, copula = indepCopula())
> U.cc.pseudo <- cCopula(U.pseudo, copula = cc, inverse = TRUE)
> ## Quasi-random sample from the Clayton copula via CDM
> set.seed(271)
> library(qrng)
> U.quasi <- ghalton(1000, d = 2) # sobol() is typically even faster
> U.cc.quasi <- cCopula(U.quasi, copula = cc, inverse = TRUE)
> plot(U.pseudo,    xlab = quote(U*"'"[1]), ylab = quote(U*"'"[2]))
> plot(U.quasi,     xlab = quote(U*"'"[1]), ylab = quote(U*"'"[2]))
> plot(U.cc.pseudo, xlab = quote(U[1]),     ylab = quote(U[2]))
> plot(U.cc.quasi,  xlab = quote(U[1]),     ylab = quote(U[2]))
```

The top row of scatter plots in Fig. 2.30 shows a pseudo-random sample (left) and a quasi-random sample (right) from the independence copula. As one can see, the top right scatter plot shows less gaps and clusters locally, and thus appears smoother than the top left scatter plot. The corresponding samples from the Clayton copula with parameter 2, obtained by applying $R^{\leftarrow}_{C^C_2}$ to the bivariate pseudo-random or quasi-random sample from Π, are displayed in the bottom row of plots in Fig. 2.30. The additional smoothness appearing in the bottom right scatter plot is due to the fact that $R^{\leftarrow}_{C^C_2}$ is one-to-one in the sense of (2.31).

Even if other transformations than $R^{\leftarrow}_C$ are used for sampling, combining them with quasi-random number generators can largely reduce the variance in simulations involving copulas. However, this may require more complicated designs (for example, it may matter which components of the quasi-random number sequence

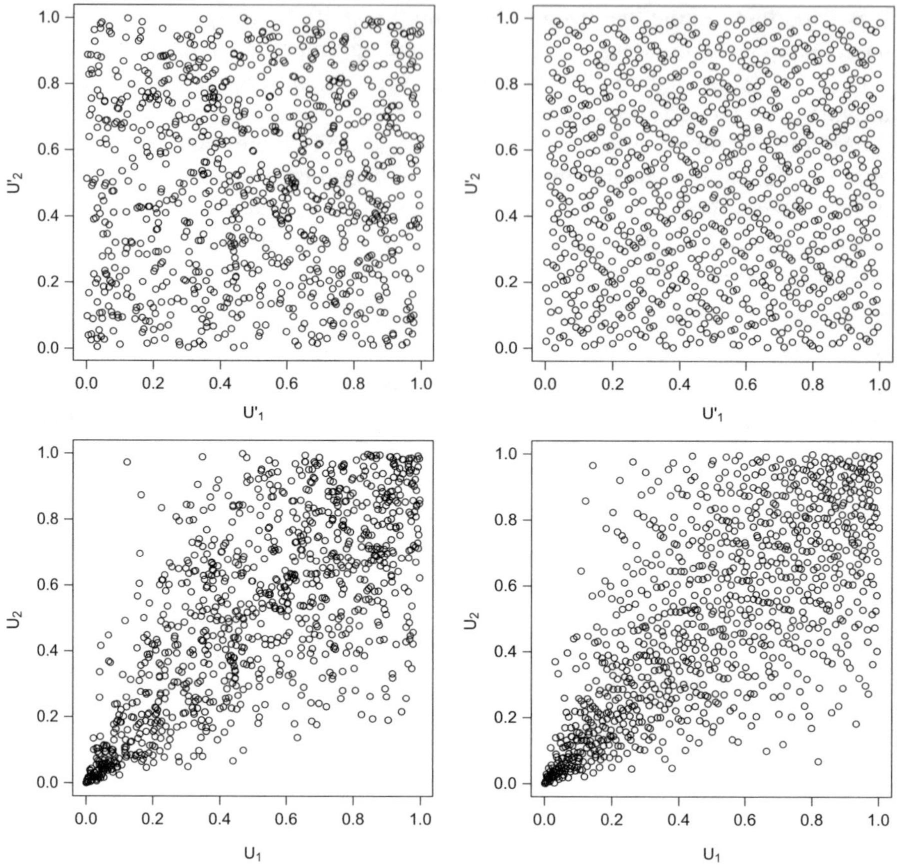

Fig. 2.30 Top row: A pseudo-random sample (left) and a quasi-random sample (right) from the independence copula. Bottom row: The corresponding samples from the Clayton copula with parameter 2 generated with the conditional distribution method

are mapped to which components of the random vector to be sampled); see Cambou et al. (2016) for more details. Besides variance reduction, an advantage of (some) quasi-random number generators (for example, a Sobol' sequence) over pseudo-random number generators (for example, the Mersenne Twister) is a shorter run time. □

Example 2.7.4 (Variance Reduction) Variance-reduction techniques are of practical interest, for example, when using Monte Carlo integration for estimating tail probabilities or when estimating risk measures in quantitative risk management. In this example, we focus on the problem of estimating, for $d = 5$, the tail probability $\mathbb{P}(\boldsymbol{U} > \boldsymbol{u})$ for $\boldsymbol{u} = (0.99, \dots, 0.99)$ and $\boldsymbol{U} \sim C^t_{\rho,\nu}$, a d-dimensional homogeneous t copula with $\nu = 3$ degrees of freedom and pairwise correlation parameter ρ such that Kendall's tau equals 0.5. In a small simulation study involving the auxiliary function `sProb()` defined below, we estimate the tail probability $\mathbb{P}(\boldsymbol{U} > \boldsymbol{u}) = \mathbb{E}(1(\boldsymbol{U} > \boldsymbol{u}))$ via Monte Carlo integration by

$$\frac{1}{n}\sum_{i=1}^{n} 1(\boldsymbol{U}_i > \boldsymbol{u}), \tag{2.32}$$

where $\boldsymbol{U}_1, \dots, \boldsymbol{U}_n$ is a random sample from $C^t_{\rho,\nu}$. As in the previous example, we use the conditional distribution method to generate $\boldsymbol{U}_1, \dots, \boldsymbol{U}_n$ from $C^t_{\rho,\nu}$ in two ways: we either use R's (default) pseudo-random number generator Mersenne Twister to obtain standard uniform random variates in $[0, 1]^d$ (see `?RNGkind`) or we utilize a quasi-random number generator (here a Sobol' sequence, see `sobol()` of the R package `qrng`), and then apply the transformation $R^{\leftarrow}_{C^t_{\rho,\nu}}$. In addition to the use of quasi-random numbers, we also consider a second variance-reduction technique known as *Latin hypercube sampling*; see Packham and Schmidt (2010). The latter efficiently extends the idea of stratified sampling to higher dimensions. It consists of replacing the sample $\boldsymbol{U}_1, \dots, \boldsymbol{U}_n$ in (2.32) (obtained from a pseudo-random number generator) by

$$\boldsymbol{U}^*_i = \left(\frac{R_{i1} - 1 + U'_{i1}}{n}, \dots, \frac{R_{id} - 1 + U'_{id}}{n}\right), \quad i \in \{1, \dots, n\},$$

where R_{ij} denotes the rank of U_{ij} among $U_{1j}, \dots, U_{nj}$ and $U'_{ij} \overset{\text{ind.}}{\sim} \mathrm{U}(0, 1)$ (also obtained from a pseudo-random number generator). In practice, given a sample $\boldsymbol{U}_1, \dots, \boldsymbol{U}_n$ from a copula, the corresponding Latin hypercube sample $\boldsymbol{U}^*_1, \dots, \boldsymbol{U}^*_n$ can easily be constructed with the function `rLatinHypercube()` of the R package `copula`.

Before reporting the results of the simulation, let us mention that variance-reduction techniques might introduce a bias when estimating quantities such as $\mathbb{P}(\boldsymbol{U} > \boldsymbol{u})$ by means of estimators such as (2.32). For Latin hypercube sampling, Packham and Schmidt (2010) show unbiasedness. For quasi-random number generators, the situation is more complicated. In principle, the estimator (2.32) based on quasi-random numbers may be biased. Unbiased estimators can be constructed by randomizing quasi-random numbers, thus leading to the so-called randomized quasi-Monte Carlo methods; see Lemieux (2009, Sect. 6.2) and Cambou et al. (2016) for more details. It is important to note that we indeed use a randomized Sobol' sequence in our small simulation below which can be obtained by setting the argument `randomize` to `TRUE` in `sobol()`.

The following code ends up producing a box plot (see Fig. 2.31) which highlights the variance-reduction effects of the two variance-reduction methods and computes the corresponding variance-reduction factors (VRFs) and percentage improvements:

```
> ##' @title Approximately computing P(U_1 > u_1,..., U_d > u_d)
> ##' @param n sample size
> ##' @param copula copula of (U_1,..., U_d)
> ##' @param u lower-left endpoint (u_1,..., u_d) of the evaluation point
> ##' @return Estimates of P(U_1 > u_1,..., U_d > u_d) by
> ##'         pseudo-random numbers, Latin hypercube sampling and
> ##'         randomized quasi-random numbers.
> sProb <- function(n, copula, u) # sample size, copula, lower-left endpoint
  {
      d <- length(u)
      stopifnot(n >= 1, inherits(copula, "Copula"), 0 < u, u < 1,
                d == dim(copula))
      umat <- rep(u, each = n)
      ## Pseudo-random numbers
      U <- rCopula(n, copula = copula)
      PRNG <- mean(rowSums(U > umat) == d)
      ## Latin hypercube sampling (based on the recycled 'U')
      U. <- rLatinHypercube(U)
      LHS <- mean(rowSums(U. > umat) == d)
      ## (Randomized) quasi-random numbers
      U.. <- cCopula(sobol(n, d = d, randomize = TRUE), copula = copula,
                     inverse = TRUE)
      QRNG <- mean(rowSums(U.. > umat) == d)
      ## Return
      c(PRNG = PRNG, LHS = LHS, QRNG = QRNG)
  }
> ## Simulate the probabilities of falling in (u_1, 1] x ... x (u_d, 1]
> library(qrng)
> N <- 500 # number of replications
> n <- 5000 # sample size
> d <- 5 # dimension
> nu <- 3 # degrees of freedom
> rho <- iTau(tCopula(df = nu), tau = 0.5) # correlation parameter
> cop <- tCopula(param = rho, dim = d, df = nu) # t copula
> u <- rep(0.99, d) # lower-left endpoint of the considered cube
> set.seed(271) # for reproducibility
> res <- replicate(N, sProb(n, copula = cop, u = u))
> ## Grab out the results and compute the sample variances
> varP <- var(PRNG <- res["PRNG",])
> varL <- var(LHS  <- res["LHS" ,])
> varQ <- var(QRNG <- res["QRNG",])
> ## Compute the VRFs and % improvements w.r.t. PRNG
> VRF.L <- varP / varL # VRF for LHS
> VRF.Q <- varP / varQ # VRF for QRNG
> PIM.L <- (varP - varL) / varP * 100 # % improvement for LHS
> PIM.Q <- (varP - varQ) / varP * 100 # % improvement for QRNG
> ## Box plot
> boxplot(list(PRNG = PRNG, LHS = LHS, QRNG = QRNG),
          sub = sprintf("N = %d replications with n = %d and d = %d", N, n, d))
> mtext(substitute("Simulated"~~P(bold(U) > bold(u))~~
                   "for a"~{C^italic(t)}[list(rho.,nu.)]~"copula",
                   list(rho. = round(rho, 2), nu. = nu)),
        side = 2, line = 4.5, las = 0)
> mtext(sprintf("VRFs (%% improvements): %.1f (%.0f%%), %.1f (%.0f%%)",
                VRF.L, PIM.L, VRF.Q, PIM.Q),
        side = 4, line = 1, adj = 0, las = 0)
```

□

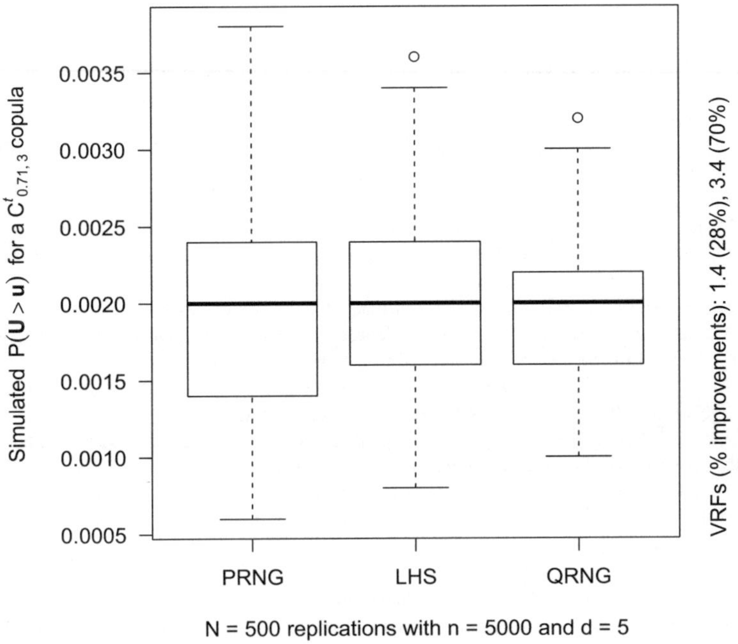

Fig. 2.31 Box plot of simulated exceedance probabilities based on a pseudo-random number generator (PRNG), Latin hypercube sampling (LHS), and a quasi-random number generator (QRNG). The given variance-reduction factors (VRFs) of LHS and QRNG, respectively, in comparison to PRNG, as well as the corresponding percentage improvements illustrate the improvement in variance of these variance-reduction techniques in this case

References

Basel Committee on Banking Supervision. (2009). *Revisions to the Basel II market risk framework.* http://www.bis.org/publ/bcbs158.pdf

Cambou, M., Hofert, M., & Lemieux, C. (2016). Quasi-random numbers for copula models. *Statistics and Computing, 27*(5), 1307–1329.

Durrett, R. (2010). *Probability: Theory and Examples* (4th ed.). London: Cambridge University Press.

Embrechts, P., Lindskog, F., & McNeil, A. J. (2003). Modelling dependence with copulas and applications to risk management. In S. Rachev (Ed.), *Handbook of heavy tailed distributions in finance* (pp. 329–384). Amsterdam: Elsevier.

Embrechts, P., McNeil, A. J., & Straumann, D. (2002). Correlation and dependency in risk management: Properties and pitfalls. In M. Dempster (Ed.), *Risk management: Value at risk and beyond* (pp. 176–223). London: Cambridge University Press.

Embrechts, P., Puccetti, G., & Rüschendorf, L. (2013). Model uncertainty and VaR aggregation. *Journal of Banking and Finance, 37*(8), 2750–2764.

Fréchet, M. (1935). Généralisations du théorème des probabilités totales. *Fundamenta Mathematica, 25*, 379–387.

Fréchet, M. (1951). Sur les tableaux de corrélations dont les marges sont données. *Annales de l'Université de Lyon, 9*, 53–77.

Genest, C. (1987). Frank's family of bivariate distributions. *Biometrika, 74*(3), 549–555.

Genest, C., & Nešlehová, J. (2007). A primer on copulas for count data. *The ASTIN Bulletin, 37*, 475–515.

Genest, C., Nešlehová, J. G., & Rémillard, B. (2013). On the estimation of Spearman's rho and related tests of independence for possibly discontinuous multivariate data. *Journal of Multivariate Analysis, 117*, 214–228.

Genz, A., Bretz, F., Miwa, T., Mi, X., Leisch, F., Scheipl, F., et al. (2017). mvtnorm: Multivariate normal and *t* distribution, R package version 1.0-6. https://CRAN.R-project.org/package=mvtnorm

Hofert, M. (2010). *Sampling nested Archimedean copulas with applications to CDO pricing*. PhD thesis, Südwestdeutscher Verlag für Hochschulschriften AG & Co. KG, ISBN 978-3-8381-1656-3.

Hofert, M. (2013). On sampling from the multivariate *t* distribution. *The R Journal, 5*(2), 129–136.

Hofert, M., & Hornik, K. (2017). qrmtools: Tools for quantitative risk management, R package version 0.0-7. https://CRAN.R-project.org/package=qrmtools

Hofert, M., & Lemieux, C. (2016). qrng: (Randomized) quasi-random number generators, R package version 0.0-3. https://CRAN.R-project.org/package=qrng

Hofert, M., & Mächler, M. (2014). A graphical goodness-of-fit test for dependence models in higher dimensions. *Journal of Computational and Graphical Statistics, 23*(3), 700–716.

Hofert, M., Memartoluie, A., Saunders, D., & Wirjanto, T. (2017). Improved algorithms for computing worst value-at-risk. *Statistics & Risk Modeling, 34*(1–2), 13–31.

Hoeffding, W. (1940). Massstabinvariante Korrelationstheorie. *Schriften des mathematischen Seminars und des Instituts für Angewandte Mathematik der Universität Berlin 5*, 181–233.

Iman, R. L., & Conover, W. J. (1982). A distribution-free approach to inducing rank correlation among input variables. *Communications in Statistics – Simulation and Computation, 11*(3), 311–334.

Jaworski, P., Durante, F., Härdle, W. K., & Rychlik, T. (Eds.) (2010). *Copula theory and its applications. Lecture notes in statistics* (Vol. 198). Berlin: Springer.

Kojadinovic, I., & Yan, J. (2010). Comparison of three semiparametric methods for estimating dependence parameters in copula models. *Insurance: Mathematics and Economics, 47*, 52–63.

Lemieux, C. (2009). *Monte Carlo and quasi-Monte Carlo sampling. Springer series in statistics*. New York: Springer.

Mächler, M. (2017), nor1mix: Normal (1-d) Mixture Models (S3 Classes and Methods), R package version 1.2-3. https://CRAN.R-project.org/package=nor1mix

Marshall, A. W., & Olkin, I. (1967). A multivariate exponential distribution. *Journal of the American Statistical Association, 62*, 30–44.

McNeil, A. J., Frey, R., & Embrechts, P. (2015). *Quantitative risk management: Concepts, techniques and tools* (2nd ed.). Princeton, NJ: Princeton University Press.

Nelsen, R. B. (2006). *An introduction to copulas*. New York: Springer.

Packham, N., & Schmidt, W. M. (2010). Latin hypercube sampling with dependence and applications in finance. *Journal of Computational Finance, 13*(3), 81–111.

Puccetti, G., & Rüschendorf, L. (2013). Sharp bounds for sums of dependent risks. *Journal of Applied Probability, 50*(1), 42–53.

Rosenblatt, M. (1952). Remarks on a multivariate transformation. *The Annals of Mathematical Statistics, 23*(3), 470–472.

Rüschendorf, L. (2009). On the distributional transform, Sklar's Theorem, and the empirical copula process. *Journal of Statistical Planning and Inference, 139*(11), 3921–3927.

Sarkar, D. (2008), *Lattice: Multivariate data visualization with R*. New York: Springer.

Scarsini, M. (1984). On measures of concordance. *Stochastica, 8*(3), 201–218.

Schmitz, V. (2003). *Copulas and stochastic processes*. PhD thesis, Rheinisch-Westfälische Technische Hochschule Aachen.

Schweizer, B., & Wolff, E. F. (1981). On nonparametric measures of dependence for random variables. *The Annals of Statistics, 9*, 879–885.

Sklar, A. (1959). Fonctions de répartition à n dimensions et leurs marges. *Publications de l'Institut de Statistique de l'Université de Paris, 8*, 229–231.

Sklar, A. (1996). Random variables, distribution functions, and copulas – A personal look backward and forward. *Distributions with Fixed Marginals and Related Topics, 28*, 1–14.

Chapter 3
Classes and Families

Several parametric copula families were already introduced in the previous chapter. As we will see, the Clayton and the Frank families, for example, share several features and properties. They belong to the same *class* of copulas, known as Archimedean copulas. Similarly, it should not come as a surprise to learn that the normal and t families belong to the same class of copulas, the so-called elliptical copulas.

The classes of elliptical copulas, Archimedean copulas, and extreme-value copulas are formally introduced in the first three sections of this chapter. The fourth section is devoted to selected copula transformations which can be used to construct more flexible copula models.

3.1 Elliptical Distributions and Copulas

Elliptical copulas are among the most widely used copulas in practice, describing, by Sklar's Theorem, the dependence of the multivariate normal distribution in terms of the normal (or Gauss) copula and the dependence of the multivariate t distribution in terms of the (Student) t copula; see Chap. 2, where these two important and widely used copula families were already introduced.

Due to their implicit construction by Sklar's Theorem, and (2.10) in particular, properties of elliptical copulas are typically derived from the properties of the corresponding elliptical distributions. It is therefore important to understand the construction and properties of elliptical distributions, which we will cover next. Elliptical copulas are addressed thereafter.

M. Hofert et al., *Elements of Copula Modeling with R*, Use R!,
https://doi.org/10.1007/978-3-319-89635-9_3

3.1.1 Elliptical Distributions

The construction of elliptical distributions is most easily grasped via stochastic representations, whereas their properties are often shown via characteristic functions. In this section, we mostly focus on stochastic representations and their implications. For detailed properties and results about elliptical distributions, see, for instance, Fang et al. (1990).

Definition 3.1.1 (Elliptical Distributions) A d-dimensional random vector $\boldsymbol{X}$ has an *elliptical distribution* with *location vector* $\boldsymbol{\mu} \in \mathbb{R}^d$, *scale* (or *dispersion*) *matrix* $\Sigma = AA'$ with $\operatorname{rank}(\Sigma) = k \le d$ for a matrix $A \in \mathbb{R}^{d\times k}$ and *radial part* $R \ge 0$ if

$$\boldsymbol{X} \overset{\text{d}}{=} \boldsymbol{\mu} + A\boldsymbol{Y}, \quad \text{for } \boldsymbol{Y} \overset{\text{d}}{=} R\boldsymbol{S},$$

where $\boldsymbol{S} \sim \mathrm{U}(\{\boldsymbol{x} \in \mathbb{R}^k : \|\boldsymbol{x}\| = 1\})$, $\|\cdot\|$ denotes the Euclidean norm (that is, $\boldsymbol{S}$ is uniformly distributed on the unit sphere in $\mathbb{R}^k$) and R and $\boldsymbol{S}$ are independent. The distribution of $\boldsymbol{Y}$ is known as *spherical distribution*. □

Some remarks concerning elliptical distributions are in order:

Remark 3.1.2

1) Since $\Sigma = AA'$, Σ is symmetric and positive semi-definite, thus a covariance matrix (for example, Σ is the covariance matrix of $A\boldsymbol{Z}$ for $\boldsymbol{Z} = (Z_1, \dots, Z_k)$ with $Z_1, \dots, Z_k \overset{\text{ind.}}{\sim} \mathrm{N}(0, 1)$). Conversely, any covariance matrix Σ can be decomposed as $\Sigma = AA'$ in terms of a lower triangular square matrix A with nonnegative entries on the diagonal (positive if Σ is positive definite). This result is known as the *Cholesky decomposition* of Σ and A as its *Cholesky factor*. Note that $\boldsymbol{\mu}$ is not necessarily the mean of $\boldsymbol{X}$ and Σ is not necessarily the covariance matrix of $\boldsymbol{X}$ (an exception being, for instance, if $\boldsymbol{X} \sim \mathrm{N}_d(\boldsymbol{\mu}, \Sigma)$, that is, if $\boldsymbol{X}$ is multivariate normal), hence the names "location vector" and "scale matrix" (or "dispersion matrix") instead of "mean vector" and "covariance matrix." If $\mathbb{E}(R) < \infty$, then $\mathbb{E}(\boldsymbol{X})$ exists and equals $\boldsymbol{\mu}$, and if $\mathbb{E}(R^2) < \infty$, then $\operatorname{Cov}(\boldsymbol{X})$ exists and equals $(\mathbb{E}(R^2)/d)\Sigma$. For instance, for the multivariate t distribution with ν degrees of freedom, $\operatorname{Cov}(\boldsymbol{X}) = (\nu/(\nu-2))\Sigma$. It follows that if $\mathbb{E}(R^2) < \infty$, the correlation matrix $\operatorname{Cor}(\boldsymbol{X})$ of $\boldsymbol{X}$ exists and equals the correlation matrix P corresponding to Σ.
2) We mostly work with scale matrices Σ which have full rank. If $\operatorname{rank}(\Sigma) = k < d$, then $\boldsymbol{X}$ is concentrated on $\mathbb{R}^k$. If $\operatorname{rank}(\Sigma) = d$, then Σ is positive definite and invertible.
3) If Σ has full rank, $\boldsymbol{X}$ has a density if and only if R has a density f_R. In this case, the density h of $\boldsymbol{X}$ can be shown to be of the form

$$h(\boldsymbol{x}) = \frac{1}{\sqrt{\det \Sigma}}\, g\big((\boldsymbol{x}-\boldsymbol{\mu})'\Sigma^{-1}(\boldsymbol{x}-\boldsymbol{\mu})\big), \quad \boldsymbol{x} \in \mathbb{R}^d, \tag{3.1}$$

in terms of a function $g : [0, \infty) \to [0, \infty)$ known as *density generator*, and it can be shown that f_R and g are related via $f_R(r) = 2\pi^{d/2} r^{d-1} g(r^2)/\Gamma(d/2)$,

$r \geq 0$, where $\Gamma(z) = \int_0^\infty x^{z-1}\exp(-x)\,\mathrm{d}x$ denotes the *gamma function*. Note that h depends on $\boldsymbol{x}$ only through $(\boldsymbol{x}-\boldsymbol{\mu})'\Sigma^{-1}(\boldsymbol{x}-\boldsymbol{\mu})$, that is, h is constant on ellipsoids, or, equivalently, h has ellipsoidal level curves (hence the name "elliptical distribution").

4) Since $\boldsymbol{S} \overset{\mathrm{d}}{=} -\boldsymbol{S}$, an elliptically distributed $\boldsymbol{X}$ satisfies $\boldsymbol{X}-\boldsymbol{\mu} \overset{\mathrm{d}}{=} RA\boldsymbol{S} \overset{\mathrm{d}}{=} -RA\boldsymbol{S} \overset{\mathrm{d}}{=} \boldsymbol{\mu}-\boldsymbol{X}$, so that $\boldsymbol{X}$ is radially symmetric about $\boldsymbol{\mu}$; see Definition 2.5.4.
5) Let $\sigma_j^2 = \Sigma_{jj}$, $j \in \{1,\dots,d\}$, and note that the location-scale transformed random variables $(X_j-\mu_j)/\sigma_j$ all have the same distribution. In other words, the marginal dfs $F_1,\dots,F_d$ of $\boldsymbol{X}$ are of the same type and may only differ in terms of location or scale.
6) There are various ways to generate $\boldsymbol{S} \sim \mathrm{U}(\{\boldsymbol{x} \in \mathbb{R}^k : \|\boldsymbol{x}\| = 1\})$. A simple one which is applicable in all dimensions follows from the fact that $\boldsymbol{S} \overset{\mathrm{d}}{=} \boldsymbol{Y}/\|\boldsymbol{Y}\|$ for any spherically distributed random vector $\boldsymbol{Y}$. One such example is given by $\boldsymbol{Y} = \boldsymbol{Z} \sim \mathrm{N}_d(\boldsymbol{0}, I_d)$, where I_d is the identity matrix in $\mathbb{R}^{d\times d}$. This amounts to generating $Z_1,\dots,Z_d \overset{\text{ind.}}{\sim} \mathrm{N}(0,1)$ and returning $\boldsymbol{S} \overset{\mathrm{d}}{=} \boldsymbol{Z}/\|\boldsymbol{Z}\|$.

To spice up the theoretical exposition, let us look at an example.

Example 3.1.3 (Construction of Elliptical Distributions and Sampling) In this example, we manually generate a sample from an elliptical distribution in R. To obtain a sample from a d-dimensional t distribution, the radial part R in Definition 3.1.1 is chosen as $R = \sqrt{dQ}$, where Q follows an F-distribution with parameters d and ν (similarly, a sample from a d-dimensional normal distribution is obtained by letting $R = \sqrt{T}$, where $T \sim \chi_d^2$, that is, T follows a chi-square distribution with d degrees of freedom).

The following code returns a sample of size $n = 1000$ from a bivariate t distribution with $\boldsymbol{\mu} = (1,3)$, $\Sigma = \left(\begin{smallmatrix}16 & 4\\ 4 & 2\end{smallmatrix}\right)$ and $\nu = 3.5$:

```
> ## Stepwise generation of a sample from a bivariate t distribution
> ## via the (general) stochastic representation of elliptical distributions
> n <- 1000 # sample size
> d <- 2 # dimension
> mu <- c(1, 3) # location vector
> Sigma <- matrix(c(16, 4,
                    4, 2), ncol = 2, byrow = TRUE) # scale matrix
> nu <- 3.5 # degrees of freedom
> set.seed(271) # set a seed (for reproducibility)
> R <- sqrt(d * rf(n, df1 = d, df2 = nu)) # sample R for a t_nu
> A <- t(chol(Sigma)) # Cholesky factor
> Z <- matrix(rnorm(n * d), ncol = d) # N_d(0,I_d)
> S <- Z/sqrt(rowSums(Z^2)) # uniform distribution on unit sphere (= Z/||Z||)
> X <- rep(mu, each = n) + R * t(A %*% t(S)) # X = mu + R A S
> plot(S, xlab = quote(S[1]), ylab = quote(S[2]))
> plot(t(A %*% t(S)), xlab = quote((AS)[1]), ylab = quote((AS)[2]))
> xlim <- range(X[,1], X[,1] - mu[1])
> ylim <- range(X[,2], X[,2] - mu[2])
> plot(R * t(A %*% t(S)), xlab = quote(R(AS)[1]), ylab = quote(R(AS)[2]),
       xlim = xlim, ylim = ylim)
> plot(X, xlab = quote(X[1]), ylab = quote(X[2]),
       xlim = xlim, ylim = ylim)
```

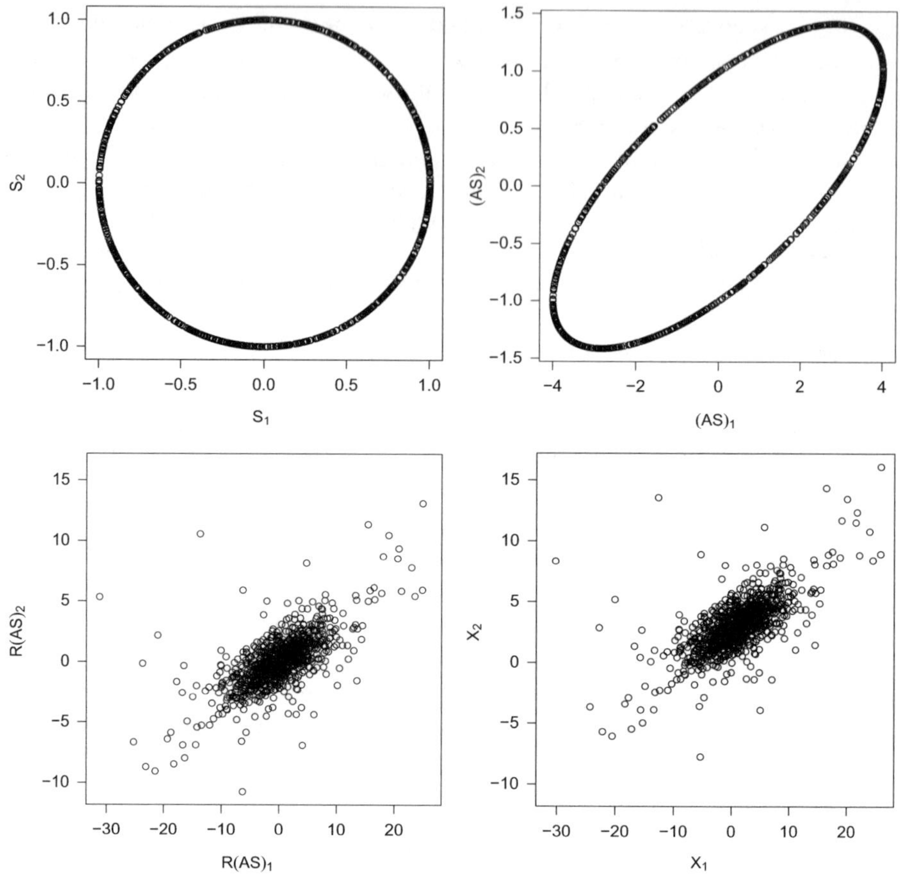

Fig. 3.1 $n = 1000$ independent observations of $\boldsymbol{S} \sim \mathrm{U}(\{\boldsymbol{x} \in \mathbb{R}^2 : \|\boldsymbol{x}\| = 1\})$ (top-left), corresponding realizations of $A\boldsymbol{S}$ (top-right) for $A = \begin{pmatrix} 4 & 0 \\ 1 & 1 \end{pmatrix}$ (the Cholesky factor of $\Sigma = \begin{pmatrix} 16 & 4 \\ 4 & 2 \end{pmatrix}$), of $RA\boldsymbol{S}$ (bottom-left) and of $\boldsymbol{X} = \boldsymbol{\mu} + RA\boldsymbol{S}$ (bottom-right)

The resulting scatter plots are displayed in Fig. 3.1. The top-left plot shows the realizations of $\boldsymbol{S}$ generated uniformly on the unit sphere. The top-right plot displays these realizations multiplied by the Cholesky factor A, which results in an ellipsoidal shaped scatter plot. When the realizations of $A\boldsymbol{S}$ are multiplied by the realizations of R generated from the radial part distribution, the bottom-left plot of Fig. 3.1 is obtained. What happens is that both coordinates of each point on the ellipse are scaled by the corresponding realization of R. This could have also been obtained by first multiplying $\boldsymbol{S}$ by R (thereby constructing a sample from the spherically distributed $\boldsymbol{Y}$ which has circular level sets) and then by multiplying the result by A. To get to the bottom-right plot representing the generated realizations of $\boldsymbol{X}$, one only needs to shift each point by $\boldsymbol{\mu}$. Note that some of the transposes `t()` or replications `rep()` in the provided R code might seem odd; the interested reader is referred to Hofert (2013).

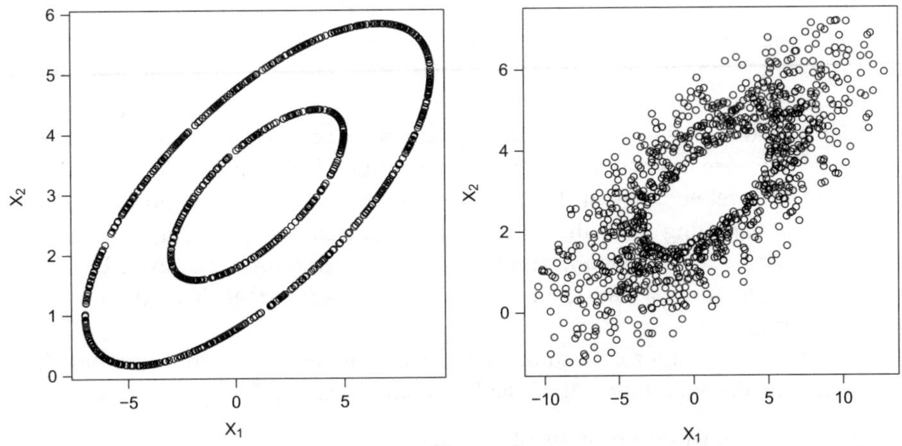

Fig. 3.2 $n = 1000$ independent observations of $\boldsymbol{X} = \boldsymbol{\mu} + RA\boldsymbol{S}$ (with the same A and $\boldsymbol{S}$ as in the bottom-right plot of Fig. 3.1), where R follows a 2-point distribution (left) and a distribution whose support is bounded (right)

Many variants of the previous construction could be considered. Starting from the same realizations of $\boldsymbol{S}$ as previously, Fig. 3.2 shows what happens to the corresponding observations of $\boldsymbol{X}$ if the distribution of the radial part is replaced by a two-point distribution or by a distribution with a bounded support. The latter is chosen as the original one leading to a multivariate t distribution but conditional on R being in $[1, 3]$:

```
> ## 2-point distribution for R and corresponding X
> R.2pt <- 1 + rbinom(n, size = 1, prob = 2/3) # prob. 1/3 to be 1, 2/3 to be 2
> X.2pt <- rep(mu, each = n) + R.2pt * t(A %*% t(S)) # compute X
> ## Bounded distribution for R and corresponding X
> pR <- function(q) pf(q^2/d, df1 = d, df2 = nu) # df of R for a t_nu
> qR <- function(p) sqrt(d * qf(p, df1 = d, df2 = nu)) # quantile function
> a <- 1; b <- 3 # [a, b]
> R.bdd <- qR(pR(a) + runif(n) * (pR(b) - pR(a))) # sample R on [a,b]
> X.bdd <- rep(mu, each = n) + R.bdd * t(A %*% t(S)) # compute X
> plot(X.2pt, xlab = quote(X[1]), ylab = quote(X[2]))
> plot(X.bdd, xlab = quote(X[1]), ylab = quote(X[2]))
```

□

3.1.2 Elliptical Copulas

Elliptical copulas are the copulas of elliptical distributions. Due to this implicit construction, elliptical copulas are of the form (2.10), where H is a multivariate elliptical df and $F_1, \ldots, F_d$ are the corresponding univariate margins. By the invariance principle, see Sect. 2.4, marginal location-scale transformations prior to the use of (2.10) will leave the copula unaffected. Hence, one might as well assume H to be such that $\boldsymbol{\mu}$ is equal to $\mathbf{0} = (0, \dots, 0)$ and Σ is a correlation matrix P. These assumptions imply that the univariate margins of H are identical, that is, $F_1 = \dots = F_d = F$.

The freedom of choosing an arbitrary df (on $[0, \infty)$) for the radial part R often comes at the price of the common univariate marginal df F of $\boldsymbol{X}$ not being tractable anymore. In a related way, if, for example, the evaluation of F requires numerical integration, sampling from the corresponding elliptical copula may be slow. For (approximate) pseudo-random number generation purposes, however, one could always sample from $\boldsymbol{X}$ and estimate F using the empirical df of the combined component samples.

The following algorithm summarizes the general recipe for sampling from elliptical copulas with radial df F_R and correlation matrix P.

Algorithm 3.1.4 (Simulation of Elliptical Copulas)

(1) Sample R from the target df F_R on $[0, \infty)$.
(2) Compute the Cholesky factor A of the correlation matrix P.
(3) Sample $Z_1, \dots, Z_d \overset{\text{ind.}}{\sim} \mathrm{N}(0, 1)$ independently of R and set $\boldsymbol{S} = \boldsymbol{Z}/\|\boldsymbol{Z}\|$.
(4) Compute $\boldsymbol{X} = RA\boldsymbol{S}$.
(5) Return $\boldsymbol{U} = (F(X_1), \dots, F(X_d))$, where F is the common univariate marginal df of $\boldsymbol{X}$.

□

Because of (2.10), the evaluation of an elliptical copula C may be difficult when the evaluation of H or its marginal quantile function $F^{\leftarrow}$ is numerically challenging. For example, for evaluating multivariate normal and t dfs in three or more dimensions, randomized quasi-Monte Carlo methods are applied. Related difficulties may also affect density evaluation. Indeed, if they exist, densities of elliptical copulas are obtained by differentiating (2.10) which leads to the generic form given in (2.11). To evaluate the latter, the quantile function $F^{\leftarrow}$ needs to be tractable.

A subclass of elliptical distributions which often lead to tractable elliptical copulas C are the so-called *normal variance mixtures* (or, equivalently, *normal scale mixtures*). A random vector $\boldsymbol{X}$ has a *normal variance mixture distribution* if

$$\boldsymbol{X} \overset{\text{d}}{=} \boldsymbol{\mu} + \sqrt{W} A \boldsymbol{Z}, \tag{3.2}$$

where $\boldsymbol{\mu}$ and A are as in Definition 3.1.1, and W is a nonnegative random variable independent of $\boldsymbol{Z} = (Z_1, \dots, Z_k)$ for $Z_1, \dots, Z_k \overset{\text{ind.}}{\sim} \mathrm{N}(0, 1)$. The conditional distribution of $\boldsymbol{X}$ given $W = w$ is then $\mathrm{N}_d(\boldsymbol{\mu}, w\Sigma)$, which implies that normal variance mixtures can be regarded as mixing (or overlaying) multivariate normal distributions with different covariance matrices (hence the name). To see that normal variance mixtures are indeed elliptical distributions, notice that $\boldsymbol{X} \stackrel{\mathrm{d}}{=} \boldsymbol{\mu} + \sqrt{W} A \boldsymbol{Z} = \boldsymbol{\mu} + \sqrt{W}\|\boldsymbol{Z}\| A \boldsymbol{Z}/\|\boldsymbol{Z}\| = \boldsymbol{\mu} + R A \boldsymbol{S}$ with $R = \sqrt{W}\|\boldsymbol{Z}\|$ and $\boldsymbol{S} = \boldsymbol{Z}/\|\boldsymbol{Z}\|$ (and one can show the possibly counterintuitive fact that R and $\boldsymbol{S}$ are independent).

The multivariate normal and Student t distributions are widely used examples of normal variance mixtures. As we already know from Chap. 2, multivariate normal and t distributions lead, through (2.10), to normal and t copulas, respectively. Recall that, in the R package `copula`, the functions `normalCopula()` and `tCopula()`, respectively, create objects representing these copulas. The next two examples provide the details.

Example 3.1.5 (Normal Copula) The d-dimensional normal copula C_P^{n}, which was already introduced in Example 2.3.3, is the copula obtained by Sklar's Theorem from the multivariate normal distribution $\mathrm{N}_d(\mathbf{0}, P)$ with correlation matrix P. If Φ_P denotes the df of the latter, $C_P^{\mathrm{n}}(\boldsymbol{u})$ is given, for any $\boldsymbol{u} \in [0,1]^d$, by

$$\begin{aligned} C_P^{\mathrm{n}}(\boldsymbol{u}) &= \Phi_P(\Phi^{-1}(u_1), \ldots, \Phi^{-1}(u_d)) \\ &= \int_{-\infty}^{\Phi^{-1}(u_d)} \cdots \int_{-\infty}^{\Phi^{-1}(u_1)} \frac{\exp(-\frac{1}{2}\boldsymbol{x}' P^{-1} \boldsymbol{x})}{(2\pi)^{\frac{d}{2}} \sqrt{\det P}} \, \mathrm{d}x_1 \ldots \mathrm{d}x_d, \end{aligned}$$

where Φ^{-1} denotes the quantile function of $\mathrm{N}(0,1)$.

The density generator of the multivariate normal distribution being the function $g(t) = \exp(-t/2)/(2\pi)^{d/2}$, $t \in [0, \infty)$, it follows from (3.1) and (2.11) that the density c_P^{n} of C_P^{n} is given by

$$c_P^{\mathrm{n}}(\boldsymbol{u}) = \frac{1}{\sqrt{\det P}} \exp\Big(-\frac{1}{2}\boldsymbol{x}'(P^{-1} - I_d)\boldsymbol{x}\Big), \quad \boldsymbol{u} \in (0,1)^d,$$

where $\boldsymbol{x} = (\Phi^{-1}(u_1), \ldots, \Phi^{-1}(u_d))$.

For $d = 2$ and P having off-diagonal entries $\rho = -1$, $C_P^{\mathrm{n}} = C_\rho^{\mathrm{n}}$ equals the lower Fréchet–Hoeffding bound W, and for $d \geq 2$ and P having off-diagonal entries equal to $\rho = 1$, the homogeneous normal copula $C_P^{\mathrm{n}} = C_\rho^{\mathrm{n}}$ equals the upper Fréchet–Hoeffding bound M. The uncorrelated case $P = I_d$ leads to $c_P^{\mathrm{n}}(\boldsymbol{u}) = 1$, $\boldsymbol{u} \in (0,1)^d$, that is, to the (density of the) independence copula.

Normal copulas C_P^{n} can be easily sampled by first sampling normal variance mixtures with a constant W. By the invariance principle, this sampling algorithm simplifies to Algorithm 3.1.6 hereafter. Note that this is simply Algorithm 2.4.4 applied to $H = \Phi_P$, the multivariate df of $\mathrm{N}_d(\mathbf{0}, P)$.

Algorithm 3.1.6 (Simulation of Normal Copulas)

(1) Compute the Cholesky factor A of the correlation matrix P.
(2) Sample $Z_1, \dots, Z_d \overset{\text{ind.}}{\sim} \mathrm{N}(0, 1)$.
(3) Compute $\boldsymbol{X} = A\boldsymbol{Z}$.
(4) Return $\boldsymbol{U} = (\Phi(X_1), \dots, \Phi(X_d))$.

□

For bivariate normal copulas C^{n}_{ρ}, one can show that Spearman's rho is given by $\rho_{\text{S}} = (6/\pi)\arcsin(\rho/2)$, as already mentioned in Sect. 2.6.2. Kendall's tau is given by $\tau = (2/\pi)\arcsin\rho$, as for all bivariate elliptically distributed random vectors $\boldsymbol{X}$ with location vector $\boldsymbol{\mu} = \boldsymbol{0}$, scale matrix P (a correlation matrix with off-diagonal entries ρ) and $\mathbb{P}(\boldsymbol{X} = \boldsymbol{0}) = 0$; see McNeil et al. (2015, Proposition 7.43). The coefficient of lower tail dependence, see Sect. 2.6.3, of C^{n}_{ρ} can be shown to be $\lambda_l = \mathbb{1}(\rho = 1)$, that is, C^{n}_{ρ} is lower tail independent ($\lambda_l = 0$) unless it is M (in which case $\lambda_l = 1$). Due to radial symmetry, see Remark 3.1.2 4), $\lambda_u = \lambda_l$.

Figure 3.3 shows wireframe and contour plots of a bivariate normal copula C^{n}_{ρ} or its density c^{n}_{ρ} for ρ such that Kendall's tau of C^{n}_{ρ} equals 0.5, as well as a scatter plot of a sample of size $n = 1000$ from C^{n}_{ρ}:

```
> nc <- normalCopula(iTau(normalCopula(), tau = 0.5))
> set.seed(271)
> U <- rCopula(1000, copula = nc) # sample from the normal copula
> wireframe2(nc, FUN = dCopula, delta = 0.025) # density
> contourplot2(nc, FUN = pCopula) # copula
> contourplot2(nc, FUN = dCopula, n.grid = 42, cuts = 33, lwd = 1/2) # density
> plot(U, xlab = quote(U[1]), ylab = quote(U[2])) # scatter plot
```

□

Example 3.1.7 (t Copula) The d-dimensional t copula $C^t_{P,\nu}$, which was already introduced in Example 2.3.3, is the copula obtained by Sklar's Theorem from the multivariate t distribution with location vector $\boldsymbol{0}$, scale matrix P and $\nu > 0$ degrees of freedom. If $t_{P,\nu}$ denotes the df of the latter, $C^t_{P,\nu}(\boldsymbol{u})$ is given, for any $\boldsymbol{u} \in [0, 1]^d$, by

$$
\begin{aligned}
C^t_{P,\nu}(\boldsymbol{u}) &= t_{P,\nu}(t_\nu^{-1}(u_1), \dots, t_\nu^{-1}(u_d)) \\
&= \int_{-\infty}^{t_\nu^{-1}(u_d)} \cdots \int_{-\infty}^{t_\nu^{-1}(u_1)} \frac{\Gamma(\frac{\nu+d}{2})}{\Gamma(\frac{\nu}{2})(\pi\nu)^{\frac{d}{2}}\sqrt{\det P}} \Big(1 + \frac{\boldsymbol{x}'P^{-1}\boldsymbol{x}}{\nu}\Big)^{-\frac{\nu+d}{2}} \,\mathrm{d}x_1 \dots \mathrm{d}x_d,
\end{aligned}
$$

where t_ν^{-1} denotes the quantile function of the df t_ν of the univariate Student t distribution with ν degrees of freedom.

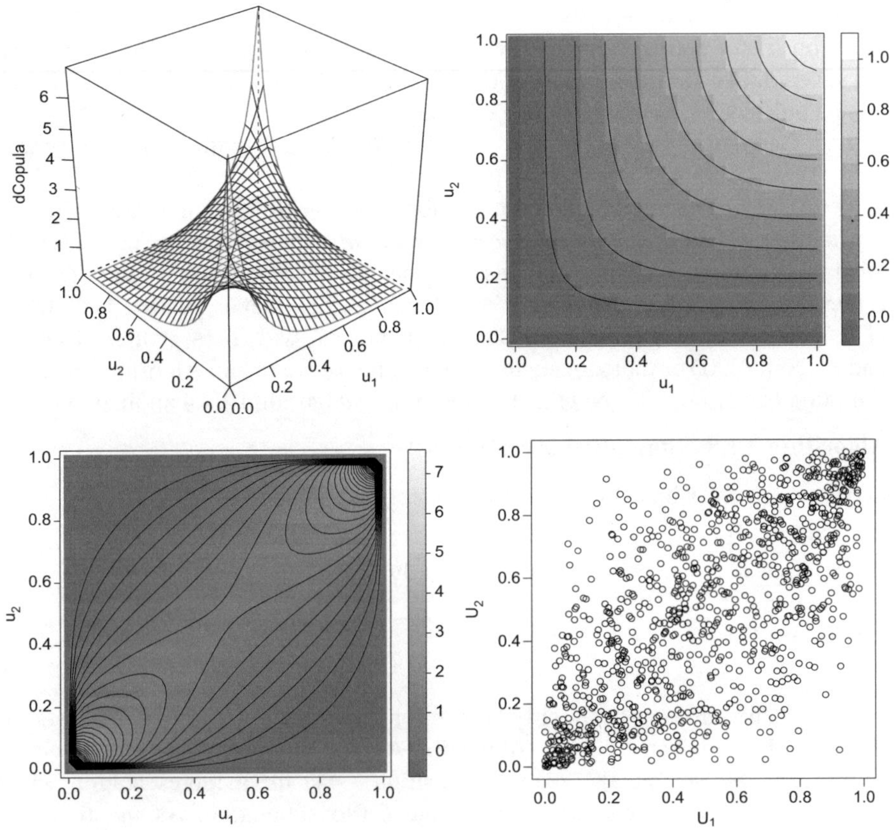

Fig. 3.3 Wireframe plot of c_ρ^{n} for $\rho \approx 0.7071$ (top left), contour plot of C_ρ^{n} (top right), of c_ρ^{n} (bottom left) and scatter plot of a sample of size $n = 1000$ from C_ρ^{n} (bottom right)

The density generator of the multivariate t distribution being

$$g(t) = \frac{\Gamma(\frac{\nu+d}{2})}{(\pi\nu)^{\frac{d}{2}}\Gamma(\frac{\nu}{2})}\left(1+\frac{t}{\nu}\right)^{-\frac{\nu+d}{2}}, \quad t \in [0, \infty),$$

we obtain from (3.1) and (2.11) that the density $c_{P,\nu}^t$ of $C_{P,\nu}^t$ is given by

$$c_{P,\nu}^t(\boldsymbol{u}) = \frac{\Gamma(\frac{\nu+d}{2})}{\Gamma(\frac{\nu}{2})\sqrt{\det P}}\left(\frac{\Gamma(\frac{\nu}{2})}{\Gamma(\frac{\nu+1}{2})}\right)^d \frac{(1+\frac{\boldsymbol{x}'P^{-1}\boldsymbol{x}}{\nu})^{-\frac{\nu+d}{2}}}{\prod_{j=1}^d (1+\frac{x_j^2}{\nu})^{-\frac{\nu+1}{2}}}, \quad \boldsymbol{u} \in (0,1)^d,$$

where $\boldsymbol{x} = (t_\nu^{-1}(u_1), \ldots, t_\nu^{-1}(u_d))$.

For $d = 2$, $C^t_{-1,\nu}$ is the lower Fréchet–Hoeffding bound W, and for $d \geq 2$, if P only consists of entries equal to 1, $C^t_{P,\nu}$ is the upper Fréchet–Hoeffding bound M. However, unlike for the normal copula, $P = I_d$ does not lead to the independence copula unless $\nu = \infty$ in which case $C^t_{P,\nu}$ equals C^{n}_P.

Let $\mathrm{Gamma}(\alpha, \beta)$ denote the *gamma distribution* with shape parameter $\alpha \in (0, \infty)$, rate parameter $\beta \in (0, \infty)$ and density $x \mapsto \beta^\alpha x^{\alpha-1} e^{-\beta x}/\Gamma(\alpha)$, $x \in (0, \infty)$. The reciprocal of a random variable following a $\mathrm{Gamma}(\alpha, \beta)$ distribution is said to follow an *inverse gamma distribution*, denoted by $\mathrm{IG}(\alpha, \beta)$. With this in mind, Student t copulas $C^t_{P,\nu}$ can be sampled by first sampling from a normal variance mixture with $W \sim \mathrm{IG}(\nu/2, \nu/2)$, that is, $W = 1/G$ where $G \sim \mathrm{Gamma}(\nu/2, \nu/2)$ (or, equivalently, $W = \nu/T$ with $T \sim \chi^2_\nu$); see Demarta and McNeil (2005). Analogously to normal copulas, we obtain Algorithm 3.1.8 for sampling from t copulas. Note that this is simply Algorithm 2.4.4 applied to $t_{P,\nu}$.

Algorithm 3.1.8 (Simulation of t Copulas)

(1) Compute the Cholesky factor A of the correlation matrix P.
(2) Sample $Z_1, \dots, Z_d \overset{\text{ind.}}{\sim} \mathrm{N}(0, 1)$.
(3) Sample $W \sim \mathrm{IG}(\nu/2, \nu/2)$ independently of $\boldsymbol{Z}$.
(4) Compute $\boldsymbol{X} = \sqrt{W} A \boldsymbol{Z}$.
(5) Return $\boldsymbol{U} = (t_\nu(X_1), \dots, t_\nu(X_d))$.

□

Spearman's rho is not explicitly known for bivariate t copulas; see, for instance, Kojadinovic and Yan (2010a, Appendix A) for an approximation based on splines and McNeil et al. (2015, Proposition 7.44) for a representation which can, for example, be evaluated by Monte Carlo simulation. As mentioned in Example 3.1.5, Kendall's tau has the same expression as for normal copulas, namely $\tau = (2/\pi) \arcsin \rho$. Furthermore, we already know from Sect. 2.6.3 that the coefficient of lower tail dependence of $C^t_{\rho,\nu}$ is given by

$$\lambda_l = 2t_{\nu+1}\left(-\sqrt{\frac{(\nu+1)(1-\rho)}{1+\rho}}\right),$$

which is increasing in ρ for fixed ν or decreasing in ν for fixed ρ; see Examples 2.6.16 and 2.6.17. Hence, t copulas are lower tail dependent (unless $\nu = \infty$ and $\rho < 1$). Again, by radial symmetry, $\lambda_u = \lambda_l$.

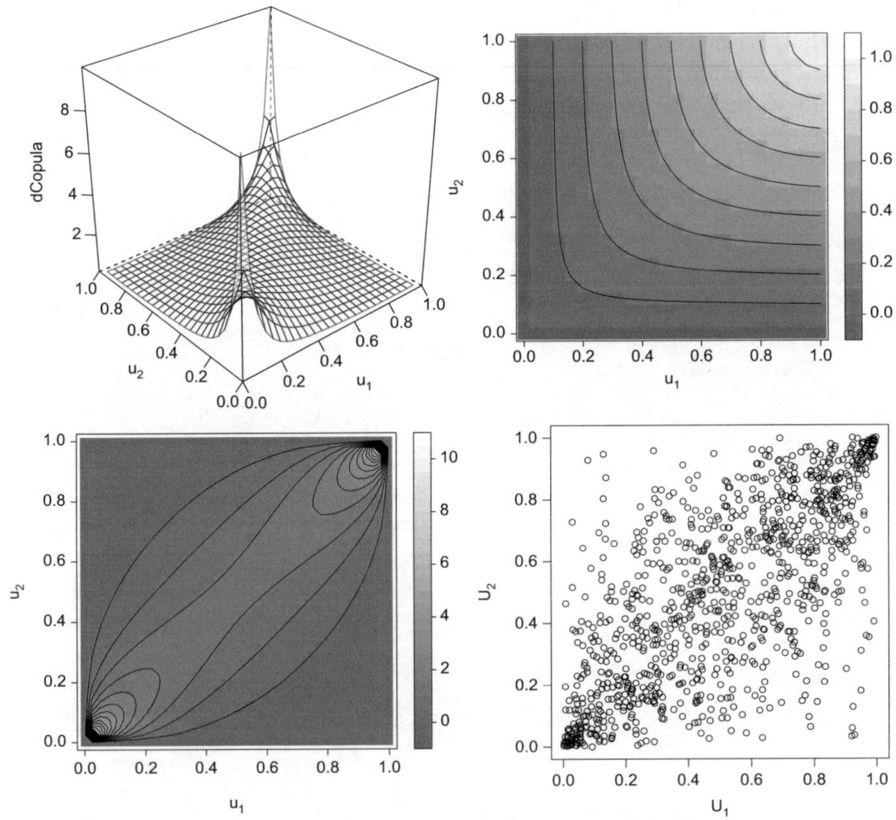

Fig. 3.4 Wireframe plot of $c^t_{\rho,\nu}$ for $\rho \approx 0.7071$ and $\nu = 4$ (top left), contour plot of $C^t_{\rho,\nu}$ (top right), of $c^t_{\rho,\nu}$ (bottom left) and scatter plot of a sample of size $n = 1000$ (bottom right)

Figure 3.4 shows wireframe and contour plots of a bivariate t copula $C^t_{\rho,\nu}$ or its density $c^t_{\rho,\nu}$ for $\nu = 4$ and ρ such that Kendall's tau of $C^t_{\rho,\nu}$ equals 0.5, as well as a scatter plot of a sample of size $n = 1000$ from $C^t_{\rho,\nu}$:

```
> nu <- 4 # needs to be an integer here (only) because of pCopula()
> tc <- tCopula(iTau(tCopula(df = nu), tau = 0.5), df = nu)
> set.seed(271)
> U <- rCopula(1000, copula = tc) # sample from the t copula
> wireframe2(tc, FUN = dCopula, delta = 0.025) # density
> contourplot2(tc, FUN = pCopula) # copula
> contourplot2(tc, FUN = dCopula, n.grid = 42, cuts = 27) # density
> plot(U, xlab = quote(U[1]), ylab = quote(U[2])) # scatter plot
```

The sample represented in the scatter plot of Fig. 3.4 was generated with the function `rCopula()`. Notice that `rCopula()` is currently based on `rmvt()` from the R package `mvtnorm` of Genz et al. (2017) and thus involves the eigenvalue decomposition of P rather than its Cholesky decomposition as in Algorithm 3.1.4 or Algorithm 3.1.8. The following code illustrates the numerical equality of t copula samples generated by directly calling `rmvt()`, generated manually (to see how `rmvt()` works), and generated using `rCopula()`:

```
> ## Setup
> n <- 1000
> d <- 5
> nu <- 3.5
> rho <- iTau(tCopula(df = nu), tau = 0.5)
> P <- matrix(rho, nrow = d, ncol = d)
> diag(P) <- 1
> ## Method 1: Directly using rmvt() from 'mvtnorm'
> library(mvtnorm)
> set.seed(271)
> X <- rmvt(n, sigma = P, df = nu)
> U <- pt(X, df = nu)
> ## Method 2: Reproducing rmvt()
> set.seed(271)
> eig <- eigen(P, symmetric = TRUE) # eigenvalue (instead of Cholesky) decomp.
> A <- t(eig$vectors %*% (t(eig$vectors) * sqrt(eig$values)))
> X.norm <- matrix(rnorm(n * d), nrow = n, byrow = TRUE) %*% A
> X.t <- X.norm / sqrt(rchisq(n, nu) / nu)
> U.rmvt <- pt(X.t, df = nu)
> stopifnot(all.equal(U.rmvt, U))
> ## Method 3: rCopula()
> set.seed(271)
> U.cop <- rCopula(n, copula = tCopula(rho, dim = d, df = nu))
> stopifnot(all.equal(U.cop, U.rmvt))
```

We conclude this example by revisiting Example 2.4.3. The plot on the left-hand side of Fig. 3.5 shows $n = 1000$ independent observations from $t_{\rho,\nu}$ with $\nu = 3.5$

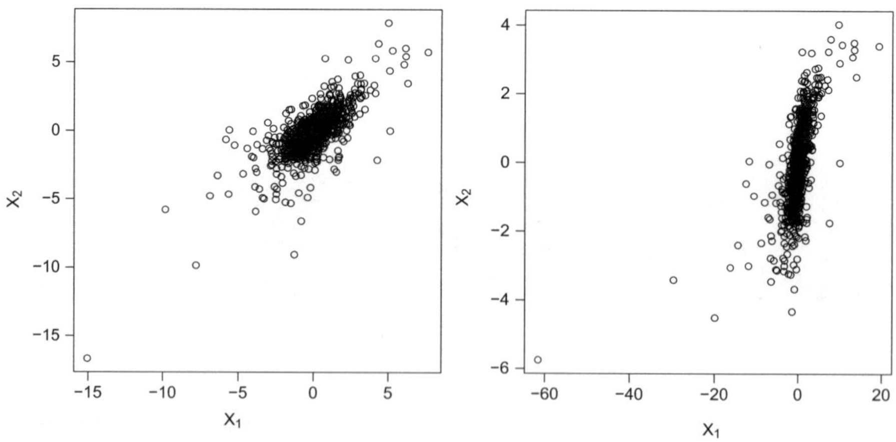

Fig. 3.5 Scatter plots of $n = 1000$ independent observations from $t_{\rho,\nu}$ with $\nu = 3.5$ and $\rho \approx 0.7071$ (left) and from the corresponding meta-t model with univariate t margins whose degrees of freedom are 2 and 10, respectively (right)

and $\rho \approx 0.7071$, while the plot on the right-hand side displays a sample of the same size from a meta-t-model with t margins whose degrees of freedom are 2 and 10, respectively. We already mentioned in Example 2.4.3 that the latter meta-t model is more flexible than the multivariate t due to the fact that the degrees of freedom of the margins can be different from the degrees of freedom of the underlying copula. In particular, the bivariate t is exchangeable whereas our meta-t is not.

```
> X.meta <- cbind(qt(U[,1], df = 2), qt(U[,2], df = 10))
> plot(X, xlab = quote(X[1]), ylab = quote(X[2]))
> plot(X.meta, xlab = quote(X[1]), ylab = quote(X[2]))
```

□

Example 3.1.9 (Marginal Inconsistency and Its Implications) Some elliptical distributions, such as the normal or the Student t, have a property known as marginal consistency. For the Student t for example, one of the consequences of this property is that the univariate margins of a d-dimensional Student t df with ν degrees of freedom are univariate Student t with ν degrees of freedom regardless of the value of d. Some elliptical distributions do not have this property, which makes their corresponding elliptical copulas less attractive to use.

Formally, following Kano (1994), a family of elliptical distributions of dimension d with density generator g_d and, without loss of generality, with location vector $\boldsymbol{\mu} = \mathbf{0}$ and scale matrix $\Sigma = I_d$, is said to be *marginally consistent* if

$$\int_{\mathbb{R}} g_{d+1}\Bigl(\sum_{j=1}^{d+1} x_j^2\Bigr) \mathrm{d}x_{d+1} = g_d\Bigl(\sum_{j=1}^{d} x_j^2\Bigr),$$

for all $d \in \mathbb{N}$ and almost every $(x_1, \dots, x_d) \in \mathbb{R}^d$. The family is said to be *marginally inconsistent* otherwise. Kano (1994) gave characterizations of marginal consistency in several forms, for example, via the density generator function, characteristic function, radial part, and normal variance mixture representation.

A popular elliptical distribution is the *exponential power family*, whose density generator in dimension d is

$$g_d(t) = K_{d,\gamma,s} \exp(-\gamma t^s), \quad t \in [0, \infty),$$

where $\gamma > 0$, $s > 0$ and $K_{d,\gamma,s} > 0$ is a constant depending on d, γ, and s. This family is a special case of the Kotz type elliptical distribution; see Fang et al. (1990, Table 3.1). It reduces to the multivariate normal family with $\gamma = 1/2$ and $s = 1$. One can also show that the radial part R is such that $R^{2s}\gamma \sim \mathrm{Gamma}(d/(2s), 1)$; see Fang et al. (1990, Section 3.2.2). The marginal inconsistency of the exponential power family is empirically illustrated by the following code based on the auxiliary function `rExpPow()` and the function `kde2d()` of the R package MASS of Ripley (2017):

```
> ##' @title Random number generator for the exponential power family
> ##' @param n sample size
> ##' @param d integer parameter d >= 1
> ##' @param s numeric parameter s > 0
> ##' @param gamma numeric parameter gamma > 0
> ##' @param mu location
> ##' @param A Cholesky factor of the scale matrix
> ##' @return Sample from an exponential power family
> rExpPow <- function(n, d, s, gamma, mu = rep(0, d), A = diag(d))
  {
      R <- (rgamma(n, d / 2 / s) / gamma)^(1 / 2 / s)
      Z <- matrix(rnorm(n * d), ncol = d)
      S <- Z / sqrt(rowSums(Z^2))
      rep(mu, each = n) + R * t(A %*% t(S))
  }
> ## Setup
> set.seed(271) # set a seed (for reproducibility)
> X.2d <- rExpPow(100000, d = 2, s = 1/2, gamma = 1) # sample for d = 2
> X.8d <- rExpPow(100000, d = 8, s = 1/2, gamma = 1) # sample for d = 8
> ## As the following plot shows, the first univariate margins differ
> plot(density(X.2d[,1]), #  est. of 1st univ. marg. of 2-dim. df
       xlim = c(-10, 10), ylim = c(0, 0.35), main = "", xlab = "")
> lines(density(X.8d[,1]), col = 2) # est. of 1st univ. marg. of 8-dim. df
> legend("topright", bty = "n",
         lty = c(1,1), lwd = c(2, 2), col = 1:2,
         legend = expression("Kernel density estimate of"~X[1]~"(d = 2)",
                             "Kernel density estimate of"~X[1]~"(d = 8)"))
> ## Empirical marg. prob. trans. followed by quant. trans. to N(0,1)
> Z.2d <- qnorm(pobs(X.2d))
> Z.8d <- qnorm(pobs(X.8d))
> ## Kernel est. (on norm. scale) of the copulas of the first bivariate marg.
> library(MASS)
> dens2d <- kde2d(Z.2d[,1], Z.2d[,2], n = 300)
> dens8d <- kde2d(Z.8d[,1], Z.8d[,2], n = 300)
> image(dens2d, xlim = c(-2, 2), ylim = c(-2, 2), xlab = quote(Z[1]),
        ylab = quote(Z[2]), col = gray(seq(1, 0, length.out = 100)))
> image(dens8d, xlim = c(-2, 2), ylim = c(-2, 2), xlab = quote(Z[1]),
        ylab = quote(Z[2]), col = gray(seq(1, 0, length.out = 100)))
```

Random variates from two exponential power distributions with $s = 1/2, \gamma = 1$, location zero, and identity scale matrix are generated, one with $d = 2$ and the other with $d = 8$. As can be seen from the left panel of Fig. 3.6 (displaying kernel density estimates of the first univariate margin of the two- and eight-dimensional dfs), the first univariate margins of the two dfs are not the same. Furthermore, as seen from the middle panel (addressing the case $d = 2$) and the right panel (addressing the case $d = 8$) of Fig. 3.6, neither are the copulas of the bivariate margin corresponding to the first two components of the dfs: in contrast to the one for $d = 8$, the one for $d = 2$ appears to be diamond-shaped on the normal scale. This is different from what one would expect from families with marginal consistency

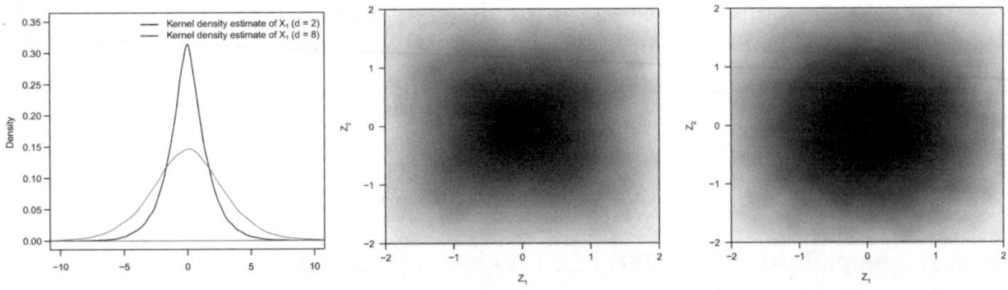

Fig. 3.6 Kernel density estimates of the first univariate margin of the underlying two- and eight-dimensional exponential power dfs (left). Kernel density estimates of the first bivariate margin of the underlying two-dimensional (middle) and eight-dimensional (right) exponential power dfs on the standard normal scale

such as the multivariate normal, Student t, and stable families. Elliptical families that are marginally inconsistent include the multivariate logistic, Pearson Type II, Pearson Type VII, Kotz type, and Bessel.

The implications of marginal inconsistency on copula modeling are discussed in Wang and Yan (2013, Section 3.2). Under marginal inconsistency, the univariate marginal dfs and quantile functions need to be derived for each d and often have no closed-form expressions; note that in the previous code we used the function `pobs()` from the R package `copula` (more formally introduced in Chap. 4) to perform marginal probability transformations empirically, thereby avoiding the evaluation of the marginal univariate dfs. The fact that the copula of a fixed multivariate margin changes as d changes is also undesirable. For example, in financial applications, this would mean that the dependence structure of a fixed number of risk-factor changes in a portfolio varies with the portfolio size. Furthermore, the dfs and densities of elliptical distributions are often hard to derive under marginal inconsistency, which further complicates numerical evaluation of their copulas and likelihood-based inference on the latter. These facts partly explain why elliptical copulas other than the popular multivariate normal and Student t are comparably rarely used in practice. □

Example 3.1.10 (Grouped Normal Variance Mixture Copulas) Grouped normal variance mixture copulas are copulas which combine a set of normal variance mixture copulas. Instead of multiplying $\boldsymbol{Y} = A\boldsymbol{Z}$ in (3.2) by the random variable $\sqrt{W}$ (that is, the same for all components of $\boldsymbol{Y}$), *grouped normal variance mixture copulas* use a fixed number $S \in \{1, \dots, d\}$ of different random variables $\sqrt{W_1}, \dots, \sqrt{W_S}$. The copula of $(\sqrt{W_1}, \dots, \sqrt{W_S})$ is often taken as the upper Fréchet–Hoeffding bound M since, otherwise, the additional randomness induced by this construction typically leads to rather weak dependence among the components of $\boldsymbol{X}$ which do not belong to the same group.

An example of the previous construction is given by the so-called grouped t copula studied in Demarta and McNeil (2005). The *grouped t copula* with S groups of dimensions $d_1, \dots, d_S$ such that $\sum_{s=1}^{S} d_s = d$ is the copula of the d-dimensional random vector

$$\boldsymbol{X} = (\sqrt{W_1}Y_1, \dots, \sqrt{W_1}Y_{d_1}, \dots, \sqrt{W_S}Y_{d_1+\dots+d_{S-1}+1}, \dots, \sqrt{W_S}Y_d)$$

where, as explained above, $(W_1, \dots, W_S)$ has copula M and $W_s \sim \mathrm{IG}(\nu_s/2, \nu_s/2)$, $s \in \{1, \dots, S\}$. Since all components are t distributed,

$$\boldsymbol{U} = (t_{\nu_1}(X_1), \dots, t_{\nu_1}(X_{d_1}), \dots, t_{\nu_S}(X_{d_1+\dots+d_{S-1}+1}), \dots, t_{\nu_S}(X_d))$$

follows a grouped t copula.

The following code generates a sample from a 10-dimensional grouped t copula with $S = 4$ groups such that $d_1 = 1$, $d_2 = 2$, $d_3 = 3$, and $d_4 = 4$:

```
> ## Sample from a grouped t copula
> d. <- 1:4 # sector dimensions
> d <- sum(d.) # total dimension
> nu <- rep(4^seq(2, -1), times = d.) # d.o.f. for each sector copula
> n <- 1000 # sample size
> set.seed(271) # set seed (for reproducibility)
> Z <- matrix(rnorm(n * d), ncol = n) # (d, n)-matrix
> P <- matrix(0.5, nrow = d, ncol = d) # correlation matrix
> diag(P) <- 1 # fix diagonal
> A <- t(chol(P)) # Cholesky factor A (s.th. AA^T = P)
> Y <- t(A %*% Z) # (n, d) matrix containing n d-vectors following N_d(0, P)
> U. <- runif(n) # n-vector of U(0,1) random variates
> W <- sapply(nu, function(nu.) 1/qgamma(U., shape = nu./2, rate = nu./2))
> X <- sqrt(W) * Y # (n, d)-matrix
> U <- sapply(1:d, function(j) pt(X[,j], df = nu[j])) # (n, d)-matrix sample
> ## Build matrix of colors
> cols <- matrix(1, nrow = d, ncol = d) # colors
> start <- c(1, cumsum(head(d., n = -1)) + 1) # block start indices
> end <- cumsum(d.) # block end indices
> for(j in seq_along(d.)) cols[start[j]:end[j], start[j]:end[j]] <- j # colors
> diag(cols) <- NA # remove colors corresponding to diagonal entries
> splom2(U, pch = ".", pscales = 0, col.mat = cols) # plot sample
```

The resulting scatter-plot matrix is shown in Fig. 3.7. Groups two, three, and four appear colorized and, by construction, $(U_2, U_3) \sim C^t_{0.5,4}$, $(U_4, U_5, U_6) \sim C^t_{0.5,1}$ and $(U_7, \dots, U_{10}) \sim C^t_{0.5,0.25}$; note the star-shaped structure appearing in the scatter plots corresponding to groups three and four due to the low values of ν. Finally, notice that while the dfs of random vectors of the form $(U_{j_1}, \dots, U_{j_s})$, $\{j_1, \dots, j_s\} \subseteq \{1, \dots, d\}$, $s \geq 2$, are t copulas when $j_1, \dots, j_s$ belong to the same group, this is not necessarily the case when $j_1, \dots, j_s$ belong to different groups.

□

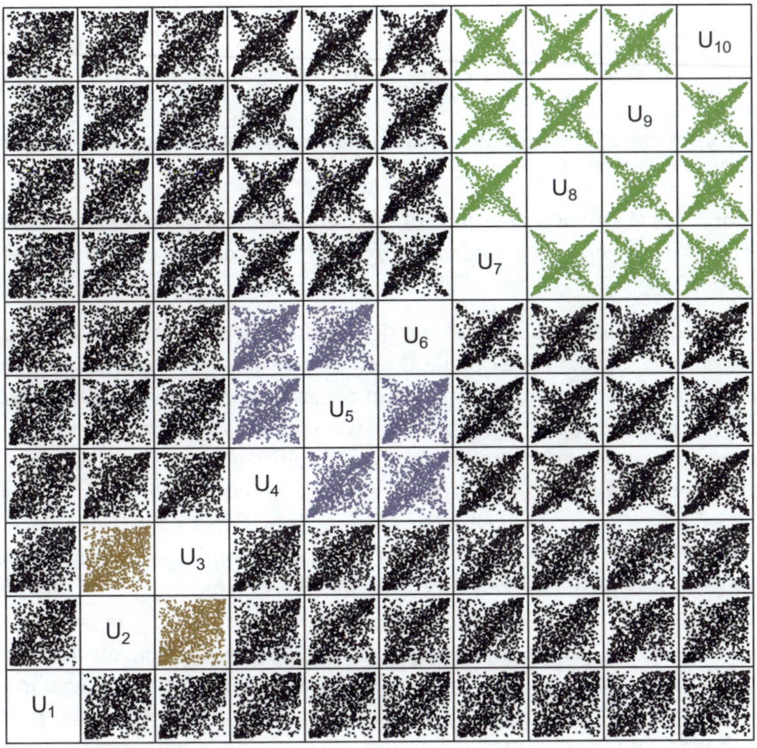

Fig. 3.7 $n = 1000$ independent realizations from a grouped t copula with four groups

3.2 Archimedean Copulas

An *Archimedean copula* is a copula of the form

$$C(\boldsymbol{u}) = \psi(\psi^{-1}(u_1) + \cdots + \psi^{-1}(u_d)), \quad \boldsymbol{u} \in [0, 1]^d, \tag{3.3}$$

for a so-called generator ψ, defined as follows.

Definition 3.2.1 (Archimedean Generator, d-Monotone, Completely Monotone) An *(Archimedean) generator* is a continuous, decreasing function $\psi : [0, \infty] \to [0, 1]$ which satisfies $\psi(0) = 1$, $\psi(\infty) = \lim_{t\to\infty} \psi(t) = 0$ and which is strictly decreasing on $[0, \inf\{t : \psi(t) = 0\}]$. Its (pseudo-)inverse $\psi^{-1} : [0, 1] \to [0, \infty]$ satisfies $\psi^{-1}(0) = \inf\{t : \psi(t) = 0\}$. The class of all generators is denoted by Ψ. If $\psi \in \Psi$ admits derivatives $\psi^{(k)}$ up to the order $k = d - 2$ satisfying $(-1)^k \psi^{(k)}(t) \geq 0$ for all $k \in \{0, \ldots, d-2\}$, $t \in (0, \infty)$, and $t \mapsto (-1)^{d-2}\psi^{(d-2)}(t)$ is decreasing and convex on $(0, \infty)$, then ψ is called *d-monotone*. If ψ is d-monotone for all d, that is, $(-1)^k \psi^{(k)}(t) \geq 0$ for all $k \in \{0, 1, 2, \ldots\}$, $t \in (0, \infty)$, then ψ is called *completely monotone*. □

It immediately follows from (3.3) that Archimedean copulas are exchangeable. As an example, $\psi(t) = \exp(-t)$, $t \in [0, \infty)$, is a completely monotone Archimedean generator and the copula of the form (3.3) it generates is the independence copula Π.

The following result provides a characterization of Archimedean copulas; see Malov (2001) and McNeil and Nešlehová (2009).

Theorem 3.2.2 (Characterization of Archimedean Copulas) *A function of the form* (3.3) *is an Archimedean copula if and only if* $\psi \in \Psi$ *and* ψ *is* d*-monotone.* □

An immediate consequence of the definition of d-monotonicity is that, in the bivariate case, C in (3.3) is a copula if and only if $\psi \in \Psi$ is convex. Furthermore, a generator $\psi \in \Psi$ generates an Archimedean copula in arbitrary dimensions if and only if ψ is completely monotone. We will mostly focus on the latter case in this section, as complete monotonicity is fulfilled by several commonly used parametric generator (and thus Archimedean copula) families, for example, the families of *Ali–Mikhail–Haq (AMH)*, *Clayton (C)*, *Frank (F)*, *Gumbel–Hougaard (GH)*, and *Joe (J)*. The expressions of the corresponding generators are given in Table 3.1; see also Hofert (2008). In the R package `copula`, the functions `amhCopula()`, `claytonCopula()`, `frankCopula()`, `gumbelCopula()` and `joeCopula()` create objects representing these Archimedean copulas. Notice that, in contrast to the definitions of the Clayton and Frank families given in Chap. 2, the parameter θ in Table 3.1 is only allowed to vary in $(0, \infty)$. The latter is the range for which the Clayton and Frank generators are completely monotone. Also note that for the Ali–Mikhail–Haq family, the range of attainable Kendall's tau is limited to $[0, 1/3)$.

Table 3.1 Well-known one-parameter completely monotone Archimedean generators ψ with corresponding frailty distributions F satisfying $\psi = \mathcal{LS}[F]$, where the transform $\mathcal{LS}$ is defined in (3.4), or, equivalently, $F = \mathcal{LS}^{-1}[\psi]$

Family	Parameter	$\psi(t)$	$V \sim F = \mathcal{LS}^{-1}[\psi]$
AMH	$\theta \in [0, 1)$	$(1-\theta)/(\exp(t)-\theta)$	$\mathrm{Geo}(1-\theta)$
C	$\theta \in (0, \infty)$	$(1+t)^{-1/\theta}$	$\mathrm{Gamma}(1/\theta, 1)$
F	$\theta \in (0, \infty)$	$-\log\big(1-(1-e^{-\theta})\exp(-t)\big)/\theta$	$\mathrm{Log}(1-e^{-\theta})$
GH	$\theta \in [1, \infty)$	$\exp(-t^{1/\theta})$	$\mathrm{S}(1/\theta, 1, \cos^{\theta}(\pi/(2\theta)), 1(\theta = 1); 1)$
J	$\theta \in [1, \infty)$	$1-(1-\exp(-t))^{1/\theta}$	$\mathrm{Sibuya}(1/\theta)$

In the last column, $\mathrm{Geo}(p)$ denotes the *geometric distribution* with probability mass function $p_k = p(1-p)^{k-1}$ at $k \in \mathbb{N} = \{1, 2, \dots\}$, $\mathrm{Gamma}(\alpha, \beta)$ denotes the *gamma distribution* as defined in Sect. 3.1.2, $\mathrm{Log}(p)$ denotes the *logarithmic distribution* with probability mass function $p_k = p^k/(-k\log(1-p))$ at $k \in \mathbb{N}$, $\mathrm{S}(1/\alpha, 1, \cos^{1/\alpha}(\alpha\pi/2), 1(\alpha = 1); 1)$ denotes the 1-parametrization of the *stable distribution* with characteristic function $\phi(t) = \exp(-(-it)^{\alpha})$ (see Nolan (2018, Section 1.3) for more details on stable distributions) and $\mathrm{Sibuya}(p)$ denotes the *Sibuya distribution* with probability mass function $p_k = \binom{p}{k}(-1)^{k-1}$ at $k \in \mathbb{N}$ (see Devroye 1993)

Example 3.2.3 (Graphs of Selected Archimedean Generators) The generators of the parametric Archimedean copula families implemented in copula, see Table 3.1, are available through specific R objects, such as copClayton for the Clayton family:

```
> copClayton@psi # generator of the Clayton family
```

```
function (t, theta)
pmax(1 + sign(theta) * t, 0)^(-1/theta)
<bytecode: 0x7ff146c00940>
<environment: namespace:copula>
```

Graphs of the generators listed in Table 3.1 (except Ali–Mikhail–Haq, which cannot attain Kendall's taus greater than or equal to 1/3) can be produced with the following code:

```
> t <- seq(0, 2, length.out = 257) # evaluation points
> tau <- 0.5 # Kendall's tau
> psi. <- cbind(Pi = exp(-t), # Pi generator
                C  = copClayton@psi(t, theta = iTau(claytonCopula(), tau)),
                F  = copFrank@psi  (t, theta = iTau(frankCopula(),   tau)),
                GH = copGumbel@psi (t, theta = iTau(gumbelCopula(),  tau)),
                J  = copJoe@psi    (t, theta = iTau(joeCopula(),     tau)))
> plot(t, psi.[,1], type = "l", lwd = 2,
       xlim = range(t), ylim = range(psi.), col = 1, ylab = "",
       xlab = quote(psi(t)~"as a function of t"))
> for(j in 2:ncol(psi.)) lines(t, psi.[,j], col = j, lwd = 2)
> legend("topright", bty = "n", lty = 1, lwd = 2, col = 1:ncol(psi.),
         legend = c("Independence", "Clayton", "Frank",
                    "Gumbel-Hougaard", "Joe"))
```

The resulting curves are displayed in Fig. 3.8.

□

Let V be a nonnegative random variable with df F. The *Laplace–Stieltjes transform* of F is defined by

$$\mathcal{LS}[F](t) = \int_0^\infty \exp(-tv)\,\mathrm{d}F(v) = \mathbb{E}(\exp(-tV)), \quad t \in [0, \infty). \tag{3.4}$$

The df F can be recovered from $\mathcal{LS}[F]$ via the corresponding inverse transform, denoted by $\mathcal{LS}^{-1}[\psi]$. The reason for considering Laplace–Stieltjes transforms at this point is due to *Bernstein's Theorem*; see Feller (1971, p. 439). This important result states that $\psi \in \Psi$ is completely monotone if and only if ψ is the Laplace–Stieltjes transform of a df F on the positive real line satisfying $F(0) = 0$, that is, $\psi(t) = \mathcal{LS}[F](t)$, $t \in [0, \infty)$. The random variable $V \sim F$ is known as *frailty* and F as *frailty distribution*. Based on (3.4), one can verify that, in terms of the

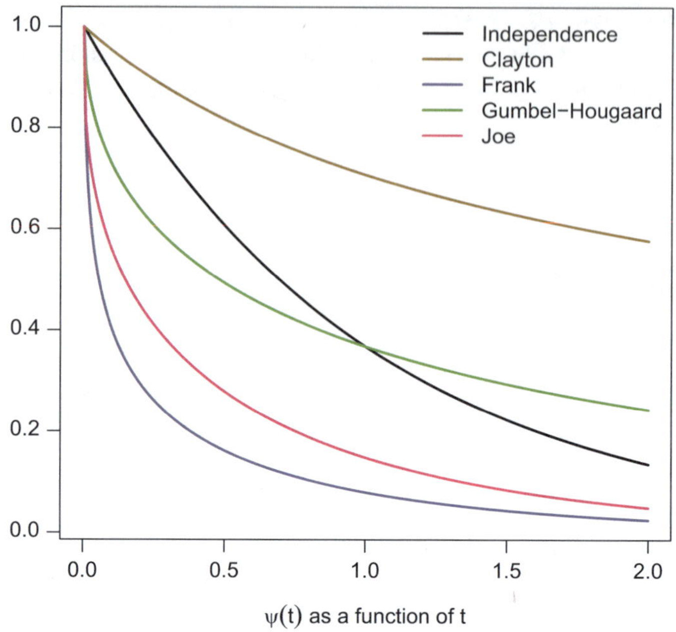

Fig. 3.8 Graphs of the Archimedean generators listed in Table 3.1

frailty V, Archimedean copulas allow for the stochastic representation

$$\boldsymbol{U} = \left(\psi\left(\frac{E_1}{V}\right), \dots, \psi\left(\frac{E_d}{V}\right)\right) \sim C, \tag{3.5}$$

where $E_1, \dots, E_d \overset{\text{ind.}}{\sim} \text{Exp}(1)$ are independent of V. This leads to the so-called *Marshall–Olkin algorithm* for sampling from Archimedean copulas generated by completely monotone $\psi \in \Psi$; see Marshall and Olkin (1988).

Algorithm 3.2.4 (Simulation of Archimedean Copulas with Completely Monotone Generators)

(1) Sample $V \sim F = \mathcal{LS}^{-1}[\psi]$.
(2) Sample $E_1, \dots, E_d \overset{\text{ind.}}{\sim} \text{Exp}(1)$, independently of V.
(3) Return $\boldsymbol{U} = (\psi(E_1/V), \dots, \psi(E_d/V))$.

□

There are two consequences of the complete monotonicity of an Archimedean generator $\psi \in \Psi$. First, such a generator is *strict*, that is, $\psi(t) > 0$ for all $t \in [0, \infty)$. Second, the corresponding Archimedean copula C given by (3.3) can be shown to be *positive quadrant dependent*, that is, $C(\boldsymbol{u}) \geq \Pi(\boldsymbol{u})$ for all $\boldsymbol{u} \in [0, 1]^d$.

For the frailty distributions of the families of Ali–Mikhail–Haq, Clayton, Frank, Gumbel–Hougaard, and Joe as listed in the last column of Table 3.1, the R package `copula` uses fast and reliable sampling algorithms; see Hofert (2011, 2012b) for details. In general, one can use numerical inversion algorithms for computing $\mathcal{LS}^{-1}$; see Hofert (2010) for some options. Note, however, that this may be prone to numerical errors (resulting for instance in an $F = \mathcal{LS}^{-1}[\psi]$ that is not necessarily increasing, and thus, not a df).

Example 3.2.5 (Sampling from a Gumbel–Hougaard Copula) The following code, utilizing the function `archmCopula()` to generate an Archimedean copula object, verifies that `rCopula()` applies Algorithm 3.2.4 for sampling from a Gumbel–Hougaard copula:

```
> ## Setup
> n <- 1000 # sample size
> d <- 5 # dimension
> family <- "Gumbel" # copula family
> th <- iTau(archmCopula(family), 0.5) # copula parameter
> ## Version 1: manually
> set.seed(271) # set seed (for reproducibility)
> cop <- getAcop(family) # define the Archimedean copula to sample from
> V <- cop@V0(n, theta = th) # generate frailties V from F
> E <- matrix(-log(runif(n*d)), ncol = d) # sample independent Exp(1)
> U.man <- cop@psi(E/V, theta = th) # construct U
> ## Version 2: via rCopula()
> cop <- archmCopula(family, param = th, dim = d) # define the copula to sample
> set.seed(271) # set seed (for reproducibility)
> U <- rCopula(n, cop) # sample
> ## Check
> stopifnot(all.equal(U.man, U))
```

Note that using `rexp()` instead of `-log(runif())` for drawing from Exp(1) would lead to different random variates here as, historically in R, exponential random variables are not generated with the inversion method. □

If the dth derivative $\psi^{(d)}$ of an Archimedean generator ψ exists and is continuous, the density c of C can be verified to be

$$c(\boldsymbol{u}) = \frac{\psi^{(d)}(\psi^{-1}(u_1) + \cdots + \psi^{-1}(u_d))}{\prod_{j=1}^{d} \psi'(\psi^{-1}(u_j))}, \quad \boldsymbol{u} \in (0,1)^d,$$

where $\psi' = \psi^{(1)}$. Evaluating $\psi^{(d)}$ can be challenging as discussed in the following example, but numerically stable evaluation procedures are available; see Hofert et al. (2012, 2013).

Example 3.2.6 (Graphs of $|\psi^{(d)}|$ for the Gumbel–Hougaard Generator) The following code displays the graphs of $|\psi^{(d)}|$ for the Gumbel–Hougaard generator given in Table 3.1 for $d \in \{2, 5, 10, 20, 50\}$:

```
> t <- 10^seq(-2, 2, by = 0.05) # evaluation points
> th <- iTau(gumbelCopula(), 0.5) # corresponding GH parameter theta
> d <- c(2, 5, 10, 20, 50) # order of the derivatives
> dPsi. <- sapply(d, function(d.) # (length(t), length(d))-mat. of derivatives
      copGumbel@absdPsi(t, theta = th, degree = d.))
> plot(t, dPsi.[,1], type = "l", log = "y", lwd = 2,
       xlim = range(t), ylim = range(dPsi.), col = 1, ylab = "",
       xlab = quote(bgroup("|",{psi^(d)}(t),"|")~"as a function of t"))
> for(j in 2:length(d)) lines(t, dPsi.[,j], col = j, lwd = 2)
> legend("topright", bty = "n", lty = 1, lwd = 2, col = 1:length(d),
         legend = as.expression(lapply(1:length(d), function(j)
             substitute(d==d., list(d.=d[j])))))
```

The resulting graphs given in Fig. 3.9 highlight the range of $|\psi^{(d)}(t)|$ as a function of t and hint at the underlying numerical difficulties which can arise when evaluating high-order derivatives.

□

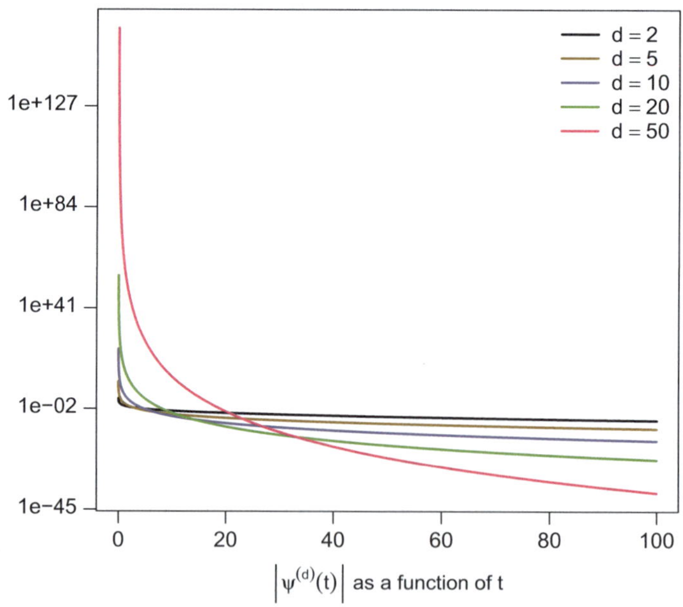

Fig. 3.9 Graphs of $|\psi^{(d)}(t)|$ as a function of t for the Gumbel–Hougaard generator $\psi(t) = \exp(-t^{1/\theta})$, where θ is chosen such that Kendall's tau equals 0.5

A consequence of the numerical stability of the evaluation of the derivatives of Archimedean generators in the R package `copula` is that density evaluation can be carried out in high dimensions as illustrated in the following example for $d = 100$.

Example 3.2.7 (Negative Log-Likelihood and Profile Likelihood of a Gumbel–Hougaard Copula) In this example, we generate a sample of size $n = 100$ from a Gumbel–Hougaard copula in dimension $d = 100$ such that its bivariate margins have a Kendall's tau of 0.5. We evaluate the negative log-likelihood with the function `loglikCopula()` of the R package `copula`. Next, we determine an initial interval and value via `initOpt()`. We then apply the optimizer `mle2()` of the R package `bbmle`, see Bolker (2017), to minimize the negative log-likelihood based on this sample, to obtain a likelihood-based 95% confidence interval for the copula parameter and to construct a profile likelihood plot for it (see the vignette `mle2` of `bbmle` for more details):

```
> ## Generate a sample from a Gumbel-Hougaard copula
> n <- 100 # sample size
> d <- 100 # dimension
> tau <- 0.5 # Kendall's tau
> family <- "Gumbel" # copula family
> th0 <- iTau(archmCopula(family), tau = tau) # true copula parameter
> cop <- archmCopula(family, param = th0, dim = d) # define copula
> set.seed(271) # set seed (for reproducibility)
> U <- rCopula(n, cop) # sample
> ## Maximum likelihood estimation (MLE)
> nLL <- function(th) -loglikCopula(th, u = U, copula = cop) # -log-likelihood
> ii <- initOpt(family) # initial interval
> iv <- initOpt(family, interval = FALSE, u = U) # initial value
> library(bbmle)
> mle <- mle2(minuslogl = nLL, optimizer = "optim", method = "L-BFGS-B",
              start = list(th = iv), lower = ii[1], upper = ii[2]) # optimize
> stopifnot(mle@details$convergence == 0) # check convergence
> th.hat <- mle@coef # estimate
> ## Confidence intervals (CIs)
> CI <- confint(mle, quietly = TRUE)
> stopifnot(CI[1] <= th0, th0 <= CI[2]) # check if theta_0 is in the CI
> prof <- profile(mle) # profile likelihood (see bbmle's vignette 'mle2')
> ## nLL plot
> th.bds <- iTau(archmCopula(family), tau = c(tau-0.015, tau+0.015)) # bounds
> th. <- seq(th.bds[1], th.bds[2], length.out = 101) # x values
> nLL. <- unlist(lapply(th., nLL)) # y values
> plot(th., nLL., type = "l", xlab = quote(theta), ylab = "")
> abline(v = th0, lty = 2)
> abline(v = CI[1], lty = 3)
> abline(v = CI[2], lty = 3)
> abline(v = th.hat, lty = 4)
```

```
> legend("bottomright", bty = "n", y.intersp = 1.1, lty = 1:4,
         legend = c(quote(- "log-likelihood"),
                    substitute("True"~theta[0]==T0, list(T0 = th0)),
                    "95% CIs", expression("Estimate"~hat(theta)[n])))
> mtext(substitute("GH copula,"~~n==N*","~~d==D, list(N = n, D = d)),
        side = 4, line = 0.6, adj = 0, las = 0) # label
> ## Profile-likelihood plot for theta
> plot(prof, xaxt = "n", main = "",
       col.prof = "black", col.minval = "black", col.conf = "black",
       xlabs = expression(theta), # a *vector*
       ylab = quote("Profile likelihood for"~theta * (abs(z)==sqrt(deviance))))
> axis(side = 1, at = c(1.96, 1.98, th.hat, th0, 2.02, 2.04),
       labels = c("1.96","1.98", quote(hat(theta)[n]),
                  expression(theta[0]), "2.02", "2.04"),
       padj = c(0.13, 0.13, 0.24, 0.35, 0.13, 0.13))
```

Figure 3.10 shows this negative log-likelihood (left) and corresponding profile likelihood (right) of the Gumbel–Hougaard copula C_2^{GH}.

□

Properties of Archimedean copulas can typically be expressed in terms of their generators. For example, given a strict and twice continuously differentiable generator ψ, Kendall's tau of the generated copula can be written as

$$\tau = 1 - 4\int_0^\infty t(\psi'(t))^2\,\mathrm{d}t = 1 + 4\int_0^1 \frac{\psi^{-1}(t)}{(\psi^{-1}(t))'}\,\mathrm{d}t.$$

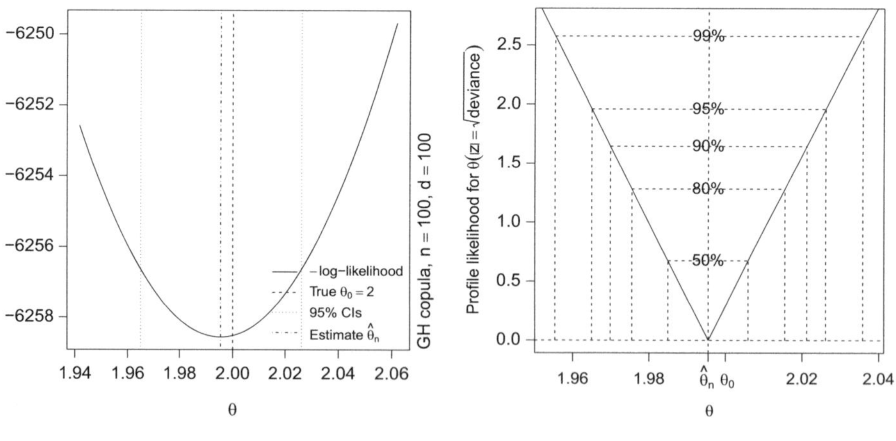

Fig. 3.10 Plot of the negative log-likelihood of a Gumbel–Hougaard copula (including a likelihood-based 95% confidence interval for the copula parameter) based on a sample of size $n = 100$ from C_2^{GH} in dimension $d = 100$ (left). Corresponding profile likelihood plot for the copula parameter (right)

Table 3.2 Kendall's tau and coefficients of tail dependence for the Archimedean copulas defined in Table 3.1

Family	τ	λ_l	λ_u
AMH	$1-2(\theta+(1-\theta)^2\log(1-\theta))/(3\theta^2)$	0	0
C	$\theta/(\theta+2)$	$2^{-1/\theta}$	0
F	$1+4(D_1(\theta)-1)/\theta$	0	0
GH	$(\theta-1)/\theta$	0	$2-2^{1/\theta}$
J	$1-4\sum_{k=1}^{\infty}1/(k(\theta k+2)(\theta(k-1)+2))$	0	$2-2^{1/\theta}$

The quantity D_1 is the *Debye function of order one*, defined by $D_1(\theta)=\int_0^{\theta}t/(\exp(t)-1)\,\mathrm{d}t/\theta$, $\theta\in(0,\infty)$

Note in passing that there is no general formula known for Spearman's rho for Archimedean copulas. The coefficients of tail dependence can be expressed in terms of the generator via

$$\lambda_l=\lim_{t\to\infty}\frac{\psi(2t)}{\psi(t)}=2\lim_{t\to\infty}\frac{\psi'(2t)}{\psi'(t)},\quad \lambda_u=2-\lim_{t\downarrow 0}\frac{1-\psi(2t)}{1-\psi(t)}=2-2\lim_{t\downarrow 0}\frac{\psi'(2t)}{\psi'(t)},$$

where the formulas involving the generator derivatives hold if ψ is differentiable; see also (2.27). Table 3.2 summarizes these properties for the families of Ali–Mikhail–Haq, Clayton, Frank, Gumbel–Hougaard, and Joe as defined in Table 3.1.

As far as symmetries are concerned, see Definition 2.5.4, Archimedean copulas are not necessarily radially symmetric; among the copulas defined in Table 3.1, only bivariate Frank copulas are. As mentioned before and as is immediate from (3.3), however, Archimedean copulas are exchangeable.

In the remaining examples in this section, we briefly discuss some construction principles and asymmetric extensions of Archimedean copulas. The interested reader is referred to the documentation in the R package `copula` for some of the functions used in the chunks and not further explained here.

Example 3.2.8 (Outer Power Archimedean Copulas) Generator transformations allow one to create more flexible, multi-parameter Archimedean generators and copulas. One construction is to consider the transformed Archimedean generator $\tilde{\psi}(t)=\psi((a^{\theta}+t)^{1/\theta}-a)$, $t\in[0,\infty)$, for $\theta\in[1,\infty)$ and $a\in[0,\infty)$; see Hofert (2011). The latter reference also shows that the corresponding frailty $\tilde{V}$ admits the stochastic representation $\tilde{V}=\tilde{S}V^{\theta}$, where V is the frailty corresponding to ψ and $\tilde{S}$ follows an *exponentially tilted stable distribution* with density $f(x)=\exp(-a^{\theta}x)f_S(x)/\psi_S(a^{\theta})$, $x\in[0,\infty)$, where f_S denotes the density of $\mathrm{S}(1/\theta,1,\cos^{\theta}(\pi/(2\theta)),\mathbf{1}(\theta=1);1)$ (a stable distribution with corresponding

Laplace–Stieltjes transform $\psi_S(t) = \exp(-t^{1/\theta})$, $t \in [0, \infty)$). For $a = 0$, one obtains the so-called *outer power Archimedean copulas* with corresponding generator $\tilde{\psi}(t) = \psi(t^{1/\theta})$, $t \in [0, \infty)$. If τ denotes Kendall's tau of the Archimedean copula generated by ψ, then Kendall's tau of the Archimedean copula generated by $\tilde{\psi}$ is given by $\tilde{\tau} = 1 - (1 - \tau)/\theta$.

Let us illustrate the previous construction when ψ is the Clayton generator with parameter θ_C. The following code based on the function `opower()` shows how one can sample from the resulting outer power Clayton copula family:

```
> ## Setup
> th.C <- copClayton@iTau(0.3) # Clayton parameter s.t. tau = 0.3
> op.C <- opower(copClayton, thetabase = th.C) # define an opC copula family
> ## Define two opC copulas (tau = 0.5 and 0.8)
> th <- sapply(c(0.5, 0.7), op.C@iTau) # choose parameter according to taus
> opC  <- onacopulaL(op.C, list(th[1], 1:2)) # define the opC copula
> opC. <- onacopulaL(op.C, list(th[2], 1:2))
> ## Sample
> set.seed(271) # set seed (for reproducibility)
> U  <- rCopula(1000, copula = opC)
> U. <- rCopula(1000, copula = opC.)
> plot(U,  xlab = quote(U[1]), ylab = quote(U[2])) # opC sample
> plot(U., xlab = quote(U[1]), ylab = quote(U[2])) # opC. sample
```

The obtained scatter plots are shown in Fig. 3.11. The coefficients of tail dependence for the resulting outer power Clayton family are given explicitly by $\tilde{\lambda}_l = 2^{-1/(\theta\theta_C)}$ and $\tilde{\lambda}_u = 2 - 2^{1/\theta}$. It follows that in contrast to Clayton copulas, outer power Clayton copulas are upper tail dependent (for $\theta > 1$) and one can

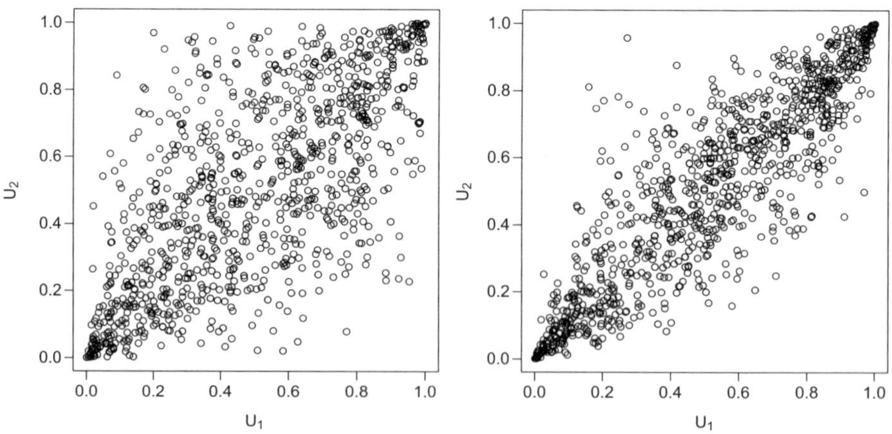

Fig. 3.11 Sample of size $n = 1000$ from an outer power Clayton copula with a Kendall's tau of 0.5 (left) and 0.7 (right)

choose θ and θ_C to obtain desired values for $\tilde{\lambda}_l$ and $\tilde{\lambda}_u$. This property is often of interest in applications in risk management, finance, and insurance.

□

Example 3.2.9 (Archimedean Copulas with d-Monotone Generators, Liouville Copulas) Following Theorem 3.2.2, d-dimensional Archimedean copulas can also be generated by (just) d-monotone generators $\psi \in \Psi$. In the same way that completely monotone generators can be characterized through the Laplace–Stieltjes transform, d-monotone generators can be characterized through the so-called Williamson d-transform. Let R be a nonnegative random variable with df F_R. The *Williamson d-transform* of F_R is defined by

$$\mathcal{W}_d[F_R](t) = \int_{(t,\infty)} (1 - t/r)^{d-1}\,\mathrm{d}F_R(r) = \begin{cases} \mathbb{E}(\max(1 - t/R, 0)^{d-1}), & \text{if } t > 0,\\ 1 - F_R(0), & \text{if } t = 0. \end{cases} \tag{3.6}$$

One can then show that a generator $\psi \in \Psi$ is d-monotone (and thus generates an Archimedean copula through (3.3) in dimension d or smaller) if and only if it is the Williamson d-transform of a df F_R on $[0, \infty)$ such that $F_R(0) = 0$ (which, using (3.6), can be denoted by $\psi(t) = \mathcal{W}_d[F_R](t)$, $t \in [0, \infty)$); see McNeil and Nešlehová (2009, Corollary 3.1). The random variable $R \sim F_R$ is known as the *radial part*. Similar to (3.5), a stochastic representation for Archimedean copulas generated by d-monotone generators is given by

$$\boldsymbol{U} = (\psi(RS_1), \ldots, \psi(RS_d)) \sim C, \tag{3.7}$$

where $R \sim F_R$ is independent of $\boldsymbol{S} \sim \mathrm{U}(\{\boldsymbol{x} \in [0, \infty)^d : \|\boldsymbol{x}\|_1 = 1\})$. Since $\boldsymbol{S}$ is uniformly distributed on the unit simplex in the first orthant, it admits the stochastic representation

$$\boldsymbol{S} = \left(\frac{E_1}{\sum_{j=1}^d E_j}, \ldots, \frac{E_d}{\sum_{j=1}^d E_j}\right), \quad \text{where } E_1, \ldots, E_d \overset{\text{ind.}}{\sim} \mathrm{Exp}(1).$$

Using $R \overset{\mathrm{d}}{=} \sum_{j=1}^d E_j/V$, where, as explained below (3.5), the frailty random variable $V > 0$ is independent of $E_1, \ldots, E_d$, it follows by comparison with (3.7) that (3.5) is indeed a special case of (3.7).

By Definition 3.2.1, an Archimedean generator is a decreasing function $\psi : [0, \infty] \to [0, 1]$ such that $\psi(0) = 1$. It can thus be interpreted as a univariate survival function. If ψ is completely monotone, then, by Bernstein's Theorem,

$\psi(t) = \mathcal{LS}[F](t)$, $t \in [0,\infty)$, where $\mathcal{LS}$ is given by (3.4). Fix $j \in \{1,\dots,d\}$. With E_j and V as in (3.5), conditioning thus implies that

$$\mathbb{P}(E_j/V > t) = \mathbb{P}(E_j > tV) = \int_0^\infty \mathbb{P}(E_j > tV \mid V = v)\,\mathrm{d}F(v)$$
$$= \int_0^\infty \exp(-tv)\,\mathrm{d}F(v) = \mathcal{LS}[F](t) = \psi(t), \quad t \in [0,\infty).$$

Therefore, the generator ψ is the common univariate marginal survival function of $(E_1/V,\dots,E_d/V)$. Combining this fact and (3.5) with Sklar's Theorem for survival functions (see Theorem 2.5.1) shows that Archimedean copulas generated by completely monotone generators are the survival copulas of $(E_1/V,\dots,E_d/V)$.

Similarly, we have that an Archimedean copula C admitting the stochastic representation (3.7) is the survival copula of $\boldsymbol{RS} = R(S_1,\dots,S_d)$; see McNeil and Nešlehová (2009, Theorem 3.1 (i)). If instead of $\boldsymbol{S} \sim \mathrm{U}(\{\boldsymbol{x} \in [0,\infty)^d : \|\boldsymbol{x}\|_1 = 1\})$, one considers $\boldsymbol{D} = (D_1,\dots,D_d)$ following a Dirichlet distribution, one arrives at Liouville copulas. Specifically, a *Liouville copula* is the survival copula of $R\boldsymbol{D}$, where

$$\boldsymbol{D} = \left(\frac{G_1}{\sum_{j=1}^d G_j},\dots,\frac{G_d}{\sum_{j=1}^d G_j}\right), \quad G_j \overset{\text{ind.}}{\sim} \mathrm{Gamma}(\alpha_j,1),\ j \in \{1,\dots,d\};$$

see McNeil and Nešlehová (2010). One advantage of this construction is that Liouville copulas are not exchangeable unless $\alpha_1,\dots,\alpha_d$ are all equal; see also Genest and Nešlehová (2013). A possible drawback is that the marginal survival functions of $R\boldsymbol{D}$ might not be available analytically; see McNeil and Nešlehová (2010, Theorem 2) for the case $\alpha_j \in \mathbb{N}$, $j \in \{1,\dots,d\}$.

Let $d = 2$. We shall first illustrate pseudo-random number generation from an Archimedean copula satisfying (3.7) for a radial part R following a Pareto distribution on $[1,\infty)$; the resulting copula is known as *Pareto–simplex copula*. In this case, the Archimedean generator appearing in (3.7) can be expressed in terms of the incomplete beta function as explained in McNeil and Nešlehová (2010, Example 3):

```
> ## Setup
> n <- 1000 # sample size
> d <- 2 # dimension
> th <- 1 # Pareto parameter
> set.seed(271) # set seed (for reproducibility)
> R <- runif(n)^(-1/th) # sample radial part (here: Pareto on [1,Inf))
> ## Sample from a so-called Pareto-simplex copula
> E <- matrix(rexp(n * d), nrow = n, ncol = d) # unit exponentials
> S <- E / matrix(rowSums(E), nrow = n, ncol = d) # S uniformly on unit simplex
> incBeta <- function(x, a, b) pbeta(x, a, b) * beta(a, b) # incomplete beta
> psi <- function(t, th) t^(-1/th) * incBeta(pmin(1,t), a = 1/th, b = d) / th
> U <- psi(R * S, th = th)
> plot(U, xlab = quote(U[1]), ylab = quote(U[2])) # Pareto-simplex sample
```

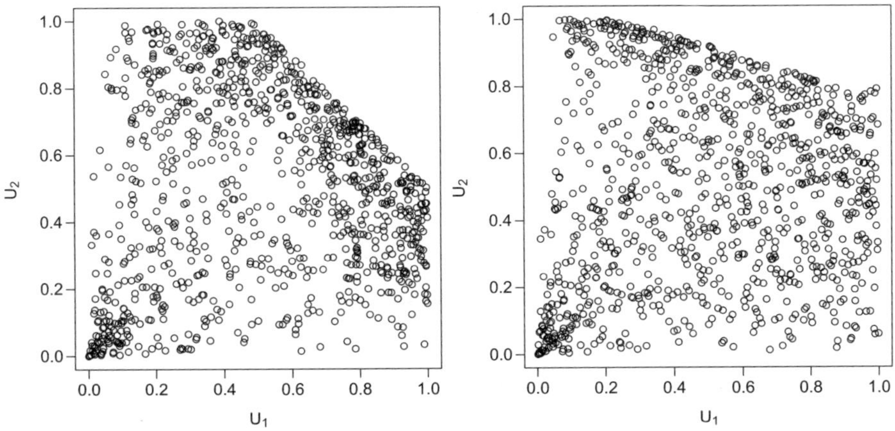

Fig. 3.12 Samples of size $n = 1000$ of a Pareto–simplex copula (left) and (approximately) of a Pareto–Liouville copula (right)

The resulting scatter plot is shown on the left-hand side of Fig. 3.12.

To obtain a sample from a Liouville copula with the same radial part and Dirichlet parameters $\alpha_1 = 0.5$ and $\alpha_2 = 2$, we need to evaluate the marginal survival functions of the random vector $R(D_1, D_2)$, as explained above. The latter are not known analytically in this case. However, for sufficiently large n, one can simply approximate them by marginal empirical survival functions. To this end, the code below conveniently invokes the function `pobs()` (see Sect. 4.1.2):

```
> ## (Approximately) sample from a so-called Pareto-Liouville copula
> alpha <- c(0.5, 2) # Dirichlet parameters
> G <- vapply(alpha, function(a) rgamma(n, a), numeric(n)) # Gamma variables
> D <- G / rowSums(G) # Dirichlet distribution on unit simplex
> U. <- 1 - pobs(R * D) # use empirical marginal survival functions
> plot(U., xlab = quote(U[1]), ylab = quote(U[2])) # Pareto-Liouville sample
```

The resulting scatter plot is shown on the right-hand side of Fig. 3.12 and, unlike the scatter on the left-hand side, it displays non-exchangeability as a consequence of considering unequal Dirichlet parameters.

Example 3.2.10 (Nested Archimedean Copulas) Besides Liouville copulas, another way of introducing (partial) asymmetry to Archimedean copulas is through nested Archimedean copulas. A *nested Archimedean copula* is an Archimedean copula with arguments replaced by other (possibly nested) Archimedean copulas. Let us consider the example

$$C(\boldsymbol{u}) = C_0\big(u_1, C_1(u_2, u_3), C_2(u_4, C_3(u_5, u_6, u_7))\big), \quad \boldsymbol{u} \in [0, 1]^7, \qquad (3.8)$$

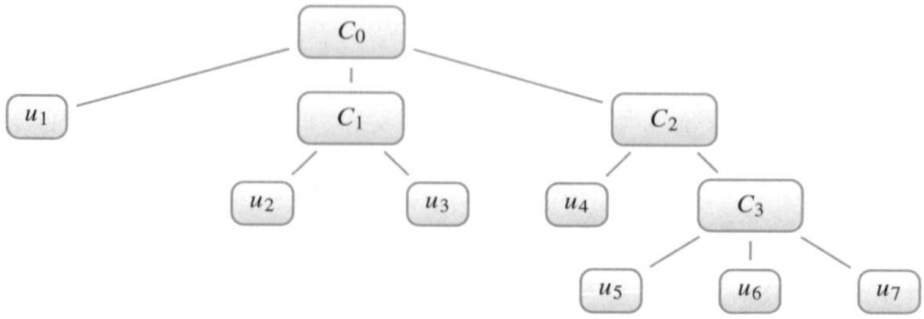

Fig. 3.13 Tree representation of the nested Archimedean copula (3.8)

with three nesting levels, where C_k is an Archimedean copula generated by ψ_k, $k \in \{0, 1, 2, 3\}$; see Fig. 3.13 for the tree representation of this nested construction.

If $\psi_0, \dots, \psi_3$ are completely monotone, C in (3.8) is indeed a copula if the so-called *sufficient nesting condition* is fulfilled, that is, in the case under consideration, $(\psi_0^{-1} \circ \psi_1)'$, $(\psi_0^{-1} \circ \psi_2)'$, and $(\psi_2^{-1} \circ \psi_3)'$ are completely monotone; see Joe (1997, pp. 87) and McNeil (2008). Under this assumption, the corresponding stochastic representation of a $\boldsymbol{U} \sim C$ is

$$\boldsymbol{U} = \left(\psi_0\left(\frac{E_1}{V_0}\right), \psi_1\left(\frac{E_2}{V_{01}}\right), \psi_1\left(\frac{E_3}{V_{01}}\right), \psi_2\left(\frac{E_4}{V_{02}}\right), \psi_3\left(\frac{E_5}{V_{23}}\right), \psi_3\left(\frac{E_6}{V_{23}}\right), \psi_3\left(\frac{E_7}{V_{23}}\right)\right),$$

where $V_0, V_{01}, V_{02}, V_{23}$ are (dependent) frailties such that V_{01}, V_{02} are conditionally independent given V_0. Therefore, when sampling from C, one first draws V_0 and then, independently of each other, V_{01}, V_{02} from a df which depends on the sampled V_0. Given V_{02}, one then draws V_{23} from a df which depends on V_{02}. Formally, if $\psi_{kl}(t; V) = \exp\big(-V\psi_k^{-1}(\psi_l(t))\big)$, $t \in [0, \infty)$, for $k \in \{0, 2\}$ and $l \in \{1, 2, 3\}$, one has the following hierarchy or nesting of conditionally drawn random variables:

1. $V_0 \sim F_0 = \mathcal{LS}^{-1}[\psi_0]$;
2. $V_{01} \mid V_0 \sim F_{01} = \mathcal{LS}^{-1}[\psi_{01}(\cdot\,; V_0)]$ and $V_{02} \mid V_0 \sim F_{02} = \mathcal{LS}^{-1}[\psi_{02}(\cdot\,; V_0)]$; and
3. $V_{23} \mid V_{02} \sim F_{23} = \mathcal{LS}^{-1}[\psi_{23}(\cdot\,; V_{02})]$.

This nesting of frailties leads to the nested copula form (3.8), hence the name of this class of copulas. Clearly, Archimedean copulas are a special case of this construction.

Figure 3.14, produced with the following code based on the function `onacopulaL()`, shows a sample of size $n = 1000$ from a nested Archimedean copula with structure as depicted in Fig. 3.13 constructed from Gumbel–Hougaard copulas $C_0, \dots, C_3$:

```
> ## Define a nested Gumbel-Hougaard copula
> family <- "Gumbel" # copula family
> tau <- c(0.2, 0.4, 0.6, 0.8) # Kendall's tau
> th <- iTau(archmCopula(family), tau = tau) # corresponding parameters
> nlist <- list(th[1], 1, list(list(th[2], 2:3), # NAC structure
                               list(th[3], 4, list(list(th[4], 5:7)))))
> NAC <- onacopulaL(family, nacList = nlist) # NAC copula
> ## Sample
> set.seed(271) # set seed (for reproducibility)
> U <- rCopula(1000, copula = NAC) # sample
> ## Build matrix of colors
> cols <- matrix(1, nrow = 7, ncol = 7)
> cols[2, 3] <- 2
> cols[4, 5:7] <- 3
> cols[5:7, 5:7] <- 4
> cols[lower.tri(cols)] <- t(cols)[lower.tri(cols)]
> diag(cols) <- NA
> splom2(U, pch = ".", pscales = 0, col.mat = cols) # plot sample
```

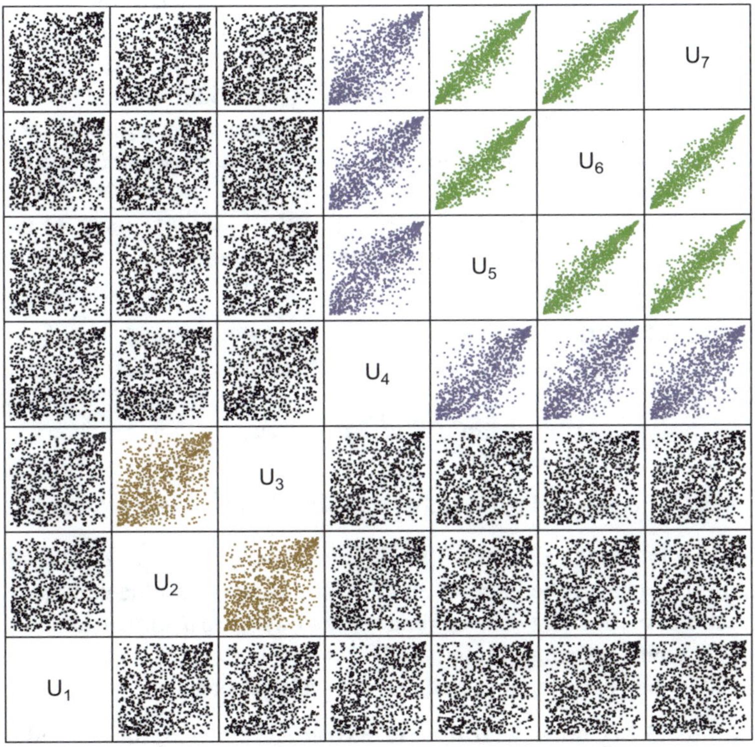

Fig. 3.14 Scatter-plot matrix of a sample of size $n = 1000$ from a nested Archimedean copula of the form (3.8). As building blocks, Gumbel–Hougaard copulas with parameters $\theta_0, \ldots, \theta_3$ (of the generators $\psi_0, \ldots, \psi_3$) with corresponding Kendall's taus equal to 0.2, 0.4, 0.6, and 0.8, respectively, are used

Extensions to trees involving different Archimedean families are available; see, for example, Hofert (2011) or Hofert (2012a). Concerning the densities of nested Archimedean copulas, see Hofert and Pham (2013). □

3.3 Extreme-Value Copulas

The copulas of random vectors distributed according to multivariate extreme-value distributions are called *extreme-value copulas*. From a practical perspective, multivariate extreme-value distributions are strongly related to the multivariate extension of the block maxima method popularized in the seminal monograph of Gumbel (1958) in the one-dimensional case; see also McNeil et al. (2015, Section 7.5). Starting from a multivariate time series (say, daily temperatures at d stations), the multivariate block maxima method consists of dividing the initial observations into blocks of equal size (say a year) and forming a multivariate sample of componentwise block maxima (so annual maxima for each of the d stations over the same years). A consequence of the extremal types theorem (see, for instance, de Haan and Ferreira 2006) is then that, under rather weak conditions, vectors of properly standardized componentwise block maxima of independent and identically distributed (iid) random vectors converge in distribution to a *multivariate extreme-value distribution* (the univariate margins of which are generalized extreme-value (GEV) distributions).

The extremal types theorem involves letting the block size increase to infinity, which explains why the copulas of multivariate extreme-value distributions are frequently formally defined as follows.

Definition 3.3.1 (Extreme-Value Copulas) A d-dimensional copula C is an *extreme-value copula* if there exists a copula C^* such that, for any $\boldsymbol{u} \in [0, 1]^d$,

$$\lim_{n\to\infty} C^*(\boldsymbol{u}^{1/n})^n = C(\boldsymbol{u}). \tag{3.9}$$

The copula C^* is then said to be in the *maximum domain of attraction* of C. □

It helps to interpret the term $C^*(\boldsymbol{u}^{1/n})^n = C^*(u_1^{1/n}, \dots, u_d^{1/n})^n$ in the previous definition: It can be verified that $C^*(\boldsymbol{u}^{1/n})^n$ is the copula, evaluated at $\boldsymbol{u} \in [0, 1]^d$, of the vector of componentwise maxima computed from a d-dimensional iid sample $\boldsymbol{Y}_1, \dots, \boldsymbol{Y}_n$ with continuous marginal dfs and copula C^*.

The following characterization of extreme-value copulas is also very useful.

Proposition 3.3.2 (Characterization of Extreme-Value Copulas Based on Max-Stability) *A copula C is an extreme-value copula if and only if it is* max-stable, *that is, if and only if, for any $\boldsymbol{u} \in [0, 1]^d$ and $r \in \mathbb{N}$,*

$$C(\boldsymbol{u}^{1/r})^r = C(\boldsymbol{u}). \tag{3.10}$$

□

This proposition can be proven relatively easily. Necessity follows by using, in combination with (3.9), the fact that, for any $\boldsymbol{u} \in [0,1]^d$ and $r \in \mathbb{N}$,

$$C(\boldsymbol{u}) = \lim_{n\to\infty} C^*(\boldsymbol{u}^{1/n})^n = \lim_{n\to\infty} C^*(\boldsymbol{u}^{1/(nr)})^{nr} = \big(\lim_{n\to\infty} C^*((\boldsymbol{u}^{1/r})^{1/n})^n\big)^r = C(\boldsymbol{u}^{1/r})^r,$$

where the first and the last equality hold since C is an extreme-value copula. Sufficiency is an immediate consequence of the fact that any max-stable copula C is in the maximum domain of attraction of itself. Interestingly enough, it can be shown that a max-stable copula satisfies (3.10) for any real $r > 0$; see, for instance, Galambos (1978, Lemma 5.4.1).

Another useful characterization is essentially due to Pickands (1981) and involves functions defined on the unit simplex

$$\Delta_{d-1} = \{(w_1, \ldots, w_{d-1}) \in [0,1]^{d-1} : w_1 + \cdots + w_{d-1} \le 1\}. \tag{3.11}$$

Proposition 3.3.3 (Characterization of Extreme-Value Copulas Based on Pickands Dependence Functions) *A copula C is an extreme-value copula if and only if there exists a function $A : \Delta_{d-1} \to [1/d, 1]$ such that, for any $\boldsymbol{u} \in (0,1]^d \setminus \{(1,\ldots,1)\}$,*

$$C(\boldsymbol{u}) = \exp\left(\left(\sum_{j=1}^{d} \log u_j\right) A\left(\frac{\log u_2}{\sum_{j=1}^d \log u_j}, \ldots, \frac{\log u_d}{\sum_{j=1}^d \log u_j}\right)\right). \tag{3.12}$$

The function A is called the Pickands dependence function *associated with C.* □

From the previous proposition, we see that Pickands dependence functions play a similar role for extreme-value copulas as Archimedean generators for Archimedean copulas.

If (3.12) holds, then A is necessarily convex and satisfies the boundary condition

$$\max\left\{w_1, \ldots, w_{d-1}, 1 - \sum_{j=1}^{d-1} w_j\right\} \le A(\boldsymbol{w}) \le 1, \quad \boldsymbol{w} = (w_1, \ldots, w_{d-1}) \in \Delta_{d-1}.$$

The latter two conditions are, however, not sufficient to characterize the class of Pickands dependence functions unless $d = 2$; see, for instance, Beirlant et al. (2004) for a counterexample.

The characterization of an extreme-value copula in terms of the corresponding Pickands dependence function is particularly attractive when $d = 2$ as the copula is then merely specified by a convex one-dimensional function $A : [0,1] \to [1/2, 1]$ satisfying $\max\{t, 1-t\} \le A(t) \le 1$ for all $t \in [0,1]$. Note that the constant function $t \mapsto 1$ is the Pickands dependence function of the independence copula Π, while the function $t \mapsto \max\{t, 1-t\}$ is the Pickands dependence function of the comonotone copula M.

Many functionals of a bivariate extreme-value copula have simple expressions in terms of A. For instance, Spearman's rho and Kendall's tau are given by

$$\rho_S = 12 \int_0^1 \frac{1}{(1 + A(t))^2}\, \mathrm{d}t - 3 \quad \text{and} \quad \tau = \int_0^1 \frac{t(1-t)}{A(t)}\, \mathrm{d}A'(t),$$

respectively. Note that the expression of τ is well defined as explained, for example, in Gudendorf and Segers (2010, Section 4), since A is a convex function on $[0, 1]$. From a practical perspective, if A is twice differentiable, the integrator in the expression of τ can be written in terms of A'' thereby resulting in a standard integral. The coefficients of tail dependence have even simpler expressions in terms of A. Starting from (2.26), it can be verified that

$$\lambda_u = 2(1 - A(1/2)) \quad \text{and} \quad \lambda_l = \mathbf{1}(A(1/2) = 1/2).$$

Since $A(1/2) = 1/2$ occurs only when the underlying bivariate extreme-value copula is the upper Fréchet–Hoeffding bound M, bivariate extreme-value copulas are independent in the lower tail, except in the case of comonotonicity.

Several other characterizations of extreme-value copulas are possible, for example using the spectral measure of C or the stable tail dependence function; see, for instance, Gudendorf and Segers (2012) for the former and Ressel (2013) and Charpentier et al. (2014) for the latter.

Example 3.3.4 (Exchangeable Extreme-Value Copulas) Several exchangeable parametric families of extreme-value copulas are implemented in the R package `copula`: the Galambos, the Gumbel–Hougaard, the Hüsler–Reiss, the Tawn and the t-EV families. Functions for creating objects representing copulas from these one-parameter families are `galambosCopula()`, `gumbelCopula()`, `huslerReissCopula()`, `tawnCopula()`, and `tevCopula()`, respectively. Notice that the independence copula Π belongs to all these families. In the bivariate case, for all these models, there is a one-to-one relationship between the parameter value and Kendall's tau (respectively, Spearman's rho). Because extreme-values copulas can model only nonnegative association, Kendall's tau or Spearman's rho can take only nonnegative values. Kendall's tau can additionally be arbitrarily close to one for all families above, except for the Tawn family for which it cannot exceed, approximately, 0.4184. Among the aforementioned models, only the Gumbel–Hougaard is implemented for $d > 2$; note that it is also Archimedean.

In the bivariate case, the Pickands dependence functions of the Gumbel–Hougaard family is $A_\theta^{\mathrm{GH}}(t) = (t^\theta + (1-t)^\theta)^{1/\theta}$, $t \in [0, 1]$. The following code based on the function `A()` produces graphs of A_θ^{GH} for several values of Kendall's tau. The resulting plot is given in Fig. 3.15. The dotted lines correspond to the constraints $\max\{1-t, t\} \le A(t) \le 1$, $t \in [0, 1]$. Recall that the functions $t \mapsto 1$ and $t \mapsto \max\{1-t, t\}$ are the Pickands dependence functions of Π and

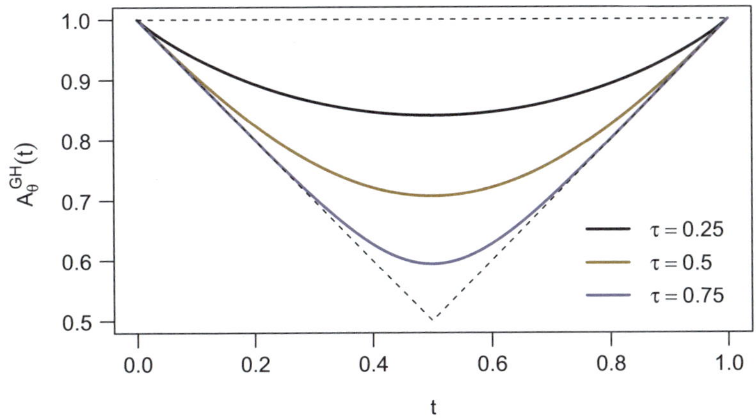

Fig. 3.15 Pickands dependence function of the bivariate Gumbel–Hougaard copula for several values of Kendall's tau. The dotted lines correspond to the constraint $\max\{1-t, t\} \leq A(t) \leq 1$, $t \in [0, 1]$

M, respectively. The plot thus illustrates the fact that the Gumbel–Hougaard family interpolates between independence and perfect positive dependence.

```
> ## Parameter values corresponding to a Kendall's tau of 0.25, 0.5 and 0.75
> th.25 <- iTau(gumbelCopula(), tau = 0.25)
> th.50 <- iTau(gumbelCopula(), tau = 0.50)
> th.75 <- iTau(gumbelCopula(), tau = 0.75)
> ## Graphs of the corresponding Pickands dependence functions A
> curve(A(gumbelCopula(th.25), w = x), from = 0, to = 1, ylim = c(0.5 ,1),
        xlab = "t", ylab = expression({A[theta]^{GH}}(t)), col = 1, lwd = 2)
> curve(A(gumbelCopula(th.50), w = x), add = TRUE, col = 2, lwd = 2)
> curve(A(gumbelCopula(th.75), w = x), add = TRUE, col = 3, lwd = 2)
> ## Every bivariate Pickands dependence function is convex
> ## and its graph is necessarily in the "triangle" plotted below
> lines(c(0, 1), c(1, 1), lty = 2)
> lines(c(0, 0.5, 1), c(1, 0.5, 1), lty = 2)
> legend("bottomright", bty = "n", lwd = 2, col = 1:3,
         legend = expression(tau == 0.25, tau == 0.5, tau == 0.75),
         inset = 0.02, y.intersp = 1.2)
```

From a practical perspective, there is hardly any difference between the aforementioned bivariate exchangeable extreme-value copulas for the same value of Kendall's tau in the sense that their Pickands dependence functions are hard to distinguish. This is illustrated by the following code which plots, for the same value of Kendall's tau, the Pickands dependence functions of all the bivariate exchangeable extreme-value copulas available in `copula`. Specifically, we begin by plotting the boundary constraints and we fix Kendall's tau:

```
> ## Draw the Pickands dependence function of the independence copula
> ## which is the same as "lines(c(0, 1), c(1, 1), lty = 2)"
> curve(A(indepCopula(), w = x), from = 0, to = 1, ylim = c(0.5, 1),
        xlab = "t", ylab = quote({A[theta]}(t)), lty = 2)
> ## Add the constraint related to the comonotone copula
> lines(c(0, 0.5, 1), c(1, 0.5, 1), lty = 2)
> ## Set Kendall's tau and color
> tau <- 0.25
> col <- adjustcolor(1, alpha.f = 0.25)
> ## And the utility A() curve plot function
> curveA <- function(x, Cop) # using  global (tau, col)
    curve(A(Cop(iTau(Cop(), tau)), w = x), add = TRUE, col = col, lwd = 2)
```

We then overlay, with alpha transparency, the graphs of the Pickands dependence functions of the available bivariate exchangeable extreme-value copulas for the above value of Kendall's tau:

```
> curveA(x, galambosCopula)
> curveA(x, gumbelCopula)
> curveA(x, huslerReissCopula)
> curveA(x, tevCopula)
> if(tau < 0.4184) curveA(x, tawnCopula) # no tawnCopula for larger tau!
```

Rerunning the above code for `tau <-0.5` and `tau <-0.75`, we obtain almost the same plot as in Fig. 3.15; see Fig. 3.16.

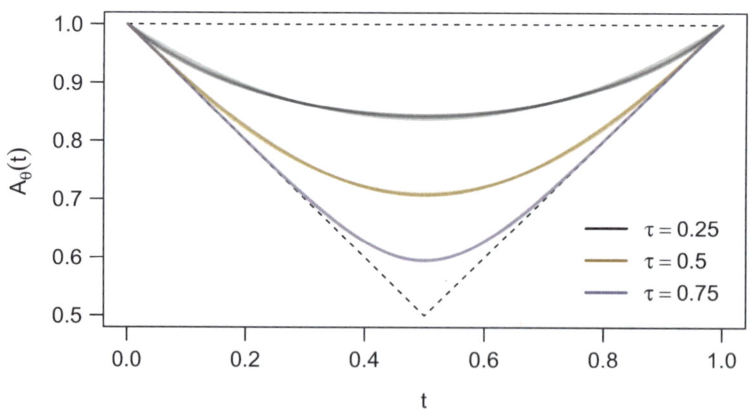

Fig. 3.16 Pickands dependence functions of the Galambos, Gumbel–Hougaard, Hüsler-Reiss, t-EV, and Tawn copula (the latter only for $\tau = 0.25$) with a Kendall's tau of $\tau \in \{0.25, 0.5, 0.75\}$. There are hardly any differences between these copulas for the same value of Kendall's tau

The strong similarity between the above bivariate exchangeable extreme-value copula families is confirmed by the numerical experiments carried out in Genest et al. (2011a), who studied goodness-of-fit tests based on the difference between a nonparametric and a parametric estimator of the Pickands dependence function (related estimators and goodness-of-fit tests will be discussed in Chaps. 4 and 5, respectively). The Monte Carlo experiments reported in the latter work highlight the fact that, unless the sample size is extremely large, there is hardly any practical difference among the aforementioned bivariate exchangeable parametric families of extreme-value copulas, and that, from a modeling perspective, a more important issue is whether an exchangeable or non-exchangeable family should be used. Note that non-exchangeable extreme-value copulas can be constructed using Khoudraji's device; see Sect. 3.4.2. □

We conclude this section with a few words on sampling. In the bivariate case, a general sampling algorithm applicable to extreme-value copulas whose Pickands dependence functions are twice differentiable is proposed in Ghoudi et al. (1998, Section 3); see also Capéraà et al. (2000, Section 5). The latter is the approach implemented in `copula` for the exchangeable copula families introduced in the previous example, with the exception of the Gumbel–Hougaard family, which, since it is also Archimedean, can be sampled as illustrated in Example 3.2.5.

3.4 Selected Copula Transformations and Constructions

3.4.1 Rotated Copulas

Let C be a d-dimensional copula. As we have seen in Proposition 2.5.2, if $\boldsymbol{U} \sim C$, then $\mathbf{1} - \boldsymbol{U} \sim \bar{C}$, where $\bar{C}$ is the survival copula corresponding to C. In dimension two, for example, it follows that a sample from $\bar{C}$ can be obtained from a sample from C by orthogonally reflecting the points in the unit square first at the line $u_1 = 1/2$ and then at the line $u_2 = 1/2$; this can be visually verified from the top row of scatter plots displayed in Fig. 2.17 when C is a bivariate Clayton copula. Clearly, the order in which the orthogonal reflections are carried out does not affect the transformation, which can also be seen as a particular rotation.

The definition of rotated copulas generalizes the above observation characterizing survival copulas. Let C be a d-dimensional copula and let $\boldsymbol{U} \sim C$. Given $\boldsymbol{r} \in \{0,1\}^d$, we call $\mathrm{rot}_{\boldsymbol{r}}(C)$ the *rotated copula* of C based on $\boldsymbol{r}$ if $\boldsymbol{U} \sim C$ is equivalent to

$$\big((1-r_1)U_1 + r_1(1-U_1), \dots, (1-r_d)U_d + r_d(1-U_d)\big) \sim \mathrm{rot}_{\boldsymbol{r}}(C). \tag{3.13}$$

It is immediate to verify that the survival copula $\bar{C}$ of C is nothing else than $\mathrm{rot}_{\mathbf{1}}(C)$. Roughly speaking, a sample from $\mathrm{rot}_{\boldsymbol{r}}(C)$ can be obtained from a sample from C by "flipping the dimensions" corresponding to those components of $\boldsymbol{r}$ which are equal to 1. The latter is sometimes equivalent to adequately rotating (hence the name) the sample from C. Formally, such a transformation is a sequence of orthogonal reflections (of arbitrary order) with respect to the hyperplanes $\{\boldsymbol{u} \in [0,1]^d : u_j = 1/2\}$ for $j \in \{1,\dots,d\}$ such that $r_j = 1$. If C admits a density c, the density $\mathrm{rot}_{\boldsymbol{r}}(c)$ of $\mathrm{rot}_{\boldsymbol{r}}(C)$ is given by

$$\mathrm{rot}_{\boldsymbol{r}}(c)(\boldsymbol{u}) = c((1-r_1)u_1 + r_1(1-u_1), \dots, (1-r_d)u_d + r_d(1-u_d)), \ \boldsymbol{u} \in (0,1)^d.$$

Example 3.4.1 (A Rotated Bivariate Clayton Copula) Let C be C_4^{C}, the Clayton copula with parameter $\theta = 4$, and let $\boldsymbol{r} = (1,0)$. Rotated copulas in the R package `copula` can be created using the function `rotCopula()`. The following code produces wireframe and contour plots of $\mathrm{rot}_{(1,0)}(C_4^{\mathrm{C}})$ and of the corresponding density $\mathrm{rot}_{(1,0)}(c_4^{\mathrm{C}})$:

```
> ## The vector r is represented by a vector of logicals
> rc <- rotCopula(claytonCopula(4), flip = c(TRUE, FALSE))
> wireframe2(rc, FUN = pCopula, # wireframe plot (copula)
             draw.4.pCoplines = FALSE)
> wireframe2(rc, FUN = dCopula, delta = 0.025) # wireframe plot (density)
> contourplot2(rc, FUN = pCopula, n.grid = 64) # contour plot (copula)
> contourplot2(rc, FUN = dCopula, n.grid = 64, cuts = 30,
               pretty = FALSE, lwd = 1/2) # contour plot (density)
```

The resulting plots are shown in Fig. 3.17.

□

Sampling from rotated copulas is straightforward via (3.13), as illustrated in the next example.

Example 3.4.2 (A Rotated Four-Dimensional Gumbel–Hougaard Copula) Let $d = 4$, C be C_3^{GH} (the Gumbel–Hougaard copula with parameter 3) and $\boldsymbol{r} = (1,0,1,0)$. The following code illustrates sampling from $\mathrm{rot}_{(1,0,1,0)}(C_3^{\mathrm{GH}})$ via (3.13):

```
> ## The logical representing vector r
> flip <- c(TRUE, FALSE, TRUE, FALSE)
> rg <- rotCopula(gumbelCopula(3, dim = 4), flip = flip)
> n <- 1000
> set.seed(2016)
> U <- rCopula(n, copula = rg)
> set.seed(2016)
> V <- rCopula(n, gumbelCopula(3, dim = 4))
> ## "Flip" the relevant components
> V[, flip] <- 1 - V[, flip]
> stopifnot(all.equal(U, V)) # check
```

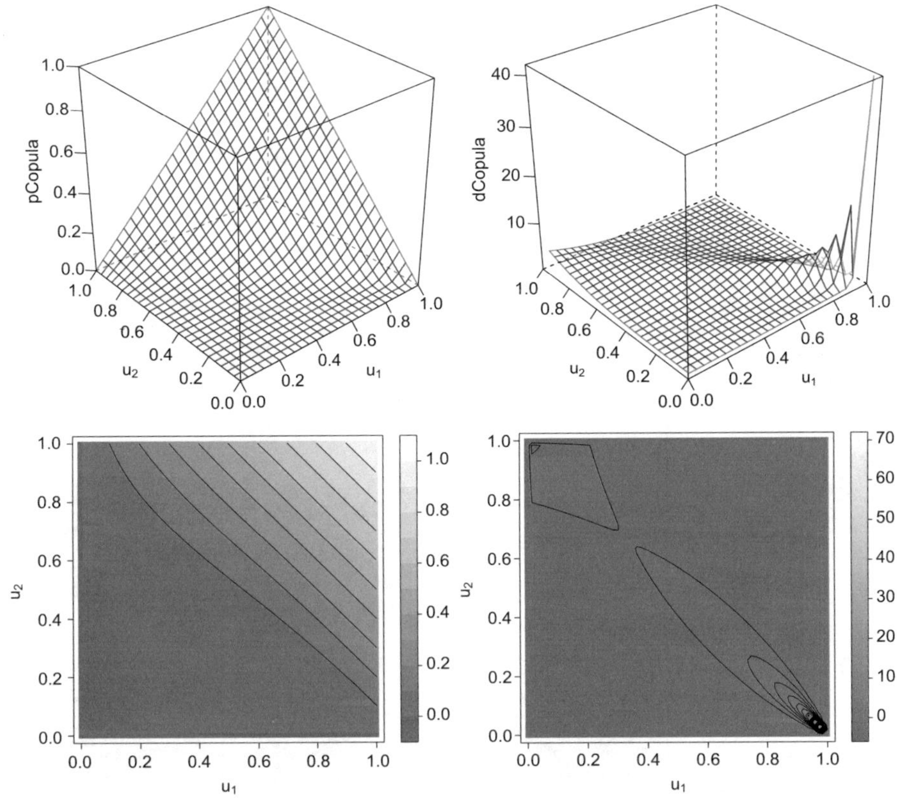

Fig. 3.17 Wireframe (top) and contour plots (bottom) of $\mathrm{rot}_{(1,0)}(C_4^{\mathrm{C}})$ (left) and its corresponding density $\mathrm{rot}_{(1,0)}(c_4^{\mathrm{C}})$ (right)

The effect of $\boldsymbol{r}$ appears clearly from the scatter-plot matrix of the generated observations shown in Fig. 3.18:

```
> splom2(U, cex = 0.3, col.mat = "black")
```

□

Finally, let us mention that, given a bivariate copula C whose Spearman's rho and Kendall's tau are denoted by ρ_{S} and τ, respectively, we immediately have that Spearman's rho and Kendall's tau of $\mathrm{rot}_{\boldsymbol{r}}(C)$ are $(-1)^{\mathbf{1}(r_1 \neq r_2)}\rho_{\mathrm{S}}$ and $(-1)^{\mathbf{1}(r_1 \neq r_2)}\tau$, respectively. For the coefficients of tail dependence of $\mathrm{rot}_{\boldsymbol{r}}(C)$ for $\boldsymbol{r} \neq \mathbf{0}$, only the case $\boldsymbol{r} = \mathbf{1}$ is trivial in general: the coefficient of upper (respectively, lower) tail dependence of $\mathrm{rot}_{\mathbf{1}}(C)$ is the coefficient of lower (respectively, upper) tail dependence of C.

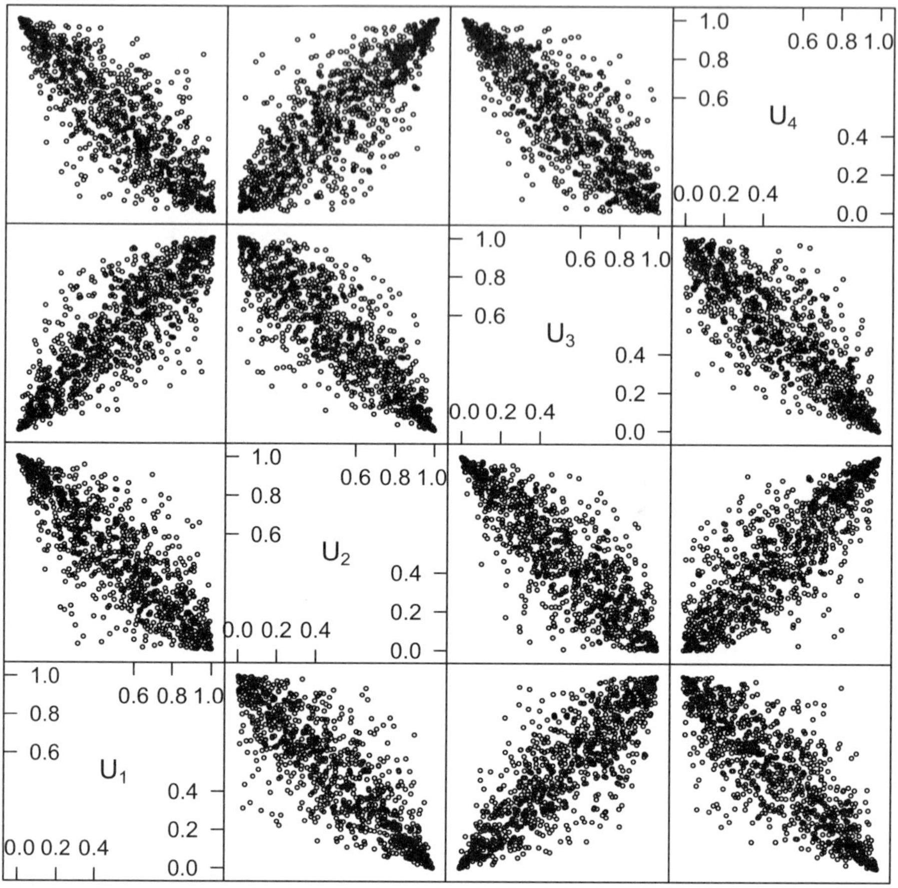

Fig. 3.18 Scatter-plot matrix of $n = 1000$ independent observations from the four-dimensional rotated Gumbel–Hougaard copula $\mathrm{rot}_{(1,0,1,0)}(C_3^{\mathrm{GH}})$

3.4.2 Khoudraji's Device

Given two d-dimensional copulas C_1, C_2 and a vector $\boldsymbol{s} \in [0, 1]^d$, *Khoudraji's device* $\mathrm{kho}_{\boldsymbol{s}}(C_1, C_2)$ with *shape vector* $\boldsymbol{s}$ is defined by

$$\mathrm{kho}_{\boldsymbol{s}}(C_1, C_2)(\boldsymbol{u}) = C_1(\boldsymbol{u}^{\mathbf{1}-\boldsymbol{s}})C_2(\boldsymbol{u}^{\boldsymbol{s}}), \quad \boldsymbol{u} \in [0, 1]^d, \tag{3.14}$$

with the convention that $\boldsymbol{u}^{\boldsymbol{v}} = (u_1^{v_1}, \dots, u_d^{v_d})$ for all $\boldsymbol{u}, \boldsymbol{v} \in [0, 1]^d$. This construction, possibly leading to a non-exchangeable d-dimensional copula, was initially proposed by Khoudraji (1995); see Genest et al. (1998) for further details. It is a special case of the general construction principle proposed by Liebscher (2008) who considers the product of an arbitrary number of copulas whose arguments are properly transformed to ensure that the product has standard uniform univariate margins.

Sampling from copulas constructed using Khoudraji's device is particularly simple given the product form of the df appearing in (3.14). Indeed, if $\boldsymbol{U} \sim C_1$ independently of $\boldsymbol{V} \sim C_2$, then it is a straightforward exercise to verify that

$$\left(\max\left\{U_1^{\frac{1}{1-s_1}}, V_1^{\frac{1}{s_1}}\right\}, \dots, \max\left\{U_d^{\frac{1}{1-s_d}}, V_d^{\frac{1}{s_d}}\right\}\right) \sim \mathrm{kho}_{\boldsymbol{s}}(C_1, C_2),$$

with the convention $u^{1/0} = 0$ for all $u \in [0, 1)$ and $1^{1/0} = 1$.

From a practical perspective, a useful subset of families constructed from Khoudraji's device is obtained by taking C_1 or C_2 to be the independence copula Π. In the bivariate case, by taking $C_1 = \Pi$ for instance, non-exchangeable versions of C_2 can be obtained by choosing s_1 or s_2 relatively close to 1. This is illustrated in the following example where C_2 is chosen to be a Clayton copula, leading to what we shall call a *Khoudraji–Clayton copula* as we continue.

Example 3.4.3 (Non-exchangeable Khoudraji–Clayton Copulas) With the R package `copula`, copulas of the form (3.14) can be created using the function `khoudrajiCopula()`. Let $d = 2$, C_2 be C_6^{C}, the Clayton copula with parameter $\theta = 6$, and let $\boldsymbol{s} = (0.2, 0.95)$. An object representing $\mathrm{kho}_{(0.2,0.95)}(\Pi, C_6^{\mathrm{C}})$ can be created as follows:

```
> (kc <- khoudrajiCopula(copula2 = claytonCopula(6), shapes = c(0.2, 0.95)))
```

```
Khoudraji copula, dim. d = 2, constructed from
 Independence copula
 Clayton copula
Dimension:  2
Parameters:
  c2.alpha    = 6.00
  shape1      = 0.20
  shape2      = 0.95
```

From the previous output, we see that $\mathrm{kho}_{(0.2,0.95)}(\Pi, C_6^{\mathrm{C}})$ has three parameters as expected: the one of $C_2 = C_6^{\mathrm{C}}$ and two shape parameters.

We now empirically investigate the influence of s_1 on the asymmetry of the copula $\mathrm{kho}_{(s_1,0.95)}(\Pi, C_6^{\mathrm{C}})$. To this end, we generate $n = 5000$ independent observations from $\mathrm{kho}_{(s_1,0.95)}(\Pi, C_6^{\mathrm{C}})$ for $s_1 \in \{0.2, 0.4, 0.6, 0.8\}$. Recall that, to change the value of the parameters of a copula, the function `setTheta()` can be used:

```
> n <- 5000
> plot(rCopula(n, copula = kc),
       cex = 0.5, xlab = quote(U[1]), ylab = quote(U[2]))
> plot(rCopula(n, copula = setTheta(kc, value = c(6, 0.4, 0.95))),
       cex = 0.5, xlab = quote(U[1]), ylab = quote(U[2]))
> plot(rCopula(n, copula = setTheta(kc, value = c(6, 0.6, 0.95))),
       cex = 0.5, xlab = quote(U[1]), ylab = quote(U[2]))
> plot(rCopula(n, copula = setTheta(kc, value = c(6, 0.8, 0.95))),
       cex = 0.5, xlab = quote(U[1]), ylab = quote(U[2]))
```

The resulting scatter plots are displayed in Fig. 3.19. As one can see, the non-exchangeability becomes clearly apparent for $s_1 \in \{0.4, 0.6, 0.8\}$. It is, however, not obvious how the scatter plots should be ranked in terms of strength of asymmetry;

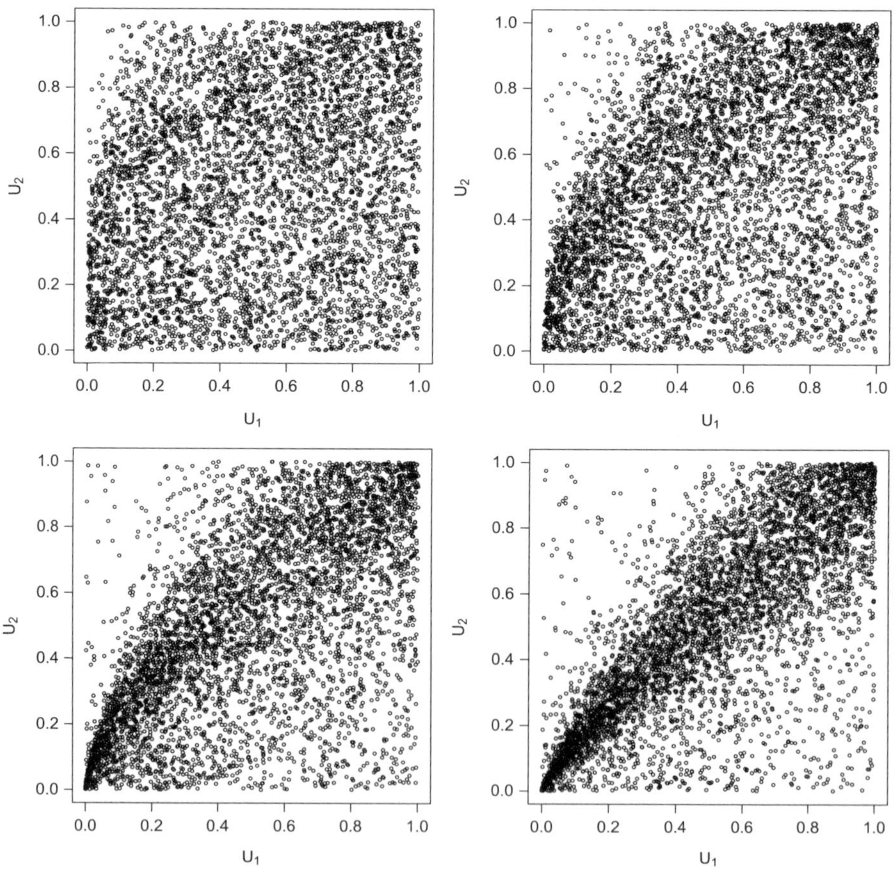

Fig. 3.19 Scatter plots of $n = 5000$ independent observations from $\text{kho}_{(0.2,0.95)}(\Pi, C_6^{\text{C}})$ (top left), $\text{kho}_{(0.4,0.95)}(\Pi, C_6^{\text{C}})$ (top right), $\text{kho}_{(0.6,0.95)}(\Pi, C_6^{\text{C}})$ (bottom left) and $\text{kho}_{(0.8,0.95)}(\Pi, C_6^{\text{C}})$ (bottom right)

see, for instance, Genest and Nešlehová (2013, Section 5.2) and the references therein for possible measures of asymmetry.

The consequence of exchanging the roles of s_1 and s_2 in (3.14) can be visualized as follows:

```
> plot(rCopula(n, copula = setTheta(kc, value = c(6, 0.95, 0.6))),
       cex = 0.5, xlab = quote(U[1]), ylab = quote(U[2]))
```

A comparison of the produced scatter plot, displayed on the left-hand side of Fig. 3.20, with the scatter plot at the bottom-left of Fig. 3.19 confirms that exchanging the roles of s_1 and s_2 simply amounts to performing an orthogonal reflection with respect to the main diagonal.

The middle scatter plot in Fig. 3.20, obtained with the following code, shows what happens when the roles of Π and C_6^{C} are exchanged:

```
> plot(rCopula(n, copula = khoudrajiCopula(copula1 = claytonCopula(6),
                                           shapes = c(0.6, 0.95))),
       cex = 0.5, xlab = quote(U[1]), ylab = quote(U[2]))
```

As one can see from the scatter plot, the resulting copula $\mathrm{kho}_{(0.6,0.95)}(C_6^{\mathrm{C}}, \Pi)$ is close to Π. Some thought and experimentation reveals that if Π is used as second copula, to recover asymmetry, either s_1 or s_2 need to be relatively close to 0. By taking $s_1 = 1 - 0.6$ and $s_2 = 1 - 0.95$ in $\mathrm{kho}_{\boldsymbol{s}}(C_6^{\mathrm{C}}, \Pi)$, unsurprisingly from (3.14), we actually recover $\mathrm{kho}_{(0.6,0.95)}(\Pi, C_6^{\mathrm{C}})$; in other words, $\mathrm{kho}_{(1-0.6,1-0.95)}(C_6^{\mathrm{C}}, \Pi) = \mathrm{kho}_{(0.6,0.95)}(\Pi, C_6^{\mathrm{C}})$. This is confirmed by the right-hand side scatter plot in Fig. 3.20, produced as follows:

```
> plot(rCopula(n, copula = khoudrajiCopula(copula1 = claytonCopula(6),
                                           shapes = c(1 - 0.6, 1 - 0.95))),
       cex = 0.5, xlab = quote(U[1]), ylab = quote(U[2]))
```

□

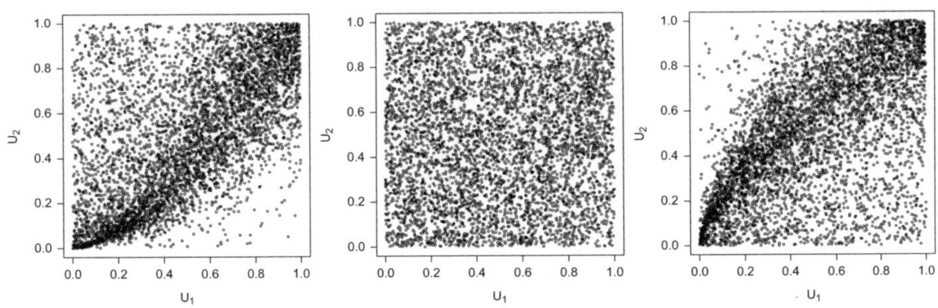

Fig. 3.20 Scatter plots of $n = 5000$ independent observations from $\mathrm{kho}_{(0.95,0.6)}(\Pi, C_6^{\mathrm{C}})$ (left), $\mathrm{kho}_{(0.6,0.95)}(C_6^{\mathrm{C}}, \Pi)$ (middle), and $\mathrm{kho}_{(1-0.6,1-0.95)}(C_6^{\mathrm{C}}, \Pi)$ (right)

An inconvenience of using Π for C_1 or C_2 in (3.14) is that the range of attainable positive association is limited. For $d = 2$, Genest et al. (2011a) showed that Kendall's tau τ of $\text{kho}_s(\Pi, C)$ with $s_1, s_2 \in (0, 1)$ satisfies

$$\tau \leq \frac{s_1 s_2}{s_1 + s_2 - s_1 s_2}. \tag{3.15}$$

For instance, if $s_1 = 0.6$ and $s_2 = 0.95$, no matter how C is chosen, τ will not be able to exceed, approximately, 0.5816. By using copulas different from Π, stronger positive association and different shapes can be obtained as illustrated in the following example.

Example 3.4.4 (Non-exchangeable Khoudraji–Gumbel–Hougaard–Clayton Copulas) Let $d = 2$, C_1 be the Gumbel–Hougaard copula $C^{\text{GH}}_{\theta_1}$ with parameter θ_1 and C_2 be the Clayton copula $C^{\text{C}}_{\theta_2}$ with parameter θ_2. In the following code, θ_1 and θ_2 are chosen such that Kendall's tau of both C_1 and C_2 is equal to $\tau \in \{0.65, 0.7, 0.95\}$. For each value of τ, $n = 5000$ independent observations from $\text{kho}_{(0.6,0.95)}(C^{\text{GH}}_{\theta_1}, C^{\text{C}}_{\theta_2})$ are generated and the corresponding scatter plot is displayed:

```
> ## The setup
> s <- c(0.6, 0.95)
> copula1 <- gumbelCopula
> copula2 <- claytonCopula
> ## A utility function to obtain the parameter values of C_1 and C_2
> param <- function(tau) c(iTau(copula1(), tau), iTau(copula2(), tau))
> ## The corresponding Khoudraji-Gumbel-Hougaard-Clayton copula
> (kho <- khoudrajiCopula(copula1 = copula1(param(0.65)[1]),
                          copula2 = copula2(param(0.65)[2]),
                          shapes = s))

Khoudraji copula, dim. d = 2, constructed from
 Gumbel copula
 Clayton copula

Dimension:  2
Parameters:
  c1.alpha    = 2.857143
  c2.alpha    = 3.714286
  shape1      = 0.600000
  shape2      = 0.950000

> n <- 5000
> U <- rCopula(n, copula = kho)
> plot(U, cex = 0.5, xlab = quote(U[1]), ylab = quote(U[2]))
> V <- rCopula(n, copula = setTheta(kho, value = c(param(0.8), s)))
> plot(V, cex = 0.5, xlab = quote(V[1]), ylab = quote(V[2]))
> W <- rCopula(n, copula = setTheta(kho, value = c(param(0.95), s)))
> plot(W, cex = 0.5, xlab = quote(W[1]), ylab = quote(W[2]))
```

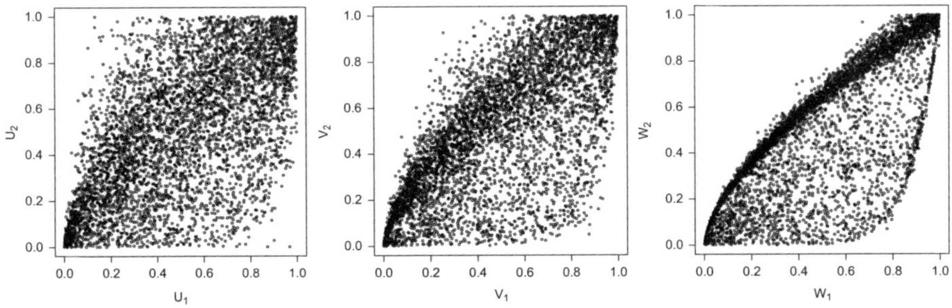

Fig. 3.21 Scatter plots of $n = 5000$ independent observations from $\mathrm{kho}_{(0.6,0.95)}(C^{\mathrm{GH}}_{\theta_1}, C^{\mathrm{C}}_{\theta_2})$ with θ_1 and θ_2 chosen so that $C^{\mathrm{GH}}_{\theta_1}$ and $C^{\mathrm{C}}_{\theta_2}$ both have a Kendall's tau of 0.65 (left), 0.8 (middle), and 0.95 (right)

The produced scatter plots are displayed in Fig. 3.21.

We can also estimate Kendall's tau from all three data sets from $\mathrm{kho}_{(0.6,0.95)}(C^{\mathrm{GH}}_{\theta_1}, C^{\mathrm{C}}_{\theta_2})$:

```
> c(cor(U, method = "kendall")[1,2], cor(V, method = "kendall")[1,2],
    cor(W, method = "kendall")[1,2])
```

```
[1] 0.4406496 0.5269185 0.6278509
```

It is thus possible to choose C_1 in (3.14) to exceed the bound (roughly 0.5816) for Kendall's tau related to (3.15). □

The previous examples highlight the variety of shapes that can arise from the use of Khoudraji's device. This construction principle can also be used to construct non-exchangeable extreme-value copula families. Indeed, if C_1 and C_2 in (3.14) are extreme-value copulas, so is $\mathrm{kho}_{\boldsymbol{s}}(C_1, C_2)$ for all $\boldsymbol{s} \in [0, 1]^d$; see Proposition 3.3.2. Taking C_1 to be the independence copula leads to relatively simple non-exchangeable versions of the extreme-value copula C_2, as illustrated in the following example in the case of a bivariate Gumbel–Hougaard copula.

Example 3.4.5 (A Non-exchangeable Extreme-Value Family) Let $d = 2$, C_2 be the Gumbel–Hougaard copula C^{GH}_4 with parameter 4, and let $s_2 = 0.95$. The following code illustrates the influence of s_1 on the shape of the Pickands dependence function of $\mathrm{kho}_{(s_1,0.95)}(\Pi, C^{\mathrm{GH}}_4)$:

```
> kg <- khoudrajiCopula(copula2 = gumbelCopula(4), shapes = c(0.2, 0.95))
> curve(A(kg, w = x), from = 0, to = 1, ylim = c(0.5, 1),
        xlab = "t", ylab = expression({A[theta]^{KGH}}(t)), col = 1, lwd = 2)
> curve(A(setTheta(kg, value = c(4, 0.4, 0.95)), w = x), add = TRUE, lwd = 2,
        col = 2)
> curve(A(setTheta(kg, value = c(4, 0.6, 0.95)), w = x), add = TRUE, lwd = 2,
        col = 3)
> curve(A(setTheta(kg, value = c(4, 0.8, 0.95)), w = x), add = TRUE, lwd = 2,
        col = 4)
> lines(c(0, 1),      c(1, 1),      lty = 2)
> lines(c(0, 0.5, 1), c(1, 0.5, 1), lty = 2)
> legend("bottomright", bty = "n", lwd = 2,, col = 1:4,
         legend = expression(s[1] == 0.2, s[1] == 0.4,
                             s[1] == 0.6, s[1] == 0.8))
```

As one can see from Fig. 3.22, the larger s_1, the closer the Pickands dependence function is to $t \mapsto \max\{1 - t, t\}$, that is, the Pickands dependence function of the comonotone copula M. Through the boundary constraints that every Pickands dependence function necessarily satisfies, it clearly appears from the previous plot that the stronger the dependence, the less room there is for strong non-exchangeability. □

Because of the product in (3.14), density evaluation for copulas constructed using Khoudraji's device becomes challenging as d increases. For implementation reasons, when $d > 2$, density evaluation in the R package `copula` is only possible when C_1 and C_2 have densities and explicit df expressions.

Example 3.4.6 (Higher-Dimensional Khoudraji Copulas) We first consider a Khoudraji–Gumbel–Hougaard–survival Clayton copula. The density expressions of

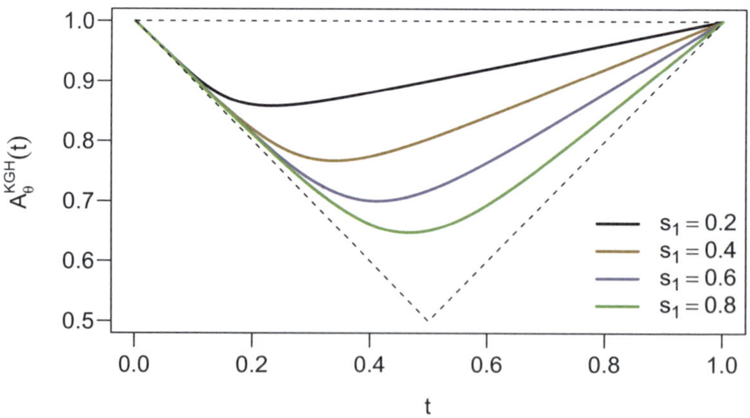

Fig. 3.22 Pickands dependence function of the bivariate Khoudraji–Gumbel–Hougaard copula $\mathrm{kho}_{(s_1,0.95)}(\Pi, C_4^{\mathrm{GH}})$ for $s_1 \in \{0.2, 0.4, 0.6, 0.8\}$

the Gumbel–Hougaard and the survival Clayton copula being explicit, the density of the corresponding Khoudraji copula can be evaluated:

```
> kgsc <- khoudrajiCopula(copula1 = gumbelCopula(2, dim = 3),
                          copula2 = rotCopula(claytonCopula(6, dim = 3)),
                          shapes = c(0.6, 0.7, 0.95))
> ## Random points in the unit hypercube where to evaluate the density
> set.seed(42)
> v <- matrix(runif(15), 5, 3)
> dCopula(v, copula = kgsc)
```

```
[1] 0.2852876148 0.7332486257 0.0007894093 0.6996430796 0.8508682765
```

However, if, for instance, a normal copula is used as second copula, density evaluation of the resulting Khoudraji copula becomes unavailable:

```
> kgn <- khoudrajiCopula(copula1 = gumbelCopula(2, dim = 3),
                         copula2 = normalCopula(0.9, dim = 3),
                         shapes = c(0.6, 0.7, 0.95))
> try(dCopula(v, copula = kgn)) # not implemented
```

```
Error in (function (classes, fdef, mtable)  :
  unable to find an inherited method for function 'dCopula' for signature '"
      matrix", "khoudrajiCopula"'
```

□

In practice, this means that the fitting of copulas constructed using Khoudraji's device can be attempted with the R package copula when C_1 and C_2 have explicit density expressions. Note that depending on the implementation of the latter, the dimension for supported Archimedean copulas is limited, and density evaluation (and thus fitting) may be affected by numerical issues, for example, near the boundary of the unit hypercube; see Yan (2007, Section 5).

3.4.3 Mixtures of Copulas

It is a simple exercise to verify that a convex combination of d-dimensional copulas is a copula. As a last construction principle, we shall thus consider mixtures of copulas. Let $C_1, \dots, C_m$ be $m \geq 2$ d-dimensional copulas and let $\boldsymbol{w} = (w_1, \dots, w_m)$ be a vector of weights such that $w_k \geq 0$ for all $k \in \{1, \dots, m\}$ with $\sum_{k=1}^m w_k = 1$. The *mixture* of $C_1, \dots, C_m$ with mixing vector $\boldsymbol{w} = (w_1, \dots, w_m)$ is the copula defined by

$$\mathrm{mix}_{\boldsymbol{w}}(C_1, \dots, C_m)(\boldsymbol{u}) = \sum_{k=1}^m w_k C_k(\boldsymbol{u}), \quad \boldsymbol{u} \in [0, 1]^d. \tag{3.16}$$

If each copula C_k, $k \in \{1, \dots, m\}$, admits a density c_k, then, as an immediate consequence of (3.16), the copula $\mathrm{mix}_{\boldsymbol{w}}(C_1, \dots, C_m)$ admits the density

$$\mathrm{mix}_{\boldsymbol{w}}(c_1, \dots, c_m)(\boldsymbol{u}) = \sum_{k=1}^{m} w_k c_k(\boldsymbol{u}), \quad \boldsymbol{u} \in (0,1)^d.$$

More generally, all linear functionals of $\mathrm{mix}_{\boldsymbol{w}}(C_1, \dots, C_m)$ can be computed as the corresponding mixture of linear functionals of the components $C_1, \dots, C_m$. For instance, in the bivariate case, Spearman's rho, the coefficient of lower tail dependence and the coefficient of upper tail dependence of $\mathrm{mix}_{\boldsymbol{w}}(C_1, \dots, C_m)$ are simply $\sum_{k=1}^{m} w_k \rho_s(C_k)$, $\sum_{k=1}^{m} w_k \lambda_l(C_k)$ and $\sum_{k=1}^{m} w_k \lambda_u(C_k)$, respectively; see (2.22) and (2.26).

Sampling from $\mathrm{mix}_{\boldsymbol{w}}(C_1, \dots, C_m)$ is immediate if one knows how to sample from the components $C_1, \dots, C_m$ because of the stochastic representation

$$M_1 \boldsymbol{U}_1 + \dots + M_m \boldsymbol{U}_m \sim \mathrm{mix}_{\boldsymbol{w}}(C_1, \dots, C_m),$$

where $\boldsymbol{U}_k \sim C_k$, $k \in \{1, \dots, m\}$, and $(M_1, \dots, M_m)$ follows a multinomial distribution with parameters 1 and $\boldsymbol{w}$.

Example 3.4.7 (A Mixture of Clayton and Gumbel–Hougaard Copulas) Objects representing mixtures of copulas can be created in the R package `copula` using the function `mixCopula()`. This is illustrated by the following code for $m = 2$, C_1 the Clayton copula $C^{\mathrm{C}}_{\theta_1}$ with parameter θ_1 and C_2 the Gumbel–Hougaard copula $C^{\mathrm{GH}}_{\theta_2}$ with parameter θ_2, both with a Kendall's tau of 0.75:

```
> cc <- claytonCopula(iTau(claytonCopula(), tau = 0.75)) # the first component
> gc <- gumbelCopula(iTau(gumbelCopula(),   tau = 0.75)) # the second component
> weights <- c(1/3, 2/3) # the corresponding weights
> (mcg <- mixCopula(list(cc, gc), w = weights)) # the mixture copula
```

```
mixCopula from 2 components
(Clyc, Gmbc)  with weights:
(0.3333333, 0.6666667)
Dimension:  2
Parameters:
  m1.alpha    = 6.0000000
  m2.alpha    = 4.0000000
  w1          = 0.3333333
  w2          = 0.6666667
```

The previous output confirms that $\mathrm{mix}_{(1/3,2/3)}(C^{\mathrm{C}}_{\theta_1}, C^{\mathrm{GH}}_{\theta_2})$ has four parameters: θ_1, θ_2, and the two weights which sum up to one.

As explained previously, computing Spearman's rho and the coefficients of tail dependence is easy if these quantities are known for the component copulas:

```
> stopifnot(all.equal(rho(mcg), weights[1] * rho(cc) + weights[2] * rho(gc)))
> lambda(mcg)
```

```
    lower     upper
0.2969662 0.5405286
```

```
> stopifnot(all.equal(lambda(mcg),
                      weights[1] * lambda(cc) + weights[2] * lambda(gc)))
```

Note that, unlike its components, the constructed mixture is both lower and upper tail dependent.

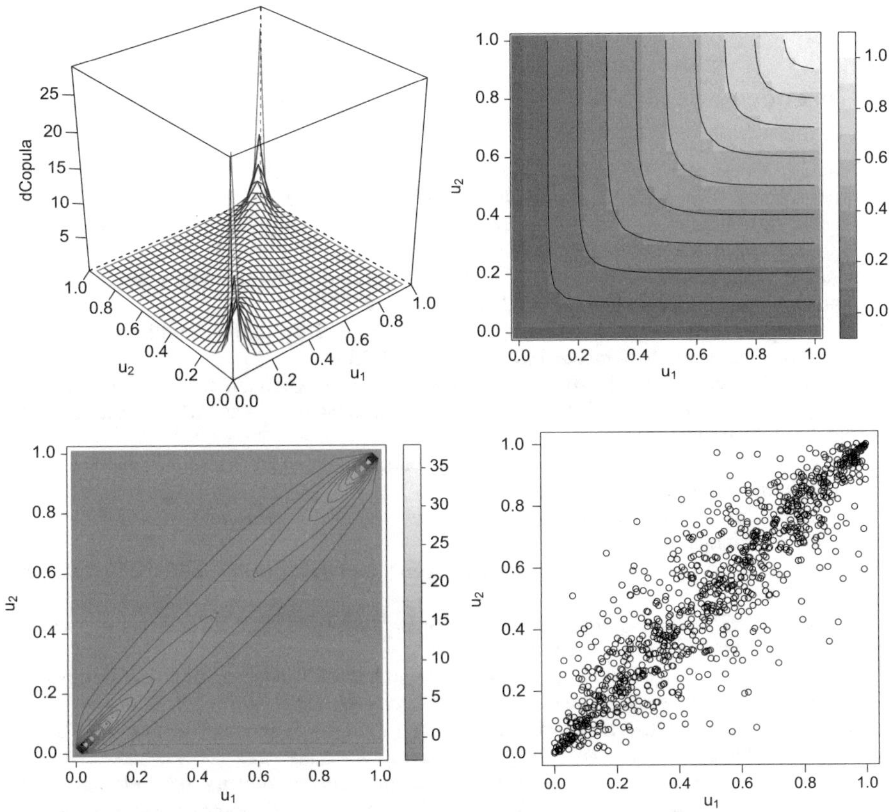

Fig. 3.23 Wireframe plot of $\mathrm{mix}_{(1/3,2/3)}(c^{\mathrm{C}}_{\theta_1}, c^{\mathrm{GH}}_{\theta_2})$ (top left), contour plot of $\mathrm{mix}_{(1/3,2/3)}(C^{\mathrm{C}}_{\theta_1}, C^{\mathrm{GH}}_{\theta_2})$ (top right), of $\mathrm{mix}_{(1/3,2/3)}(c^{\mathrm{C}}_{\theta_1}, c^{\mathrm{GH}}_{\theta_2})$ (bottom left) and scatter plot of a sample of size $n = 1000$ from $\mathrm{mix}_{(1/3,2/3)}(C^{\mathrm{C}}_{\theta_1}, C^{\mathrm{GH}}_{\theta_2})$ (bottom right), where θ_1 and θ_2 are chosen so that $C^{\mathrm{C}}_{\theta_1}$ and $C^{\mathrm{GH}}_{\theta_2}$ both have a Kendall's tau of 0.75 □

We end this example by producing wireframe and contour plots of this mixture or its density, as well as a scatter plot of a sample of size $n = 1000$; see Fig. 3.23.

```
> set.seed(127)
> U <- rCopula(1000, copula = mcg) # sample from the mixture
> wireframe2(mcg, FUN = dCopula, delta = 0.025) # density
> contourplot2(mcg, FUN = pCopula) # copula
> contourplot2(mcg, FUN = dCopula, cuts = 32, # density
               n.grid = 50, pretty = FALSE,
               col = adjustcolor(1, 1/3), alpha.regions = 3/4)
> plot(U, xlab = quote(U[1]), ylab = quote(U[2])) # scatter plot
```

□

References

Beirlant, J., Goegebeur, Y., Segers, J., & Teugels, J. (2004). *Statistics of extremes: Theory and applications. Wiley series in probability and statistics*. Chichester: Wiley.

Bolker, B. (2017). bbmle: Tools for general maximum likelihood estimation. R package version 1.0.20. https://CRAN.R-project.org/package=bbmle.

Capéraà, P., Fougères, A.-L., & Genest, C. (2000). Bivariate distributions with given extreme value attractor. *Journal of Multivariate Analysis, 72*, 30–49.

Charpentier, A., Fougères, A.-L., Genest, C., & Nešlehová, J. G. (2014). Multivariate Archimax copulas. *Journal of Multivariate Analysis, 126*, 118–136.

De Haan, L., & Ferreira, A. (2006). *Extreme value theory: An introduction*. New York: Springer.

Demarta, S., & McNeil, A. J. (2005). The *t* copula and related copulas. *International Statistical Review, 73*(1), 111–129.

Devroye, L. (1993). A triptych of discrete distributions related to the stable law. *Statistics & Probability Letters, 18*(5), 349–351.

Fang, K.-T., Kotz, S., & Ng, K.-W. (1990). *Symmetric multivariate and related distributions*. London: Chapman & Hall.

Feller, W. (1971). *An introduction to probability theory and its applications* (2nd ed.), vol. 2. New York: Wiley.

Galambos, J. (1978). *The asymptotic theory of extreme order statistics. Wiley series in probability and mathematical statistics*. New York: Wiley.

Genest, C., Ghoudi, K., & Rivest, L.-P. (1998). Discussion of "Understanding relationships using copulas", by E. Frees and E. Valdez. *North American Actuarial Journal, 3*, 143–149.

Genest, C., Kojadinovic, I., Nešlehová, J. G., & Yan, J. (2011a). A goodness-of-fit test for bivariate extreme-value copulas. *Bernoulli, 17*(1), 253–275.

Genest, C., & Nešlehová, J. G. (2013). Assessing and modeling asymmetry in bivariate continuous data. In P. Jaworski, F. Durante & W. K. Härdle (Eds.), *Copulae in mathematical and quantitative finance* (pp. 91–114). *Lecture notes in statistics*. Berlin: Springer.

Genz, A., Bretz, F., Miwa, T., Mi, X., Leisch, F., & Scheipl, F. (2017). mvtnorm: Multivariate normal and *t* distribution. R package version 1.0-6. https://CRAN.R-project.org/package=mvtnorm.

Ghoudi, K., Khoudraji, A., & Rivest, L.-P. (1998). Propriétés statistiques des copules de valeurs extrêmes bidimensionnelles. *The Canadian Journal of Statistics, 26*(1), 187–197.

Gudendorf, G., & Segers, J. (2010), Extreme-value copulas. In P. Jaworski, F. Durante, W. K. Härdle & W. Rychlik (Eds.), *Copula theory and its applications (Warsaw, 2009)* (pp. 127–146). *Lecture notes in statistics*. Berlin: Springer.

Gudendorf, G., & Segers, J. (2012). Nonparametric estimation of multivariate extreme-value copulas. *Journal of Statistical Planning and Inference, 143*, 3073–3085.

Gumbel, E. J. (1958). *Statistics of extremes*. New York: Columbia University Press.

Hofert, M. (2008). Sampling Archimedean copulas. *Computational Statistics & Data Analysis, 52*, 5163–5174.

Hofert, M. (2010). *Sampling nested Archimedean copulas with applications to CDO pricing*. PhD thesis, Südwestdeutscher Verlag für Hochschulschriften AG & Co. KG. ISBN 978-3-8381-1656-3.

Hofert, M. (2011). Effciently sampling nested Archimedean copulas. *Computational Statistics & Data Analysis, 55*, 57–70.

Hofert, M. (2012a). A stochastic representation and sampling algorithm for nested Archimedean copulas. *Journal of Statistical Computation and Simulation, 82*(9), 1239–1255.

Hofert, M. (2012b). Sampling exponentially tilted stable distributions. *ACM Transactions on Modeling and Computer Simulation, 22*(1). https://doi.org/10.1145/2043635.2043638.

Hofert, M. (2013). On sampling from the multivariate t distribution. *The R Journal, 5*(2), 129–136.

Hofert, M., Mächler, M., & McNeil, A. J. (2012). Likelihood inference for Archimedean copulas in high dimensions under known margins. *Journal of Multivariate Analysis, 110*, 133–150.

Hofert, M., Mächler, M., & McNeil, A. J. (2013). Archimedean copulas in high dimensions: Estimators and numerical challenges motivated by financial applications. *Journal de la Société Française de Statistique, 154*(1), 25–63.

Hofert, M., & Pham, D. (2013). Densities of nested Archimedean copulas. *Journal of Multivariate Analysis, 118*, 37–52.

Joe, H. (1997). *Multivariate models and dependence concepts*. London: Chapman & Hall.

Kano, Y. (1994). Consistency property of elliptical probability density functions. *Journal of Multivariate Analysis, 51*, 139–147.

Khoudraji, A. (1995). *Contributions à l'étude des copules et à la modélisation des valeurs extrêmes bivariées*. PhD thesis, Québec, Canada: Université Laval.

Kojadinovic, I., & Yan, J. (2010a). Comparison of three semiparametric methods for estimating dependence parameters in copula models. *Insurance: Mathematics and Economics, 47*, 52–63.

Liebscher, E. (2008). Construction of asymmetric multivariate copulas. *Journal of Multivariate Analysis, 99*, 2234–2250.

Malov, S. V. (2001). On finite-dimensional Archimedean copulas. In N. Balakrishnan, I. A. Ibragimov & V. B. Nevzorov (Eds.), *Asymptotic methods in probability and statistics with applications* (pp. 19–35). Basel: Birkhäuser.

Marshall, A. W., & Olkin, I. (1988). Families of multivariate distributions. *Journal of the American Statistical Association, 83*(403), 834–841.

McNeil, A. J. (2008). Sampling nested Archimedean copulas. *Journal of Statistical Computation and Simulation, 78*(6), 567–581.

McNeil, A. J., Frey, R., & Embrechts, P. (2015). *Quantitative risk management: Concepts, techniques and tools* (2nd ed.). Princeton, NJ: Princeton University Press.

McNeil, A. J., & Nešlehová, J. (2009). Multivariate Archimedean copulas, d-monotone functions and l_1-norm symmetric distributions. *The Annals of Statistics, 37*(5b), 3059–3097.

McNeil, A. J., & Nešlehová, J. (2010). From Archimedean to Liouville copulas. *Journal of Multivariate Analysis, 101*, 1772–1790.

Nolan, J. P. (2018). *Stable distributions—Models for heavy tailed data*. In progress, Chapter 1 online at http://fs2.american.edu/jpnolan/www/stable/stable.html. Boston: Birkhauser.

Pickands, J. (1981). Multivariate extreme value distributions. With a discussion. *Bulletin de l'Institut international de statistique, 49*, 859–878, 894–902. *Proceedings of the 43rd session of the Internatinal Statistical Institute*.

Ressel, P. (2013). Homogeneous distributions — And a spectral representation of classical mean values and stable tail dependence functions. *Journal of Multivariate Analysis, 117*, 246–256.

Ripley, B. D. (2017). MASS: Support functions and datasets for Venables and Ripley's MASS. R package version 7.3–47. http://CRAN.R-project.org/package=MASS.
Wang, X., & Yan, J. (2013). Practical notes on multivariate modeling based on elliptical copulas. *Journal de la Société Française de Statistique, 154*(1), 102–115.
Yan, J. (2007). Enjoy the joy of copulas: With a package copula. *Journal of Statistical Software, 21*(4), 1–21.

Chapter 4
Estimation

Assume that one has a *random sample* $\boldsymbol{X}_1, \dots, \boldsymbol{X}_n$ (that is, independent copies) of a d-variate random vector of interest $\boldsymbol{X}$ with df H and that the univariate margins $F_1, \dots, F_d$ of H are continuous. From Sklar's Theorem, see Theorem 2.3.1, there exists a unique copula C on $[0, 1]^d$, such that

$$H(\boldsymbol{x}) = C\big(F_1(x_1), \dots, F_d(x_d)\big), \quad \boldsymbol{x} \in \mathbb{R}^d. \tag{4.1}$$

Consider furthermore the realistic situation in which $F_1, \dots, F_d$ are all unknown. The aim of this chapter is to provide some answers to the question about how C can be estimated from $\boldsymbol{X}_1, \dots, \boldsymbol{X}_n$ in this setting. We shall discuss the estimation of the unknown copula C first, under the assumption that it belongs to a parametric family of copulas such as those introduced in Chaps. 2 and 3, and then in a nonparametric setting.

4.1 Estimation Under a Parametric Assumption on the Copula

Assume that C in (4.1) belongs to an absolutely continuous parametric family of copulas

$$\mathcal{C} = \{C_{\boldsymbol{\theta}} : \boldsymbol{\theta} \in \Theta\},$$

where Θ, the parameter space, is a subset of $\mathbb{R}^p$ for some integer $p \geq 1$.

First of all, notice that assuming that $C \in \mathcal{C}$ is equivalent to assuming that there exists $\boldsymbol{\theta}_0 \in \Theta$ such that $C = C_{\boldsymbol{\theta}_0}$. In other words, estimating C in the current context amounts to estimating the unknown parameter vector $\boldsymbol{\theta}_0$. Next, let us discuss the role of the unknown margins $F_1, \dots, F_d$ of H in the estimation of $\boldsymbol{\theta}_0$. If these were

M. Hofert et al., *Elements of Copula Modeling with R*, Use R!,
https://doi.org/10.1007/978-3-319-89635-9_4

known, the sample

$$\boldsymbol{U}_i = \big(F_1(X_{i1}), \ldots, F_d(X_{id})\big), \quad i \in \{1, \ldots, n\}, \tag{4.2}$$

would be observable, iid, and, from Lemma 2.4.1, it would be a sample from C. It follows that $\boldsymbol{\theta}_0$ could be estimated using classical techniques for iid data such as maximum likelihood estimation; see also Sect. 4.1.1. However, under the realistic assumption that the margins of H are unknown, the random sample (4.2) is not observable, and the margins are nuisance parameters that have to be estimated so that $\boldsymbol{\theta}_0$ can be estimated.

With primary interest in the estimation of $\boldsymbol{\theta}_0$, one can distinguish between two approaches to estimating $F_1, \ldots, F_d$: parametric estimation and nonparametric estimation. In the former case, the resulting estimate of $C_{\boldsymbol{\theta}_0}$ will be *fully parametric*, while in the latter case, it will be a *semi-parametric* estimate.

4.1.1 Parametrically Estimated Margins

Let us assume that there are good reasons to believe that $F_1, \ldots, F_d$ belong to absolutely continuous parametric families of univariate dfs $\mathcal{F}_1, \ldots, \mathcal{F}_d$, respectively, where, for any $j \in \{1, \ldots, d\}$,

$$\mathcal{F}_j = \{F_{j,\boldsymbol{\gamma}_j} : \boldsymbol{\gamma}_j \in \Gamma_j\},$$

and Γ_j is a subset of $\mathbb{R}^{p_j}$ for some integer $p_j \geq 1$. In other words, we assume that, for any $j \in \{1, \ldots, d\}$, there exists $\boldsymbol{\gamma}_{0,j} \in \Gamma_j$ such that $F_j = F_{j,\boldsymbol{\gamma}_{0,j}}$.

The estimation problem of interest then consists of estimating the unknown copula $C_{\boldsymbol{\theta}_0}$ from $\boldsymbol{X}_1, \ldots, \boldsymbol{X}_n$ under the assumption that the unknown df H belongs to the parametric family of dfs

$$\mathcal{H} = \big\{C\big(F_1(\cdot), \ldots, F_d(\cdot)\big) : C \in \mathcal{C} \text{ and } F_j \in \mathcal{F}_j \text{ for all } j \in \{1, \ldots, d\}\big\}. \tag{4.3}$$

Maximum Likelihood Estimator

As the aforementioned setting is a classical setting in parametric statistics, a straightforward approach is to estimate the vector of unknown parameters $(\boldsymbol{\gamma}_{0,1}, \ldots, \boldsymbol{\gamma}_{0,d}, \boldsymbol{\theta}_0)$ and obtain an estimate of $\boldsymbol{\theta}_0$ as a by-product. Notice that, as a consequence of the absolute continuity of $\mathcal{C}$ and $\mathcal{F}_1, \ldots, \mathcal{F}_d$, the family of parametric dfs $\mathcal{H}$ is absolutely continuous. Differentiating Sklar's representation (4.1) of $H \in \mathcal{H}$ implies that the corresponding density h is given by

$$h(\boldsymbol{x}) = c_{\boldsymbol{\theta}}\big(F_{1,\boldsymbol{\gamma}_1}(x_1), \ldots, F_{d,\boldsymbol{\gamma}_d}(x_d)\big) \prod_{j=1}^{d} f_{j,\boldsymbol{\gamma}_j}(x_j),$$

where $c_{\boldsymbol{\theta}}$ is the density of $C_{\boldsymbol{\theta}}$ and $f_{j,\boldsymbol{\gamma}_j}$ is the density of $F_{j,\boldsymbol{\gamma}_j}$, $j \in \{1, \ldots, d\}$; see also 4) in Remark 2.3.2. The parameter vector $(\boldsymbol{\gamma}_{0,1}, \ldots, \boldsymbol{\gamma}_{0,d}, \boldsymbol{\theta}_0)$ can then be estimated by the *maximum likelihood estimator (MLE)*, that is, the maximizer $(\boldsymbol{\gamma}_{n,1}, \ldots, \boldsymbol{\gamma}_{n,d}, \boldsymbol{\theta}_n)$ of the log-likelihood function

$$\begin{aligned} \ell_n(\boldsymbol{\gamma}_1, \ldots, \boldsymbol{\gamma}_d, \boldsymbol{\theta}) = &\sum_{i=1}^{n} \log c_{\boldsymbol{\theta}}\big(F_{1,\boldsymbol{\gamma}_1}(X_{i1}), \ldots, F_{d,\boldsymbol{\gamma}_d}(X_{id})\big) \\ &+ \sum_{j=1}^{d} \sum_{i=1}^{n} \log f_{j,\boldsymbol{\gamma}_j}(X_{ij}). \end{aligned} \tag{4.4}$$

The above approach is theoretically well grounded; see, for instance, Lehmann and Casella (1998). In addition, it has the advantage of providing not only the estimate $C_{\boldsymbol{\theta}_n}$ of C but also the estimate $F_{j,\boldsymbol{\gamma}_{n,j}}$ of $F_{j,\boldsymbol{\gamma}_j}$, $j \in \{1, \ldots, d\}$, and thus the estimate $C_{\boldsymbol{\theta}_n}\big(F_{1,\boldsymbol{\gamma}_{n,1}}(\cdot), \ldots, F_{d,\boldsymbol{\gamma}_{n,d}}(\cdot)\big)$ of H.

Example 4.1.1 (Estimation of Copula Parameters via the MLE) Assume that $d = 2$, $\mathcal{C}$ is the Clayton copula family, $\mathcal{F}_1$ is the family of univariate $\mathrm{N}(\mu, \sigma^2)$ dfs, and $\mathcal{F}_2$ is the family of $\mathrm{Exp}(\lambda)$ dfs. As we have seen in Chap. 2, objects representing multivariate distributions belonging to $\mathcal{H}$ in (4.3) can be created using the R function `mvdc()`. The following code illustrates the estimation of the Clayton parameter θ_0 from a sample of $n = 1000$ independent observations from such a distribution with marginal parameters $\boldsymbol{\gamma}_{0,1} = (\mu_0, \sigma_0) = (0, 1)$ and $\boldsymbol{\gamma}_{0,2} = \lambda_0 = 1$, and copula parameter $\theta_0 = 3$. The key R function here is `fitMvdc()`:

```
> ## The "unknown" copula (a 2-dim. Clayton copula with parameter 3)
> cc <- claytonCopula(3)
> ## The "unknown" distribution (N(0,1), Exp(1) margins)
> mcc <- mvdc(cc, margins = c("norm", "exp"),
              paramMargins = list(list(mean = 0, sd = 1),
                                  list(rate = 1)))
> ## Generate the "observed" sample
> set.seed(712)
> n <- 1000
> X <- rMvdc(n, mvdc = mcc)
> ## The function fitMvdc() estimates all the parameters of the mvdc object
> ## mcc (whose parameter values are not used). Starting values need to be
> ## provided.
> start <- c(mu0 = mean(X[,1]), sig0 = sd(X[,1]), lam0 = 1 / mean(X[,2]),
             th0 = 2)
> (mle <- fitMvdc(X, mvdc = mcc, start = start))
```

```
Call: fitMvdc(data = X, mvdc = mcc, start = start)
Maximum Likelihood estimation based on 1000 2-dimensional observations.
Copula:  claytonCopula
Margin 1 :
 m1.mean    m1.sd
-0.02231  0.99149
Margin 2 :
m2.rate
  1.014
Copula:
alpha
3.117
The maximized loglikelihood is -1760
Optimization converged
```

A richer output containing standard errors for the marginal and copula parameter estimates can be obtained by calling `summary()` on the object returned by `fitMvdc()`:

```
> summary(mle)
```

```
Call: fitMvdc(data = X, mvdc = mcc, start = start)
Maximum Likelihood estimation based on 1000 2-dimensional observations.
Copula:  claytonCopula
Margin 1 :
        Estimate Std. Error
m1.mean -0.02231      0.026
m1.sd    0.99149      0.013
Margin 2 :
        Estimate Std. Error
m2.rate    1.014      0.031
Clayton copula, dim. d = 2
      Estimate Std. Error
alpha    3.117      0.125
The maximized loglikelihood is -1760
Optimization converged
Number of loglikelihood evaluations:
function gradient
      49       10
```

□

In the context under consideration, maximum likelihood estimation suffers from two drawbacks. From (4.4), we see that the estimation of $\boldsymbol{\theta}_0$ will be affected by the parametric assumptions made on $F_1, \ldots, F_d$. If the margins are partially misspecified, that is, if, for some $j \in \{1, \ldots, d\}$, $F_j \notin \mathcal{F}_j$, the estimation of $\boldsymbol{\theta}_0$ will be biased. The second drawback is computational as the log-likelihood function given in (4.4) has to be maximized over a potentially high-dimensional parameter space, and high-dimensional optimization is known to be challenging. For instance, if $d = 10$, $\mathcal{C}$ is the normal copula family and $\mathcal{F}_1 = \cdots = \mathcal{F}_{10}$ is the family of univariate gamma dfs, the dimension of the parameter space of $\mathcal{H}$ is $\binom{10}{2} + 10 \cdot 2 = 65$.

Inference Functions for Margins Estimator

The computational burden potentially associated with maximum likelihood estimation can be decreased by employing a two-stage estimator known as the *inference functions for margins estimator (IFME)*; see Joe and Xu (1996), Joe (1997), and Joe (2005). It consists of first estimating the unknown marginal parameter vectors $\boldsymbol{\gamma}_{0,1}, \dots, \boldsymbol{\gamma}_{0,d}$ by computing, for each $j \in \{1, \dots, d\}$,

$$\boldsymbol{\gamma}_{n,j} = \operatorname*{argsup}_{\boldsymbol{\gamma}_j \in \Gamma_j} \sum_{i=1}^{n} \log f_{j,\boldsymbol{\gamma}_j}(X_{ij}), \tag{4.5}$$

and then estimating the unknown copula parameter vector $\boldsymbol{\theta}_0$ by

$$\boldsymbol{\theta}_n = \operatorname*{argsup}_{\boldsymbol{\theta} \in \Theta} \sum_{i=1}^{n} \log c_{\boldsymbol{\theta}}\big(F_{1,\boldsymbol{\gamma}_{n,1}}(X_{i1}), \dots, F_{d,\boldsymbol{\gamma}_{n,d}}(X_{id})\big).$$

The computational advantage of the latter approach generally comes at the price of an efficiency loss relative to the maximum likelihood estimator; see Joe (2005) for the theoretical properties of the IFME.

From a more intuitive perspective, one can view the IFME as first forming the sample

$$\boldsymbol{U}_{i,\boldsymbol{\gamma}_n} = \big(F_{1,\boldsymbol{\gamma}_{n,1}}(X_{i1}), \dots, F_{d,\boldsymbol{\gamma}_{n,d}}(X_{id})\big), \quad i \in \{1, \dots, n\}, \tag{4.6}$$

where $\boldsymbol{\gamma}_n = (\boldsymbol{\gamma}_{n,1}, \dots, \boldsymbol{\gamma}_{n,d})$, and then estimating $\boldsymbol{\theta}_0$ by maximizing a log-likelihood-like function, that is,

$$\boldsymbol{\theta}_n = \operatorname*{argsup}_{\boldsymbol{\theta} \in \Theta} \sum_{i=1}^{n} \log c_{\boldsymbol{\theta}}(\boldsymbol{U}_{i,\boldsymbol{\gamma}_n}). \tag{4.7}$$

Under the assumption that the margins are correctly specified, the sample (4.6) can be viewed as a consistently estimated version of the unobservable iid sample (4.2). For reasons that will become clear in Sect. 4.1.2, we could refer to $\boldsymbol{U}_{1,\boldsymbol{\gamma}_n}, \dots, \boldsymbol{U}_{n,\boldsymbol{\gamma}_n}$ as *(parametric) pseudo-observations* from C. As the latter sample is not a true sample from C, $\boldsymbol{\theta}_n$ in (4.7) is not a proper maximum likelihood estimator.

Example 4.1.2 (Estimation of Copula Parameters via the IFME) Consider the same setting as in Example 4.1.1 and notice that the marginal MLEs in (4.5) have simple closed forms in this case. The key R function here is `fitCopula()` with argument `method = "ml"`:

```
> ## Parametric pseudo-observations obtained from X by marginal MLE
> U <- cbind(pnorm(X[,1], mean = mean(X[,1]),
                     sd = sqrt((n - 1) / n) * sd(X[,1])),
             pexp(X[,2], rate = 1 / mean(X[,2])))
> ifme <- fitCopula(claytonCopula(), data = U, method = "ml")
> summary(ifme)
```

```
Call: fitCopula(copula, data = data, method = "ml")
Fit based on "maximum likelihood" and 1000 2-dimensional observations.
Clayton copula, dim. d = 2
      Estimate Std. Error
alpha    3.062      0.117
The maximized loglikelihood is 637.7
Optimization converged
Number of loglikelihood evaluations:
function gradient
       5        5
```

The standard error(s) in the result of `summary(fitCopula(, method = "ml"))` will be consistently estimated if the data passed as argument are a true sample from the unknown copula. In the context of the above example (and typically in practice), this is clearly not the case as they are parametric pseudo-observations. As a consequence, since the estimation errors for the marginal parameters are not taken into account, the standard error computed in the previous code will underestimate on average the true unknown standard error and should thus not be used in inferences.

From a numerical perspective, notice that the optimization method for computing (4.7) is controlled by the argument `optim.method` of `fitCopula()`. The latter is specified, by default, through the function `optimMeth()`, which suggests a numerical optimization method for each of the available copula families given the dimension d. For instance, in the previous code, `optim.method` was automatically set to:

```
> optimMeth(claytonCopula(), method = "ml", dim = 2)
```

```
[1] "L-BFGS-B"
```

However, in practice it can happen that the numerical optimization based on the suggestion of `optimMeth()` fails, in which case the user can try some of the other methods listed on the help page of the R function `optim()`. □

Finally, let us stress the fact that the computational advantage of the inference functions for margins method does not prevent it from suffering from the same drawback as maximum likelihood estimation: If the margins are partially misspecified, the estimation of $\boldsymbol{\theta}_0$ will be biased. Empirical evidence of this phenomenon is provided in Fermanian and Scaillet (2005, Section 2) and Kim et al. (2007).

4.1.2 Nonparametrically Estimated Margins

A natural way to avoid the issues mentioned in the previous section related to any misspecification of the margins consists of estimating $F_1, \dots, F_d$ nonparametrically. In the copula literature, the unknown univariate marginal dfs are typically estimated by the (rescaled) empirical dfs of the component samples of $\boldsymbol{X}_1, \dots, \boldsymbol{X}_n$, that is, for any $j \in \{1, \dots, d\}$, F_j is estimated by

$$F_{n,j}(x) = \frac{1}{n+1} \sum_{i=1}^{n} \mathbf{1}(X_{ij} \leq x), \quad x \in \mathbb{R}. \tag{4.8}$$

The estimated margins are then typically used to form the sample

$$\boldsymbol{U}_{i,n} = \big(F_{n,1}(X_{i1}), \dots, F_{n,d}(X_{id})\big), \quad i \in \{1, \dots, n\}, \tag{4.9}$$

from which $\boldsymbol{\theta}_0$ will be estimated. This sample can be regarded as a consistently estimated version of the unobservable iid sample (4.2) and is therefore frequently referred to as a sample of *pseudo-observations* from C. Notice that not only $\boldsymbol{U}_{1,n}, \dots, \boldsymbol{U}_{n,n}$ are not true observations from C but $\boldsymbol{U}_{1,n}, \dots, \boldsymbol{U}_{n,n}$ are not independent (because $F_{n,j}$ depends on the jth component sample $X_{1j}, \dots, X_{nj}$, $j \in \{1, \dots, d\}$).

The reason for using $n+1$ instead of n in (4.8) is to ensure that the sample (4.9) lies in the interior $(0, 1)^d$ of the unit hypercube. As will become clear later in this section, this asymptotically negligible rescaling is of particular importance when carrying out maximum pseudo-likelihood estimation, see (4.11) below, since several copula families have nonfinite density on the edges of $[0, 1]^d$ ending at $(1, \dots, 1)$.

For any $j \in \{1, \dots, d\}$, let R_{ij} be the rank of X_{ij} among $X_{1j}, \dots, X_{nj}$. It is then easy to verify that $F_{n,j}(X_{ij}) = R_{ij}/(n+1)$, that is, the sample (4.9) is merely a sample of multivariate scaled ranks:

$$\boldsymbol{U}_{i,n} = \frac{1}{n+1}(R_{i1}, \dots, R_{id}), \quad i \in \{1, \dots, n\}. \tag{4.10}$$

Example 4.1.3 (Pseudo-Observations of Daily Log-Returns) We consider the daily log-return data for the period 1996–2000 for the Intel, Microsoft, and General Electric stocks analyzed in McNeil et al. (2015, Chapter 7). Notice that the resulting trivariate observations contain a small number of ties (which is incompatible with the assumption of continuous margins considered in this chapter) and are weakly serially dependent (which is a stylized fact of financial time series). The practical handling of such aspects is postponed to Chap. 6. For simplicity, these important issues will be ignored in the remaining part of this chapter and in the next one.

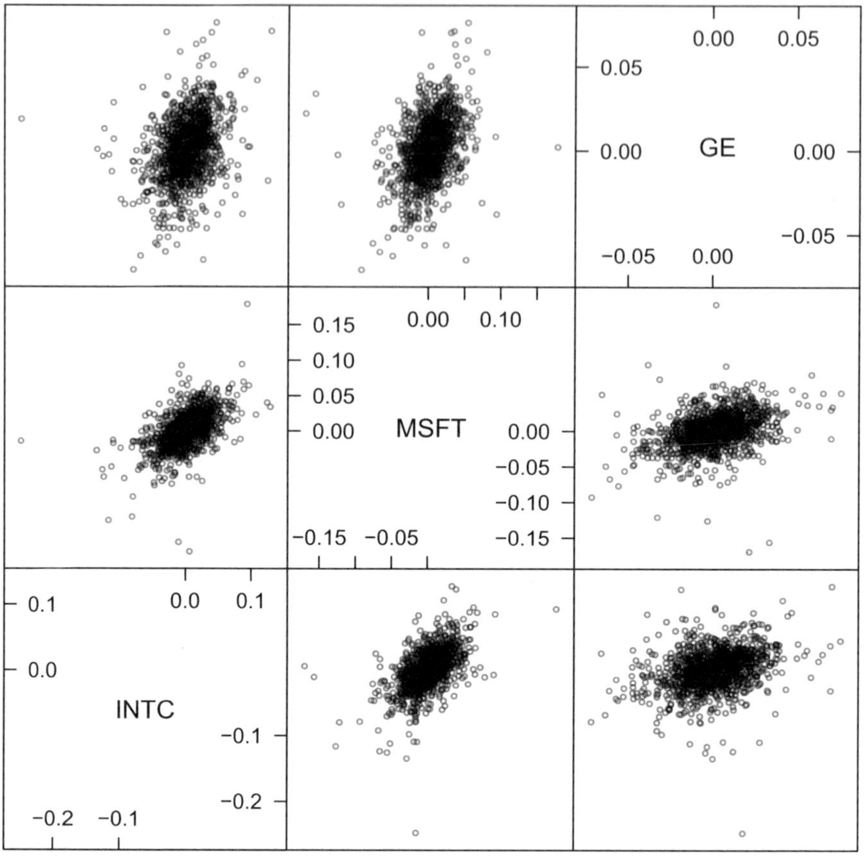

Fig. 4.1 Scatter-plot matrix of the `rdj` data set

We start by loading and plotting the data (see Fig. 4.1):

```
> data(rdj) # 'head(rdj)' for looking at the first six observations
> splom2(rdj[,2:4], cex = 0.4, col.mat = adjustcolor("black", 0.5))
```

The pseudo-observations in (4.9) can be computed using the R function `pobs()`:

```
> U <- pobs(rdj[,2:4])
> splom2(U, cex = 0.4, col.mat = "black")
```

The resulting plot is shown in Fig. 4.2.

□

Two of the most popular estimation methods among those relying on non-parametrically estimated margins are the so-called method-of-moments approaches

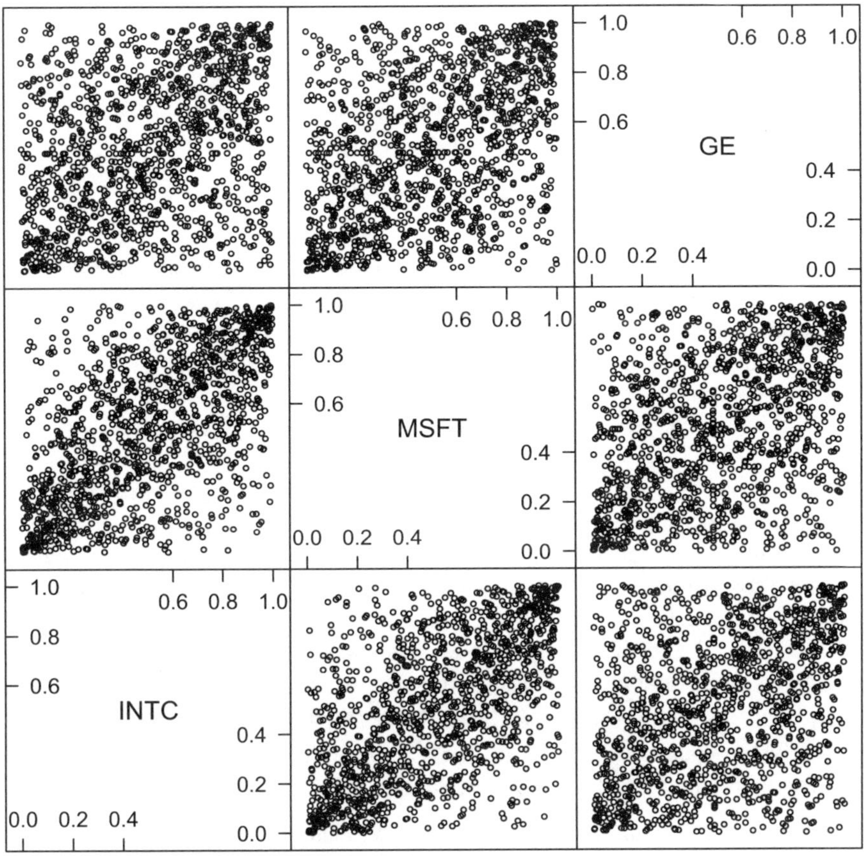

Fig. 4.2 Pseudo-observations corresponding to Fig. 4.1

based on Kendall's tau or Spearman's rho (presented next), and maximum pseudo-likelihood estimation (presented thereafter). Because they are solely based on the pseudo-observations in (4.9), that is, on the multivariate scaled ranks (4.10), these estimation methods are invariant under strictly increasing transformations of the margins, or more simply put, they are *margin-free*.

Method-of-Moments Estimators

Method-of-moments estimators in the copula context are direct transpositions of well-known method-of-moments estimators used in many areas of statistics. Early references on the former in a copula setting are, among others, Oakes (1982), Genest (1987), and Genest and Rivest (1993). Moments of random variables are replaced by moments of the copula such as Kendall's tau or Spearman's rho. Although it is possible to use method-of-moments estimators when $d \geq 3$ or $p \geq 2$ (for instance, for fitting elliptical copulas, as shall be explained later), we shall first only consider the most frequently encountered case in the literature, namely that of a bivariate sample and a one-parameter copula family $\mathcal{C} = \{C_\theta : \theta \in \Theta\}$.

Given $\mathcal{C}$, let g_τ and g_{ρ_s} be the functions defined by

$$g_\tau(\theta) = \tau(C_\theta) \quad \text{and} \quad g_{\rho_s}(\theta) = \rho_s(C_\theta), \qquad \theta \in \Theta \subseteq \mathbb{R},$$

where $\rho_s(C_\theta)$ and $\tau(C_\theta)$ are defined in (2.22) and (2.23), respectively. Method-of-moments estimators based on Kendall's tau (respectively, Spearman's rho) can be used for the family $\mathcal{C}$ if the function g_τ (respectively, g_{ρ_s}) is one-to-one. In that case, the estimator θ_n of θ_0 is simply given by

$$\theta_n = g_\tau^{-1}(\tau_n) \quad \big(\text{respectively, } \theta_n = g_{\rho_s}^{-1}(\rho_{s,n})\big),$$

where τ_n (respectively, $\rho_{s,n}$) is the sample version of Kendall's tau (respectively, Spearman's rho) defined in (2.25) (respectively (2.24)) and computed from the original sample, or, equivalently, from (4.9). Standard errors can be computed as explained for instance in Genest and Favre (2007, p. 353) or Kojadinovic and Yan (2010a, Section 2.1).

When $d > 2$ and the family $\mathcal{C}$ is exchangeable but the number of parameters p remains equal to one, a natural extension of the previous approach consists of applying g_τ^{-1} (respectively, $g_{\rho_s}^{-1}$) to the average of the sample Kendall's taus (respectively, Spearman's rhos) of the $\binom{d}{2}$ different bivariate margins; see, for instance, Kojadinovic and Yan (2010a, Section 2.2).

Example 4.1.4 (Estimation of Copula Parameters via the Method of Moments Based on Kendall's Tau) Assume that $\mathcal{C}$ is the Gumbel–Hougaard copula family. Then, $g_\tau(\theta) = 1 - 1/\theta$ and $\theta_n = g_\tau^{-1}(\tau_n) = 1/(1-\tau_n)$. The name of the R function corresponding to g_τ^{-1} is `iTau()` as already seen in Sect. 2.6.2. The following code illustrates the estimation of θ_0 from $n = 1000$ independent observations from a bivariate distribution with Gumbel–Hougaard copula with parameter $\theta_0 = 3$ and standard normal margins:

```
> ## The "unknown" copula (a 2-dim. Gumbel-Hougaard copula with parameter 3)
> gc <- gumbelCopula(3)
> ## The "unknown" distribution (N(0,1) margins)
> mgc <- mvdc(gc, margins = c("norm", "norm"),
              paramMargins = list(list(mean = 0, sd = 1),
                                  list(mean = 0, sd = 1)))
> ## Generate the "observed" sample
> set.seed(49)
> X <- rMvdc(1000, mvdc = mgc)
> ## The sample version of Kendall's tau
> tau.n <- cor(X[,1], X[,2], method = "kendall")
> ## The corresponding copula parameter estimate
> (itau <- iTau(gc, tau = tau.n))
```

```
[1] 2.976439
```

```
> stopifnot(all.equal(itau, 1 / (1 - tau.n))) # the same
> ## The same but with a standard error
> summary(fitCopula(gumbelCopula(), data = pobs(X), method = "itau"))
```

```
Call: fitCopula(copula, data = data, method = "itau")
Fit based on "inversion of Kendall's tau" and 1000 2-dimensional observations.
Gumbel copula, dim. d = 2
      Estimate Std. Error
alpha    2.976      0.114
```

Notice that the choice of the margins is irrelevant in the above setting since Kendall's correlation coefficient (being rank-based) is invariant under strictly increasing marginal transformations, making this estimation method indeed margin-free as mentioned previously. Furthermore, note that the standard error returned by `summary(fitCopula(, method = "itau"))` is computed from a formula assuming that the input data are pseudo-observations; see Kojadinovic and Yan (2010a, Section 2.1). Consequently, the argument `data` of `fitCopula()` needs to be set equal to pseudo-observations for the standard error to be consistently estimated. □

Example 4.1.5 (Estimation of Copula Parameters via the Method of Moments Based on Spearman's Rho) Assume that C is the normal copula family. Then, $g_{\rho_s}(\theta) = 6/\pi \arcsin(\theta/2)$ and $\theta_n = g_{\rho_s}^{-1}(\rho_{s,n}) = 2\sin(\pi\rho_{s,n}/6)$. The name of the R function corresponding to $g_{\rho_s}^{-1}$ is `iRho()` as seen in Sect. 2.6.2. The following code illustrates the estimation of θ_0 from $n = 1000$ independent observations from any meta-normal distribution with copula parameter $\theta_0 = 0.5$ and continuous margins:

```
> ## The "unknown" copula (a 2-dim. normal copula with parameter 0.5)
> nc <- normalCopula(0.5)
> ## Generate the "observed" sample
> set.seed(314)
> X <- rCopula(1000, nc)
> ## The sample estimate of Spearman's rho
> rho.n <- cor(X[,1], X[,2], method = "spearman")
> ## The corresponding copula parameter estimate
> (irho <- iRho(nc, rho = rho.n))
```

```
[1] 0.5498842
```

```
> stopifnot(all.equal(irho, 2 * sin(pi * rho.n / 6))) # the same
> ## The same but with a standard error
> summary(fitCopula(normalCopula(), data = pobs(X), method = "irho"))
```

```
Call: fitCopula(copula, data = data, method = "irho")
Fit based on "inversion of Spearman's rho" and 1000 2-dimensional observations.
Normal copula, dim. d = 2
      Estimate Std. Error
rho.1   0.5499      0.024
```

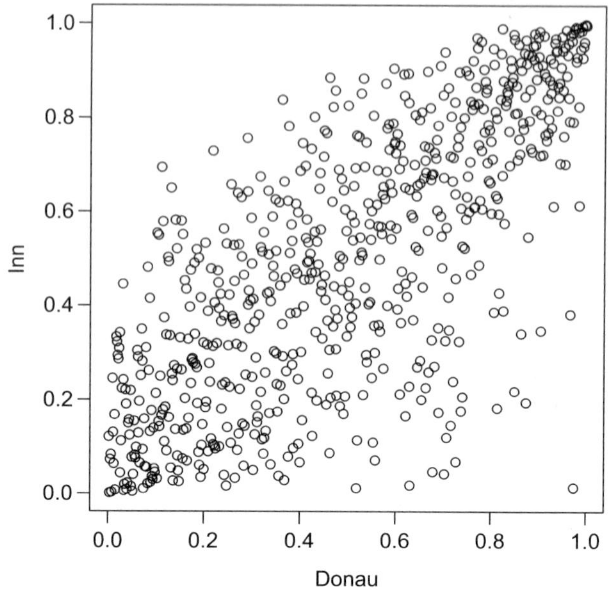

Fig. 4.3 Scatter plot of the pseudo-observations of the `danube` data set

As previously, for the standard error returned by `summary(fitCopula(, method = "irho"))` to be estimated consistently, `fitCopula()` needs to be called on pseudo-observations. □

Example 4.1.6 (Application of the Estimation via the Method of Moments Based on Kendall's Tau) As a last illustration of method-of-moments estimation, we consider the `danube` data set provided by the R package `lcopula` of Belzile and Genest (2017). It consists of bivariate pseudo-observations of preprocessed flow measurements from the Global River Discharge project of the Oak Ridge National Laboratory Distributed Active Archive Center. The initial observations are monthly average flow rates for two stations, one situated at Scharding (Austria) on the Inn river and the other at Nagymaros (Hungary) on the Danube. More details can be found on the help page describing the data set.

```
> data(danube, package = "lcopula") # already pseudo-observations
> U <- as.matrix(danube)
> plot(U, xlab = "Donau", ylab = "Inn")
```

The resulting pseudo-observations are displayed in Fig. 4.3. Method-of-moments estimators can be used to fit the Gumbel–Hougaard, the Plackett, or the normal copula family to these bivariate data:

```
> summary(fitCopula(gumbelCopula(), data = U, method = "itau"))
```

```
Call: fitCopula(copula, data = data, method = "itau")
Fit based on "inversion of Kendall's tau" and 659 2-dimensional observations.
Gumbel copula, dim. d = 2
      Estimate Std. Error
alpha    2.215      0.091

> summary(fitCopula(plackettCopula(), data = U, method = "itau"))

Call: fitCopula(copula, data = data, method = "itau")
Fit based on "inversion of Kendall's tau" and 659 2-dimensional observations.
plackettCopula copula, dim. d = 2
      Estimate Std. Error
alpha    15.21      1.729

> summary(fitCopula(normalCopula(), data = U, method = "itau"))

Call: fitCopula(copula, data = data, method = "itau")
Fit based on "inversion of Kendall's tau" and 659 2-dimensional observations.
Normal copula, dim. d = 2
      Estimate Std. Error
rho.1   0.7588      0.019
```

Diagnostics and tests that can be used to guide the choice of the hypothesized copula family will be discussed in Chap. 5. □

Maximum Pseudo-Likelihood Estimator

If the margins $F_1, \dots, F_d$ were known, the maximum likelihood estimator of $\boldsymbol{\theta}_0$ would be given by

$$\boldsymbol{\theta}_n = \operatorname*{argsup}_{\boldsymbol{\theta} \in \Theta} \sum_{i=1}^{n} \log c_{\boldsymbol{\theta}}(\boldsymbol{U}_i)$$

in terms of the iid sample $\boldsymbol{U}_1, \dots, \boldsymbol{U}_n$ defined in (4.2). As the margins are unknown, it seems natural to replace the unobservable sample $\boldsymbol{U}_1, \dots, \boldsymbol{U}_n$ in the previous display by the sample of pseudo-observations (4.9), thereby leading to the *maximum pseudo-likelihood estimator (MPLE)*

$$\boldsymbol{\theta}_n = \operatorname*{argsup}_{\boldsymbol{\theta} \in \Theta} \sum_{i=1}^{n} \log c_{\boldsymbol{\theta}}(\boldsymbol{U}_{i,n}). \tag{4.11}$$

For copula families whose parameters can be estimated by method-of-moments estimators, the MPLE was observed to be more efficient both in finite-samples and asymptotically; see for instance Genest et al. (1995), Tsukahara (2005), and Kojadinovic and Yan (2010a). Another key advantage is that the range of applicability of the MPLE is in principle far greater than the one of the method-of-moments estimators, especially when multiparameter copulas are considered.

Indeed, as long as the density c_{θ} can be evaluated, one can attempt to maximize the logarithmic pseudo-likelihood. In practice, this is generally carried out using numerical optimization routines such as those implemented in the R function `optim()`. The underlying optimization can typically be made more efficient when an initial value for $\boldsymbol{\theta}_0$ can be provided. When possible, a natural choice is to use method-of-moments estimators, possibly based on averages of pairwise sample versions of Kendall's tau or Spearman's rho. Numerical optimization can, however, become unstable when the dimension d of the data is large, the number p of parameters is large, or the sample size n is small.

Standard errors for the MPLE can be computed as explained in Genest et al. (1995); see also Kojadinovic and Yan (2010a, Section 3). Their estimation requires the computation of certain partial derivatives of the (logarithmic) density. For those families in `copula` for which these partial derivatives are not implemented, numerical differentiation based on the R package `numDeriv` of Gilbert and Varadhan (2016) is used (and a warning message is displayed).

Example 4.1.7 (Estimation of Copula Parameters via the MPLE) Assume that C is the Frank copula family. The following code illustrates the estimation of θ_0 by its MPLE from $n = 1000$ independent observations from any meta-Frank distribution with copula parameter $\theta_0 = 3$ and continuous margins:

```
> ## The "unknown" copula (a 2-dim. Frank copula with parameter 3)
> fc <- frankCopula(3)
> ## Generate the "observed" sample
> set.seed(271)
> U <- rCopula(1000, fc)
> ## Compute the MPLE and its standard error
> summary(fitCopula(frankCopula(), data = pobs(U), method = "mpl"))

Call: fitCopula(copula, data = data, method = "mpl")
Fit based on "maximum pseudo-likelihood" and 1000 2-dimensional observations.
Frank copula, dim. d = 2
      Estimate Std. Error
alpha    2.907        0.2
The maximized loglikelihood is 101.5
Optimization converged
Number of loglikelihood evaluations:
function gradient
       4        4
```

Similar to method-of-moments estimators, for the standard errors returned by `summary(fitCopula(, method = "mpl"))` to be estimated consistently, `fitCopula()` needs to be called on pseudo-observations; see, for instance, Kojadinovic and Yan (2010a, Equation (11)).

From a numerical perspective, the argument `optim.method` of `fitCopula()` specifies which optimization method is used to carry out the maximization in (4.11). As explained in Example 4.1.2, the default optimization method is determined by `optimMeth()`. For the bivariate Frank family, we have:

```
> optimMeth(frankCopula(), method = "mpl", dim = 2)
```

```
[1] "L-BFGS-B"
```

As already mentioned, if the optimization carried out in `fitCopula()` fails, the user can try to set `optim.method` to some of the other methods listed on the help page of `optim()`. □

Example 4.1.8 (Application of the Estimation via the MPLE) As a second example of maximum pseudo-likelihood estimation, we consider again the trivariate daily log-return data set (Intel, Microsoft, General Electric) introduced in Example 4.1.3 and fit two trivariate copula families to these data: the (nonhomogeneous) normal and t families. The first has three parameters while the second one has four. When carrying out the fitting, it is thus necessary to specify that the correlation matrices P of C^{n}_{P} and $C^{t}_{P,\nu}$ should be considered as parametrized by $\binom{d}{2}$ elements with $d = 3$ (instead of just one correlation parameter as in the homogeneous case). This can be done by changing the default value of the argument `dispstr` of `normalCopula()` and `tCopula()` from `"ex"` (for "exchangeable") to `"un"` (for "unstructured"):

```
> U <- pobs(rdj[,2:4]) # compute the pseudo-observations
> ## MPLE for the normal copula
> summary(fitCopula(normalCopula(dim = 3, dispstr = "un"), data = U))

Call: fitCopula(copula, data = data)
Fit based on "maximum pseudo-likelihood" and 1262 3-dimensional observations.
Normal copula, dim. d = 3
      Estimate Std. Error
rho.1   0.5781      0.016
rho.2   0.3400      0.020
rho.3   0.4017      0.018
The maximized loglikelihood is 375.7
Optimization converged
Number of loglikelihood evaluations:
function gradient
      29        6
```

```
> ## MPLE for the t copula
> summary(fitCopula(tCopula(dim = 3, dispstr = "un"), data = U))

Call: fitCopula(copula, data = data)
Fit based on "maximum pseudo-likelihood" and 1262 3-dimensional observations.
t-copula, dim. d = 3
      Estimate Std. Error
rho.1   0.5877      0.020
rho.2   0.3593      0.026
rho.3   0.4225      0.025
df      6.5025         NA
The maximized loglikelihood is 419.3
Optimization converged
Number of loglikelihood evaluations:
function gradient
      31       10
```

As one can see from the last output, `summary(fitCopula(, method = "mpl"))` (`"mpl"` being the default value of the argument `method`) does not return a standard error for the estimate of the degrees of freedom of a t copula. In addition, the standard errors of the correlation parameters are estimated as if the underlying t copula had its degree of freedom parameter fixed (to its estimated value); see Sect. 4.1.5 below. As a consequence, the computed standard errors underestimate, on average, the true unknown standard errors (simulations indicate that the bias might not be of practical importance, though). These shortcomings are, among other things, due to the fact that the t copula cannot be evaluated for non-integer degrees of freedom for the moment, as already mentioned in Example 2.3.3.

Diagnostics and tests that can be used to choose between the normal and the t families will be presented in Chap. 5. □

The asymptotic properties of the MPLE are investigated in Genest et al. (1995), Shih and Louis (1995), and Tsukahara (2005). Historical details on this approach can be found in Genest et al. (1995). Conditions for the asymptotic semi-parametric efficiency of the MPLE are investigated in Genest and Werker (2002). It is argued on intuitive grounds that efficiency should actually be the exception rather than the norm, as, in practice, only under independence or under the normal copula family, is the MPLE identified to be asymptotically semi-parametrically efficient; see Klassen and Wellner (1997).

4.1.3 Estimators of Elliptical Copula Parameters

As the dimension d of the data increases, the MPLE for elliptical copulas becomes prohibitive and numerically unstable because of the number of parameters (which is $\binom{d}{2}$ for the correlation matrix alone). In order to fit a normal copula C_P^{n}, method-of-moments estimators based on Spearman's rho or Kendall's tau remain a feasible alternative. The idea simply consists of first computing the d by d matrix of sample Spearman's rhos (respectively, Kendall's taus) and then applying $g_{\rho_{\mathrm{s}}}^{-1}$ (respectively, g_{τ}^{-1}) elementwise to obtain an estimate of P. To guarantee that the estimate of P is indeed a proper correlation matrix, the method of Higham (2002) is applied by default, as illustrated in the next example.

Example 4.1.9 (Estimation of Normal Copula Parameters via Method-of-Moments)
We consider pseudo-observations computed from log-returns of all $d = 20$ constituents of the Swiss Market Index (SMI) from 2011-09-09 to 2012-03-28. Let us first load the data, compute the log-returns (based on `returns()` from the R package `qrmtools` of Hofert and Hornik 2017) and the corresponding pseudo-observations:

```
> data(SMI.12) # load the SMI constituent data
> library(qrmtools)
> X <- returns(SMI.12) # compute log-returns
> U <- pobs(X) # compute pseudo-observations
> d <- ncol(U) # 20 dimensions
```

We wish to fit a (nonhomogeneous) normal copula C_P^{n} to these pseudo-observations using method-of-moments estimators based on Spearman's rho and Kendall's tau:

```
> f.irho <- fitCopula(normalCopula(dim = d, dispstr = "un"), data = U,
                      method = "irho")
> f.itau <- fitCopula(normalCopula(dim = d, dispstr = "un"), data = U,
                      method = "itau")
```

An inspection of `coef(f.irho)` and `coef(f.itau)` reveals that, internally, correlation matrices are represented as vectors of $\binom{d}{2}$ elements. They can be converted to correlation matrices using the function `p2P()`:

```
> P.irho <- p2P(coef(f.irho), d = d)
> P.itau <- p2P(coef(f.itau), d = d)
```

Notice that the inverse operation is available via the function `P2p()`.

Finally, to guarantee that an estimate of the correlation matrix P is indeed a proper correlation matrix, the argument `posDef` of `fitCopula()` is `TRUE` by default when called on an object representing an elliptical copula. This results in internally applying `nearPD()` of the R package `Matrix` of Bates and Maechler (2017), which computes the nearest positive definite matrix (with respect to a weighted Frobenius norm; see Higham (2002) for more details) to the initial correlation matrix estimate. □

Because of the degrees of freedom parameter, the method-of-moments estimator based on Spearman's rho or Kendall's tau cannot be directly applied to the t family. For estimating the parameters of a $C_{P,\nu}^t$ copula, there exists, however, a convenient alternative which combines method-of-moments estimation based on Kendall's tau for the estimation of P and maximum pseudo-likelihood-like estimation for the estimation of ν. The procedure is presented in Mashal and Zeevi (2002) based on the idea of inverting Kendall's tau for estimating the correlation matrix as introduced in a RiskLab report in 2001 later published as Embrechts et al. (2003); see also Demarta and McNeil (2005). The idea is to first use pairwise inversion of Kendall's tau for estimating the $\binom{d}{2}$ correlation parameters in P and then maximum pseudo-likelihood-like estimation based on the estimated P and the available pseudo-observations to estimate the degrees of freedom parameter ν.

Example 4.1.10 (Estimation of t Copula Parameters Using the Method of Mashal and Zeevi) We briefly demonstrate how to fit a t family using the method of Mashal and Zeevi (2002) to the pseudo-observations computed in Example 4.1.9:

```
> fit <- fitCopula(tCopula(dim = d, dispstr = "un"), data = U,
                   method = "itau.mpl")
```

For `method = "itau.mpl"`, the estimation of the standard errors is not implemented yet. An analog estimation with `method = "mpl"` fails for numerical reasons. □

As mentioned in Sect. 2.6, using sample (linear) correlation is truly meaningful only when the observations arise from an elliptical distribution. In case of the multivariate t distribution with scale matrix Σ and $\nu > 2$ degrees of freedom, for instance, the sample correlation matrix is a natural estimator of the correlation matrix P corresponding to Σ, see Remark 3.1.2 (1), and P is a parameter of the underlying t copula. The following example compares this estimator of P with the method-of-moments estimator based on inverting Kendall's tau through simulations.

Example 4.1.11 (Linear Correlation vs Kendall's Tau for Estimating t Distributions) We focus on the bivariate t distribution with fixed degree of freedom $\nu > 2$. We start by implementing an auxiliary function to draw B samples of size n from this distribution with correlation parameter ρ. The auxiliary function then computes, for each of the B samples, an estimate of the correlation parameter ρ once via the sample linear correlation and once via the method-of-moments estimator based on inverting Kendall's tau:

```
> ##' @title Compute (repeated) parameter estimates for the bivariate t
> ##' @param nu degrees of freedom parameter
> ##' @param rho correlation parameter
> ##' @param B number of replications
> ##' @param n sample size
> ##' @return (B, 2)-matrix containing the B estimates computed
> ##'           via Pearson's rho and Kendall's tau
> estimates <- function(nu, rho, B, n)
  {
      ## Generate data (B-list of (n, 2)-matrices)
      tc <- tCopula(rho, df = nu) # define the corresponding t copula
      X <- lapply(1:B, function(i) qt(rCopula(n, copula = tc), df = nu))
      ## For each of the B data sets, estimate the correlation parameter of the
      ## t distribution via Pearson's rho and Kendall's tau.
      pea <- vapply(X, function(x) cor(x[,1], x[,2]), numeric(1))
      ken <- iTau(tCopula(df = nu),
                  tau = sapply(X, function(x) cor(x, method = "kendall")[2,1]))
      ## Return
      cbind(Pearson = pea, Kendall = ken) # (B, 2)-matrix
  }
```

Similar to McNeil et al. (2015, Example 6.30), we use the previous function to generate $B = 3000$ samples of size $n = 90$ from a bivariate t distribution with correlation $\rho = 0.5$ and $\nu = 3$ degrees of freedom. We display box plots of the two samples of estimates of ρ:

```
> set.seed(271) # for reproducibility
> nu <- 3 # degrees of freedom
> rho <- 0.5 # correlation parameter
> r <- estimates(nu, rho = rho, B = 3000, n = 90) # (B, 2)-matrix
> varP <- var(r[,"Pearson"]) # variance of sample linear correlation
> varK <- var(r[,"Kendall"]) # variance of inverting Kendall's tau
> VRF <- varP / varK # variance reduction factor
> PIM <- (varP - varK) / varP * 100 # % improvement
> boxplot(r, names = c("Sample linear correlation", "Inverting Kendall's tau"),
          ylab = substitute("Estimates of"~rho~"of a"
                            ~t[nu.]~"distribution with"~rho==rho.,
                            list(nu. = nu, rho. = rho)))
> mtext(substitute("VRF (% improvement):"~~v~"("*i*"%)",
                   list(v = round(VRF, 2), i = round(PIM))),
        side = 4, line = 1, adj = 0, las = 0)
```

As the box plots on the left-hand side of Fig. 4.4 indicate, the method-of-moments estimator based on inverting Kendall's tau has a smaller variance than the sample correlation in this case.

With the help of our auxiliary function, we can also investigate, by simulation, how the ratio of the variances of the two estimators changes as a function of the degrees of freedom ν and the correlation parameter ρ of a bivariate t distribution:

```
> ## Estimate variance reduction factor for various nu and rho
> nu. <- 2 + 2^seq(-2, 7, by = 0.5) # degrees of freedom to consider
> rho. <- c(-0.99, -0.8, 0, 0.8, 0.99) # correlations to consider
> res <- lapply(nu., function(nu..) lapply(rho., function(rho..)
                estimates(nu.., rho = rho.., B = 3000, n = 90)))
> vars <- lapply(res, function(r) lapply(r, function(r.)
                 var(r.[,"Pearson"])/var(r.[,"Kendall"])))
> VRF. <- matrix(unlist(vars), nrow = length(rho.), ncol = length(nu.),
                 dimnames = list(rho = rho., nu = nu.))
> ylim <- range(VRF.)
> lrho <- length(rho.)
> plot(nu., VRF.[1,], type = "l", log = "xy", ylim = ylim, col = 1,
       xlab = "Degrees of freedom",
       ylab = "VRF (correlation over inverting Kendall's tau)")
> for(k in 2:lrho) lines(nu., VRF.[k,], col = k)
> abline(h = 1, lty = 2)
> legend("topright", bty = "n", lty = rep(1, lrho), col = 1:lrho,
         legend = as.expression(lapply(1:lrho, function(k)
             substitute(rho == rho., list(rho. = rho.[k])))))
```

As the right-hand side of Fig. 4.4 reveals, the estimator of ρ based on inverting of Kendall's tau has a smaller variance than sample correlation if $\nu > 2$ is small. For moderate ν, inversion of Kendall's tau is only preferred for $|\rho|$ not too large. Furthermore, for moderate and large ν, sample correlation seems to have the smallest variance when $|\rho|$ is large. For large ν, sample correlation appears to outperform inversion of Kendall's tau for all investigated ρ. □

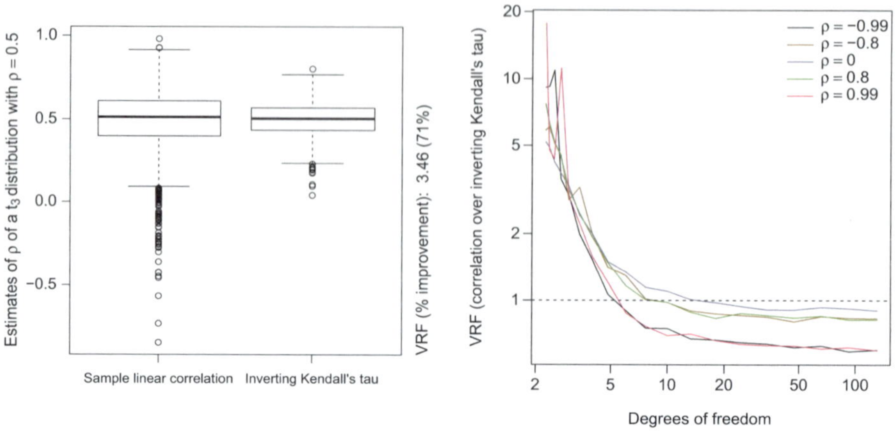

Fig. 4.4 Box plots of sample linear correlation and method-of-moments estimates (based on inverting Kendall's tau) of ρ for $B = 3000$ samples of size $n = 90$ from a bivariate t distribution with $\nu = 3$ degrees of freedom and correlation parameter $\rho = 0.5$ (left); note the smaller variance of the method-of-moments estimator in this case. Estimated variance-reduction factor (VRF) as a function of the degrees of freedom ν and the correlation parameter ρ (right)

4.1.4 *Other Semi-parametric Estimators*

Method-of-moments and maximum pseudo-likelihood estimators are not the only semi-parametric estimators available in the literature. For instance, two other semi-parametric estimation methods are investigated in Tsukahara (2005), namely rank approximate Z-estimators and minimum-distance estimators. In a simulation study, both methods are found overall to lead to a higher estimated mean squared error than the MPLE, though.

A version of the MPLE where the unknown marginal densities are approximated by linear combinations of finite-dimensional known basis functions with increasing complexity, called *sieves*, is investigated in Chen et al. (2006). The authors show that the resulting estimator of $\boldsymbol{\theta}_0$ is asymptotically semi-parametrically efficient provided additional smoothness conditions are satisfied. The study of the finite-sample performance of the method for sample size $n = 400$ reveals that this approach performs significantly better than the standard MPLE when one marginal df is known. This advantage does not seem to hold anymore when all marginals are unknown. The finite-sample performance of this more complex approach would need to be studied for sample sizes smaller than 400 to better judge the overall performance of this estimator.

4.1.5 Estimation of Copula Models with Partly Fixed Parameters

When dealing with a multiparameter copula family $\mathcal{C}$ (so when $p > 1$), it is sometimes of interest to fix some components of the parameter vector $\boldsymbol{\theta}$ to predetermined values. This is typically done to reduce the complexity of the estimation (but can also be relevant in the context of stress testing or carried out to incorporate expert opinion into the model). The key functions for fixing parameter values in the `copula` package are `fixParam()` and `fixedParam()`. As it shall be illustrated below, the former controls which component of a parameter vector should be treated as fixed when creating an R object representing a copula while the latter can be used to modify an existing R object in terms of which parameters are fixed.

Example 4.1.12 (Estimation of Elliptical Copulas with Partly Fixed Parameters) We first consider the estimation of the parameters of a trivariate normal copula whose first parameter is fixed to 0.6:

```
> ## The "unknown" copula (a 3-dim. normal copula)
> nc  <- normalCopula(param = c(0.6, 0.3, 0.2), dim = 3, dispstr = "un")
> ## Generate the "observed" sample and compute corresponding pobs
> set.seed(819)
> U <- pobs(rCopula(1000, nc))
> ## A trivariate normal copula whose first parameter is fixed to 0.6
> (ncf <- normalCopula(param = fixParam(c(0.6, NA_real_, NA_real_),
                                        c(TRUE, FALSE, FALSE)),
                      dim = 3, dispstr = "un"))
```

```
Normal copula, dim. d = 3
Dimension:  3
Parameters (partly fixed, see ':='):
  rho.1   := 0.6
  rho.2    =  NA
  rho.3    =  NA
dispstr:  un
```

```
> fitCopula(ncf, data = U) # MPLE
```

```
Call: fitCopula(copula, data = data)
Fit based on "maximum pseudo-likelihood" and 1000 3-dimensional observations.
Copula: normalCopula
 rho.2  rho.3
0.2882 0.1834
The maximized loglikelihood is 283.3
Optimization converged
```

```
> fitCopula(ncf, data = U, method = "itau")
```

```
Call: fitCopula(copula, data = data, method = "itau")
Fit based on "inversion of Kendall's tau" and 1000 3-dimensional observations.
Copula: normalCopula
 rho.2  rho.3
0.2861 0.1952
```

```
> fitCopula(ncf, data = U, method = "irho")
```

```
Call: fitCopula(copula, data = data, method = "irho")
Fit based on "inversion of Spearman's rho" and 1000 3-dimensional observations.
Copula: normalCopula
 rho.2  rho.3
0.2863 0.1948
```

As an alternative to using the function `fixParam()` when creating a copula object, one can fix the parameter values of an existing copula object using the function `fixedParam()`. For instance:

```
> fixedParam(nc) <- c(TRUE, FALSE, FALSE)
> nc
```

```
Normal copula, dim. d = 3
Dimension:  3
Parameters (partly fixed, see ':='):
  rho.1   := 0.6
  rho.2    = 0.3
  rho.3    = 0.2
dispstr:  un
```

As a consequence, instead of considering calls of the form `fitCopula(ncf,)`, one could alternatively consider calls of the form `fitCopula(nc,)` to attempt to fit the normal copula family with first correlation parameter fixed to 0.6.

A similar example for the t copula family follows, this time with two fixed parameters (which leaves two free parameters, a correlation parameter and the degrees of freedom):

```
> ## The "unknown" copula is a 3-dim. t copula (with 4 d.o.f. by default)
> tc  <- tCopula(param = c(0.6, 0.3, 0.2), dim = 3, dispstr = "un")
> ## Generate the "observed" sample and compute corresponding pobs
> set.seed(314)
> U <- pobs(rCopula(1000, tc))
> ## A trivariate t copula whose first two parameters are fixed to 0.6 and 0.3
> (tcf <- tCopula(param = fixParam(c(0.6, 0.3, NA_real_),
                                   c(TRUE, TRUE, FALSE)),
                  dim = 3, dispstr = "un"))
```

```
t-copula, dim. d = 3
Dimension:  3
Parameters (partly fixed, see ':='):
  rho.1   := 0.6
  rho.2   := 0.3
  rho.3    =  NA
  df       = 4.0
dispstr:  un
```

```
> fitCopula(tcf, data = U) # MPLE
```

```
Call: fitCopula(copula, data = data)
Fit based on "maximum pseudo-likelihood" and 1000 3-dimensional observations.
Copula: tCopula
 rho.3     df
0.2208 4.2502
The maximized loglikelihood is 366.1
Optimization converged
```

```
> fitCopula(tcf, data = U, method = "itau.mpl")
```

```
Call: fitCopula(copula, data = data, method = "itau.mpl")
Fit based on "itau for dispersion matrix P and maximum likelihood for df" and
    1000 3-dimensional observations.
Copula: tCopula
rho.3    df
0.246 4.257
The maximized loglikelihood is 365.6
Optimization converged
```

To keep the degrees of freedom fixed, set `df.fixed = TRUE` in the function `tCopula()`:

```
> ## A trivariate t copula whose first correlation parameter is fixed to 0.6
> ## and whose number of degrees of freedom is fixed to 4 (default value)
> (tcf2 <- tCopula(param = fixParam(c(0.6, NA_real_, NA_real_),
                                    c(TRUE, FALSE, FALSE)),
                   dim = 3, dispstr = "un", df.fixed = TRUE))
```

```
t-copula, dim. d = 3
Dimension:  3
Parameters (partly fixed, see ':='):
  rho.1   := 0.6
  rho.2    =  NA
  rho.3    =  NA
  df      := 4.0
dispstr:  un
```

```
> fitCopula(tcf2, data = U) # MPLE
```

```
Call: fitCopula(copula, data = data)
Fit based on "maximum pseudo-likelihood" and 1000 3-dimensional observations.
Copula: tCopula
 rho.2  rho.3
0.3241 0.2337
The maximized loglikelihood is 366.2
Optimization converged
```

```
> fitCopula(tcf2, data = U, method = "itau")
```

```
Call: fitCopula(copula, data = data, method = "itau")
Fit based on "inversion of Kendall's tau" and 1000 3-dimensional observations.
Copula: tCopula
 rho.2  rho.3
0.3393 0.2460
```

□

Example 4.1.13 (Estimation of Khoudraji–Clayton Copulas with Partly Fixed Parameters) As explained in Sect. 3.4.2, Khoudraji's device can be used to construct potentially non-exchangeable copula families from exchangeable ones. Recall that, in the bivariate case, if C_1 in (3.14) is chosen to be the independence copula Π, a non-exchangeable version of C_2 can be obtained by setting one of the shape parameters, s_1 or s_2, relatively close to 1. This is illustrated in Example 3.4.3 when C_2 is a Clayton copula.

In a modeling framework, as a simpler alternative to the three-parameter copula $\mathrm{kho}_{s}(\Pi, C_{\theta}^{\mathrm{C}})$, one could consider one of the two-parameter models $\mathrm{kho}_{(s_1,1)}(\Pi, C_{\theta}^{\mathrm{C}})$ or $\mathrm{kho}_{(1,s_2)}(\Pi, C_{\theta}^{\mathrm{C}})$, according to the direction of the asymmetry observed in the scatter plot of pseudo-observations. As we shall see from the following code, the two-parameter copulas turn out to be easier to fit (see also Example 5.2.16 for goodness-of-fit testing):

```
> ## The "unknown" copula (a 2-dim. Khoudraji-Clayton copula)
> kc <- khoudrajiCopula(copula2 = claytonCopula(6), shapes = c(0.4, 1))
> set.seed(1307)
> U <- pobs(rCopula(1000, kc))
> ## The default optimization method for fitting bivariate copulas
> ## constructed with Khoudraji's device by MPLE
> optimMeth(khoudrajiCopula(), method = "mpl", dim = 2)
```

```
[1] "BFGS"
```

Fitting the three-parameter Khoudraji–Clayton copula using the BFGS optimization method (which requires starting values for all the components of the parameter vector) fails:

```
> try(fitCopula(khoudrajiCopula(copula2 = claytonCopula()),
                start = c(1.1, 0.5, 0.5), data = U))
```

```
Error in optim(start, logL, lower = lower, upper = upper, method = optim.method
    , :
  non-finite finite-difference value [3]
```

The same task based on the more robust Nelder–Mead optimization method is successful:

```
> fitCopula(khoudrajiCopula(copula2 = claytonCopula()),
            start = c(1.1, 0.5, 0.5), data = U,
            optim.method = "Nelder-Mead")
```

```
Call: fitCopula(copula, data = data, start = ..1, optim.method = "Nelder-Mead")
Fit based on "maximum pseudo-likelihood" and 1000 2-dimensional observations.
Copula: khoudrajiExplicitCopula
c2.alpha    shape1    shape2
  5.9257    0.3929    1.0000
The maximized loglikelihood is 166.5
Optimization converged
```

Fitting a Khoudraji–Clayton copula whose second shape parameter is fixed to `1` works even with the default BFGS method:

```
> kcf <- khoudrajiCopula(copula2 = claytonCopula(),
                         shapes = fixParam(c(NA_real_, 1),
                                           c(FALSE, TRUE)))
> fitCopula(kcf, start = c(1.1, 0.5), data = U)
```

```
Call: fitCopula(copula, data = data, start = ..1)
Fit based on "maximum pseudo-likelihood" and 1000 2-dimensional observations.
Copula: khoudrajiExplicitCopula
c2.alpha    shape1
  5.9249    0.3929
The maximized loglikelihood is 166.5
Optimization converged
```

□

4.2 Nonparametric Estimation of the Copula

A smooth nonparametric estimate of the copula or its density can be obtained using the kernel method or wavelets; see, for example, Charpentier et al. (2007), Omelka et al. (2009), and Genest et al. (2009). With Chap. 5 in mind, we mostly focus on the simplest and most studied nonparametric estimator of C, namely the so-called empirical copula, introduced below, and briefly mention two smoothed versions of the latter which are expected to behave better in small samples. Furthermore, we succinctly introduce a class of nonparametric estimators of interest under the additional assumption that C is an extreme-value copula.

4.2.1 The Empirical Copula

In the same way that the empirical df is a natural nonparametric estimator of a df, a sensible nonparametric estimator of C is

$$C_n(\boldsymbol{u}) = \frac{1}{n}\sum_{i=1}^{n} \mathbf{1}(\boldsymbol{U}_{i,n} \le \boldsymbol{u}) = \frac{1}{n}\sum_{i=1}^{n}\prod_{j=1}^{d} \mathbf{1}(U_{ij,n} \le u_j), \quad \boldsymbol{u} \in [0,1]^d, \tag{4.12}$$

where $\boldsymbol{U}_{i,n} = (U_{i1,n}, \ldots, U_{id,n})$, $i \in \{1,\ldots,n\}$, are the pseudo-observations defined in (4.9) and inequalities $\boldsymbol{U}_{i,n} \le \boldsymbol{u}$, $i \in \{1,\ldots,n\}$, are to be understood componentwise. The quantity C_n, which is merely the empirical df of the pseudo-observations, is referred to as the *empirical copula* of $\boldsymbol{X}_1, \ldots, \boldsymbol{X}_n$ and seems to first appear (as defined in (4.12) or under an asymptotically equivalent form) in Rüschendorf (1976), Ruymgaart (1978), and Deheuvels (1979, 1981). It is a consistent estimator of C whose asymptotics follow from those of the so-called *empirical copula process*

$$\sqrt{n}\big(C_n(\boldsymbol{u}) - C(\boldsymbol{u})\big), \quad \boldsymbol{u} \in [0,1]^d. \tag{4.13}$$

The latter was studied in Stute (1984), Gänssler and Stute (1987), Fermanian et al. (2004), Tsukahara (2005), Van der Vaart and Wellner (2007), Segers (2012), and Berghaus et al. (2017), among others. The empirical copula process arises, in particular, as a key theoretical tool for showing the asymptotic validity of many inference procedures on the unknown copula C, such as those to be presented in Chap. 5.

Example 4.2.1 (Nonparametric Estimation by the Empirical Copula) Assume that C is the Clayton copula C_3^{C} with parameter 3. The following code illustrates the nonparametric estimation of C by the empirical copula of a random sample of size n from any meta-Clayton distribution with continuous margins constructed from C_3^{C}. As the empirical copula is based on the multivariate scaled ranks (4.10), it is margin-free. In the R package `copula`, the empirical copula can be evaluated using the function `C.n()`:

```
> ## The "unknown" copula (a 3-dim. Clayton copula with parameter 3)
> d <- 3
> cc <- claytonCopula(3, dim = d)
> ## Generate a sample from the copula, which will be transformed
> ## to pseudo-observations in 'C.n()'
> n <- 1000
> set.seed(65)
> U <- rCopula(n, copula = cc)
> ## Generate random points where to evaluate the empirical copula
> v <- matrix(runif(n * d), nrow = n, ncol = d)
> ec <- C.n(v, X = U)
> ## Compare with the true copula; increase n to decrease the error
> true <- pCopula(v, copula = cc)
> round(mean(abs(true - ec) / true) * 100, 2) # mean relative error (in %)

[1] 1.55
```

□

The empirical copula, being a particular multivariate empirical df, often exhibits a large bias when the sample size is small. A promising class of smooth nonparametric estimators of C are the so-called empirical Bernstein copulas introduced in Sancetta and Satchell (2004) and further studied in Janssen et al. (2012). A member of the class of empirical Bernstein copulas of particular practical interest is the *empirical beta copula* of Segers et al. (2017) which is the empirical Bernstein copula obtained by setting the degrees of all Bernstein polynomials equal to the sample size. The empirical beta copula of $\boldsymbol{X}_1, \dots, \boldsymbol{X}_n$ can be conveniently computed as

$$C_n^{\beta}(\boldsymbol{u}) = \frac{1}{n}\sum_{i=1}^{n}\prod_{j=1}^{d} F_{n,R_{ij}}(u_j), \quad \boldsymbol{u} \in [0,1]^d,$$

where, for any $r \in \{1, \dots, n\}$, $F_{n,r}$ denotes the df of the beta distribution with parameters r and $n+1-r$ and where, for any $j \in \{1, \dots, d\}$, R_{ij} is the rank of X_{ij} among $X_{1j}, \dots, X_{nj}$. The smoothness of C_n^{β} is a consequence of the replacement of the indicator function in (4.12) by a product of particular beta dfs. Unlike the empirical copula C_n, the empirical beta copula C_n^{β} has standard uniform univariate margins as long as there are no ties in the component samples of $\boldsymbol{X}_1, \dots, \boldsymbol{X}_n$ and is therefore a genuine copula in that case.

An even simpler smooth version of the empirical copula is the so-called *empirical checkerboard copula* of $\boldsymbol{X}_1, \dots, \boldsymbol{X}_n$ defined by

$$C_n^{\#}(\boldsymbol{u}) = \frac{1}{n}\sum_{i=1}^{n}\prod_{j=1}^{d} \min\{\max\{nu_j - R_{ij} + 1, 0\}, 1\}, \quad \boldsymbol{u} \in [0,1]^d;$$

see, for instance, Carley and Taylor (2002) and the references therein. Being a multilinear extension of (an asymptotically equivalent version of) the empirical copula C_n, it is again a genuine copula as long as there are no ties in the components samples of $\boldsymbol{X}_1, \ldots, \boldsymbol{X}_n$. Note that an adequate version of the empirical checkerboard copula is an important tool when studying copulas in the context of discontinuous data; see, for instance, Genest and Nešlehová (2007) and Genest et al. (2014, 2017).

Example 4.2.2 (The Empirical Beta and Checkerboard Copulas) To evaluate the empirical beta and checkerboard copulas, one can use the function `C.n()` with argument `smoothing = "beta"` and `smoothing = "checkerboard"`, respectively. To illustrate the improved small-sample behavior of these smooth nonparametric estimators compared to the empirical copula, we use a Gumbel–Hougaard copula as data-generating model. We consider a function which generates a sample of size n from this "unknown" copula, evaluates the three empirical copulas at the same random points, and returns the corresponding mean relative errors as percentages:

```
> gc <- gumbelCopula(4, dim = 3) # the 'unknown' copula
> ##' @title Mean relative error in % (for empirical copula, empirical beta
> ##'           copula and empirical checkerboard copula)
> ##' @param n sample size
> ##' @param cop copula object
> ##' @return mean relative errors in %
> compareEmpCops <- function(n, cop)
  {
      d <- dim(cop)
      U <- rCopula(n, copula = cop) # a sample from the true copula
      v <- matrix(runif(n * d), nrow = n, ncol = d) # random evaluation points
      ec     <- C.n(v, X = U) # the empirical copula values
      beta  <- C.n(v, X = U, smoothing = "beta") # the emp. beta cop. values
      check <- C.n(v, X = U, smoothing = "checkerboard") # emp. check. cop. val
      true <- pCopula(v, copula = cop) # the true copula values
      c(ec    = mean(abs(true - ec) / true),
        beta  = mean(abs(true - beta) / true),
        check = mean(abs(true - check) / true)) * 100 # mean rel. error in %
  }
```

We execute the above function 100 times and return the average of the errors:

```
> set.seed(2013)
> round(rowMeans(replicate(100, compareEmpCops(30, cop = gc))), 2)

   ec  beta check
20.68  6.52  7.52
```

Hence, in the above setting, the empirical beta and checkerboard copulas are better in terms of mean relative error than the empirical copula. As one can expect, this advantage progressively vanishes as the sample size increases:

```
> set.seed(2013)
> round(rowMeans(replicate(100, compareEmpCops(300, cop = gc))), 2)

   ec  beta check
 4.40  2.07  2.45
```

We end this example by empirically verifying that the two smooth estimators under consideration have standard uniform univariate margins:

```
> set.seed(2008)
> U <- rCopula(30, copula = gc) # a sample from the true copula
> m <- 100 # number of evaluation points
> v <- runif(m) # random points where to evaluate the margins of the estimators
> w1 <- cbind(v, 1, 1) # evaluations points margin 1
> w2 <- cbind(1, v, 1) # evaluations points margin 2
> w3 <- cbind(1, 1, v) # evaluations points margin 3
> ## Checks
> stopifnot(all.equal(C.n(w1, X = U, smoothing = "beta"), v))
> stopifnot(all.equal(C.n(w2, X = U, smoothing = "beta"), v))
> stopifnot(all.equal(C.n(w3, X = U, smoothing = "beta"), v))
> stopifnot(all.equal(C.n(w1, X = U, smoothing = "checkerboard"), v))
> stopifnot(all.equal(C.n(w2, X = U, smoothing = "checkerboard"), v))
> stopifnot(all.equal(C.n(w3, X = U, smoothing = "checkerboard"), v))
```

□

4.2.2 *Under Extreme-Value Dependence*

In this section, let us assume that the unknown copula C is of extreme-value type; see Sect. 3.3. Notice that such an assumption can be particularly meaningful in fields such as hydrology, meteorology, or climatology when $\boldsymbol{X}_1, \ldots, \boldsymbol{X}_n$ are actually componentwise block maxima (for instance, componentwise seasonal or annual maxima) obtained from some high frequency multivariate observations. Then, we know that C can be expressed as in (3.12) in terms of the Pickands dependence function A. Given a nonparametric estimator A_n of A computed from $\boldsymbol{X}_1, \ldots, \boldsymbol{X}_n$, a sensible nonparametric estimator of C is then simply obtained by plugging A_n into (3.12).

Two particularly easy-to-compute nonparametric estimators of A are derived in Gudendorf and Segers (2012) as extensions of those proposed in Genest and Segers (2009) in the bivariate case. They are the rank-based versions of two well-known estimators of A, namely the *Pickands estimator* studied in Pickands (1981) and the *Capéraà–Fougères–Genest estimator* proposed in Capéraà et al. (1997).

Let

$$\zeta_{i,n}(\boldsymbol{w}) = \min_{j \in \{1,\ldots,d\}} \frac{-\log U_{ij,n}}{w_j}, \qquad \boldsymbol{w} \in \Delta_{d-1},\ i \in \{1, \ldots, n\},$$

where $U_{ij,n}$ is the jth component of $\boldsymbol{U}_{i,n}$ in (4.9) and Δ_{d-1} is defined by (3.11). The rank-based version of the Pickands (P) and Capéraà–Fougères–Genest (CFG) estimators are then defined by

$$A_n^{\mathrm{P}}(\boldsymbol{w}) = 1 \Big/ \frac{1}{n} \sum_{i=1}^{n} \zeta_{i,n}(\boldsymbol{w}), \qquad \boldsymbol{w} \in \Delta_{d-1},$$

and

$$A_n^{\mathrm{CFG}}(\boldsymbol{w}) = \exp\left(-\gamma - \frac{1}{n}\sum_{i=1}^{n} \log \zeta_{i,n}(\boldsymbol{w})\right), \quad \boldsymbol{w} \in \Delta_{d-1},$$

respectively, where $\gamma = -\int_0^\infty \log(x)e^{-x}\,\mathrm{d}x \approx 0.577$ is the Euler–Mascheroni constant.

From the above definitions, it is easy to verify that $A_n^{\mathrm{P}}(\boldsymbol{e}_1) = \cdots = A_n^{\mathrm{P}}(\boldsymbol{e}_d)$ and that $A_n^{\mathrm{CFG}}(\boldsymbol{e}_1) = \cdots = A_n^{\mathrm{CFG}}(\boldsymbol{e}_d)$, where $\boldsymbol{e}_1, \ldots, \boldsymbol{e}_d$ are the standard basis vectors of $\mathbb{R}^d$. To ensure that the endpoint constraints $A_n^{\mathrm{P}}(\boldsymbol{e}_j) = A_n^{\mathrm{CFG}}(\boldsymbol{e}_j) = 1$, $j \in \{1, \ldots, d\}$, are satisfied (see Sect. 3.3), the previous estimators can be corrected as follows:

$$A_{n,c}^{\mathrm{P}}(\boldsymbol{w}) = (1/A_n^{\mathrm{P}}(\boldsymbol{w}) - 1/A_n^{\mathrm{P}}(\boldsymbol{e}_1) + 1)^{-1}, \quad \boldsymbol{w} \in \Delta_{d-1},$$

and

$$A_{n,c}^{\mathrm{CFG}}(\boldsymbol{w}) = \exp(\log A_n^{\mathrm{CFG}}(\boldsymbol{w}) - \log A_n^{\mathrm{CFG}}(\boldsymbol{e}_1)), \quad \boldsymbol{w} \in \Delta_{d-1},$$

respectively. These corrections are suggested in Gudendorf and Segers (2012) as natural extensions of those proposed in the case of known margins for $d = 2$ by Deheuvels (1991) and Capéraà et al. (1997), respectively.

In the bivariate case, the above corrected versions are found to behave better than the uncorrected versions in small samples in Genest and Segers (2009) and Genest et al. (2011). As verified in Gudendorf and Segers (2012), A_n^{P} and $A_{n,c}^{\mathrm{P}}$ (respectively, A_n^{CFG} and $A_{n,c}^{\mathrm{CFG}}$) become indistinguishable as n tends to infinity. Also, $A_{n,c}^{\mathrm{CFG}}$ is found, overall, to outperform $A_{n,c}^{\mathrm{P}}$ in several Monte Carlo experiments in the bivariate case; see, for example, Genest and Segers (2009) and Kojadinovic and Yan (2010b). The same empirical conclusion is obtained by Gudendorf and Segers (2012) in dimension three.

Example 4.2.3 (Nonparametric Estimation of the Pickands Dependence Function) Assume that C is the bivariate Khoudraji–Gumbel–Hougaard copula $\mathrm{kho}_{(0.6,0.95)}(\Pi, C_3^{\mathrm{GH}})$ given by (3.14). The following code illustrates the nonparametric estimation of the associated Pickands dependence function $A_{\boldsymbol{\theta}}^{\mathrm{KGH}}$ by $A_{n,c}^{\mathrm{P}}$ and $A_{n,c}^{\mathrm{CFG}}$. In `copula`, the latter estimators can be evaluated using the function `An.biv()`. Note that its multivariate analog (which is not used in this example) is `An()`.

```
> ## The "unknown" copula (a 2-dim. extreme-value copula)
> kg <- khoudrajiCopula(copula1 = indepCopula(),
                        copula2 = gumbelCopula(3),
                        shapes = c(0.6, 0.95))
> ## Generate a sample from this copula transformed to pseudo-observations
> set.seed(172)
> U <- pobs(rCopula(100, copula = kg))
```

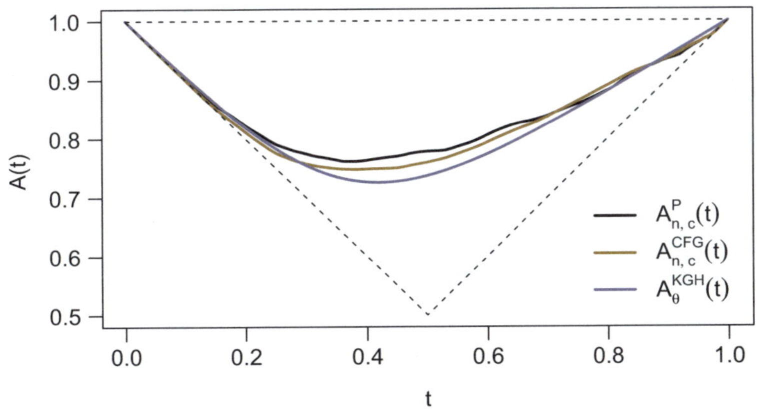

Fig. 4.5 Pickands dependence function $A_{\boldsymbol{\theta}}^{\mathrm{KGH}}$ of the bivariate Khoudraji–Gumbel–Hougaard copula $\mathrm{kho}_{(0.6,0.95)}(\Pi, C_3^{\mathrm{GH}})$ and its estimates $A_{n,c}^{\mathrm{P}}$ and $A_{n,c}^{\mathrm{CFG}}$

The graphs of $A_{\boldsymbol{\theta}}^{\mathrm{KGH}}$ and its two estimates $A_{n,c}^{\mathrm{P}}$ and $A_{n,c}^{\mathrm{CFG}}$ can be produced as follows:

```
> ## Graphs of the Pickands dependence function A and of its two estimates
> curve(An.biv(U, x, estimator = "Pickands"), from = 0, to = 1, col = 1,
        lwd = 2, ylim = c(0.5, 1), xlab = "t", ylab = "A(t)")
> curve(An.biv(U, x, estimator = "CFG"), 0, 1, add = TRUE, lwd = 2, col = 2)
> curve(A(kg, w = x), 0, 1, add = TRUE, lwd = 2, col = 3)
> lines(c(0, 0.5, 1), c(1, 0.5, 1), lty = 2)
> lines(c(0, 1),      c(1, 1),      lty = 2)
> legend("bottomright", bty = "n", lwd = 2, col = 1:3,
         legend = expression({A[list(n,c)]^{P}}(t), {A[list(n,c)]^{CFG}}(t),
                             {A[bold(theta)]^{KGH}}(t)),
         inset = 0.02, y.intersp = 1.2)
```

The resulting plot is shown in Fig. 4.5.

□

References

Bates, D., & Maechler, M. (2017). Matrix: Sparse and dense matrix classes and methods, R package version 1.2-12. http://CRAN.R-project.org/package=Matrix

Belzile, L., & Genest, C. (2017). lcopula: Liouville Copulas, R package version 1.0. https://CRAN.R-project.org/package=lcopula

Berghaus, B., Bücher, A., & Volgushev, S. (2017). Weak convergence of the empirical copula process with respect to weighted metrics. *Bernoulli, 23*(1), 743–772.

Capéraà, P., Fougères, A.-L., & Genest, C. (1997). A nonparametric estimation procedure for bivariate extreme value copulas. *Biometrika, 84*, 567–577.

Carley, H., & Taylor, M. D. (2002). A new proof of Sklar's theorem. In C. M. Cuadras, J. Fortiana, & J. A. Rodríguez-Lallena (Eds.), *Distributions with given marginals and statistical modelling* (pp. 29–34). Dordrecht: Kluwer Academic Publishers.

Charpentier, A., Fermanian, J.-D., & Scaillet, O. (2007). The estimation of copulas: Theory and practice. In J. Rank (Ed.), *Copulas: From theory to application in finance* (pp. 35–60). London: Risk Books.

Chen, X., Fan, Y., & Tsyrennikov, V. (2006). Effcient estimation of semiparametric multivariate copula models. *Journal of the American Statistical Association, 101*, 1228–1240.

Deheuvels, P. (1979). La fonction de dépendance empirique et ses propriétés: un test non paramétrique d'indépendance. *Académie Royale de Belgique Bulletin de la Classe des Sciences 5th Series, 65*, 274–292.

Deheuvels, P. (1981). A non parametric test for independence. *Publications de l'Institut de Statistique de l'Université de Paris, 26*, 29–50.

Deheuvels, P. (1991). On the limiting behavior of the Pickands estimator for bivariate extreme-value distributions. *Statistics & Probability Letters, 12*, 429–439.

Demarta, S., & McNeil, A. J. (2005). The *t* copula and related copulas. *International Statistical Review, 73*(1), 111–129.

Embrechts, P., Lindskog, F., & McNeil, A. J. (2003). Modelling dependence with copulas and applications to risk management. In S. Rachev (Ed.), *Handbook of heavy tailed distributions in finance* (pp. 329–384). Amsterdam: Elsevier.

Fermanian, J.-D., Radulovic, D., & Wegkamp, M. (2004). Weak convergence of empirical copula processes. *Bernoulli, 10*(5), 847–860.

Fermanian, J.-D., & Scaillet, O. (2005). Some statistical pitfalls in copula modelling for financial applications. In E. Klein (Ed.), *Capital formation, governance and banking* (pp. 59–74). Hauppauge, NY: Nova Science.

Gänssler, P., & Stute, W. (1987). *Seminar on empirical processes*, DMV Seminar 9. Basel: Birkhäuser.

Genest, C. (1987). Frank's family of bivariate distributions. *Biometrika, 74*(3), 549–555.

Genest, C., & Favre, A.-C. (2007). Everything you always wanted to know about copula modeling but were afraid to ask. *Journal of Hydrological Engineering, 12*, 347–368.

Genest, C., Ghoudi, K., & Rivest, L.-P. (1995). A semiparametric estimation procedure of dependence parameters in multivariate families of distributions. *Biometrika, 82*, 543–552.

Genest, C., Kojadinovic, I., Nešlehová, J. G., & Yan, J. (2011). A goodness-of-fit test for bivariate extreme-value copulas. *Bernoulli, 17*(1), 253–275.

Genest, C., Masiello, E., & Tribouley, K. (2009). Estimating copula densities through wavelets. *Insurance: Mathematics and Economics, 44*, 170–181.

Genest, C., & Nešlehová, J. (2007). A primer on copulas for count data. *The ASTIN Bulletin, 37*, 475–515.

Genest, C., Nešlehová, J. G., & Rémillard, B. (2014). On the empirical multilinear copula process for count data. *Bernoulli, 20*, 1344–1371.

Genest, C., Nešlehová, J. G., & Rémillard, B. (2017). Asymptotic behavior of the empirical multilinear copula process under broad conditions. *Journal of Multivariate Analysis, 20*, 82–110.

Genest, C., & Rivest, L.-P. (1993). Statistical inference procedures for bivariate Archimedean copulas. *Journal of the American Statistical Association, 88*, 1034–1043.

Genest, C., & Segers, J. (2009). Rank-based inference for bivariate extreme-value copulas. *The Annals of Statistics, 37*, 2990–3022.

Genest, C., & Werker, B. J. M. (2002). Conditions for the asymptotic semiparametric efficiency of an omnibus estimator of dependence parameters in copula models. In C. M. Cuadras, J. Fortiana, & J. A. Rodríguez-Lallena (Eds.), *Distributions with given marginals and statistical modelling* (pp. 103–112). London: Kluwer Academic Publishers.

Gilbert, P., & Varadhan, R. (2016). numDeriv: Accurate numerical derivatives, R package version 2016.8-27. https://CRAN.R-project.org/package=numDeriv

Gudendorf, G., & Segers, J. (2012). Nonparametric estimation of multivariate extreme-value copulas. *Journal of Statistical Planning and Inference, 143*, 3073–3085.

Higham, N. (2002). Computing the nearest correlation matrix – A problem from finance, *IMA Journal of Numerical Analysis, 22*, 329–343.

Hofert, M., & Hornik, K. (2017). qrmtools: Tools for quantitative risk management, R package version 0.0-7. https://CRAN.R-project.org/package=qrmtools

Janssen, P., Swanepoel, J., & Veraverbeke, N. (2012). Large sample behavior of the Bernstein copula estimator. *Journal of Statistical Planning and Inference, 142*, 1189–1197.

Joe, H. (1997). *Multivariate models and dependence concepts*. London: Chapman & Hall.

Joe, H. (2005). Asymptotic efficiency of the two-stage estimation method for copula-based models. *Journal of Multivariate Analysis, 94*, 401–419.

Joe, H., & Xu, J. J. (1996). *The Estimation Method of Inference Functions for Margins for Multivariate Models*. Technical Report, Department of Statistics, University of British Columbia.

Kim, G., Silvapulle, M. J., & Silvapulle, P. (2007). Comparison of semiparametric and parametric methods for estimating copulas. *Computational Statistics & Data Analysis, 51*(6), 2836–2850.

Klassen, C. A. J., & Wellner, J. A. (1997). Effcient estimation in the bivariate normal copula model: Normal marginals are least favourable. *Bernoulli, 3*, 55–77.

Kojadinovic, I., & Yan, J. (2010a). Comparison of three semiparametric methods for estimating dependence parameters in copula models. *Insurance: Mathematics and Economics, 47*, 52–63.

Kojadinovic, I., & Yan, J. (2010b). Nonparametric rank-based tests of bivariate extreme-value dependence. *Journal of Multivariate Analysis, 101*(9), 2234–2249.

Lehmann, E. L., & Casella, G. (1998). *Theory of point estimation*. New York: Springer.

Mashal, R., & Zeevi, A. (2002). *Beyond correlation: Extreme co-movements between financial assets*. Technical Report, Columbia School of Business, https://www0.gsb.columbia.edu/faculty/azeevi/PAPERS/BeyondCorrelation.pdf

McNeil, A. J., Frey, R., & Embrechts, P. (2015). *Quantitative risk management: Concepts, techniques and tools* (2nd ed.). Princeton, NJ: Princeton University Press.

Oakes, D. (1982). A model for association in bivariate survival data. *Journal of the Royal Statistical Society Series B, 44*, 414–422.

Omelka, M., Gijbels, I., & Veraverbeke, N. (2009). Improved kernel estimation of copulas: Weak convergence and goodness-of-fit testing. *The Annals of Statistics, 37*(5B), 3023–3058.

Pickands, J. (1981). Multivariate extreme value distributions. With a discussion. Proceedings of the 43rd session of the International Statistical Institute. *Bulletin de l'Institut international de statistique, 49*, 859–878, 894–902.

Rüschendorf, L. (1976). Asymptotic distributions of multivariate rank order statistics. *The Annals of Statistics, 4*, 912–923.

Ruymgaart, F. H. (1978). *Asymptotic theory of ranks tests for independence*. Amsterdam: Mathematisch Centrum.

Sancetta, A., & Satchell, S. (2004). The Bernstein copula and its applications to modeling and approximations of multivariate distributions. *Econometric Theory, 20*, 535–562.

Segers, J. (2012). Asymptotics of empirical copula processes under nonrestrictive smoothness assumptions. *Bernoulli, 18*, 764–782.

Segers, J., Sibuya, M., & Tsukahara, H. (2017). The empirical beta copula. *Journal of Multivariate Analysis, 155*, 35–51.

Shih, J. H., & Louis, T. A. (1995). Inferences on the association parameter in copula models for bivariate survival data. *Biometrics, 51*(4), 1384–1399.

Stute, W. (1984). The oscillation behavior of empirical processes: The multivariate case. *The Annals of Probability, 12*(2), 361–379.

Tsukahara, H. (2005). Semiparametric estimation in copula models. *The Canadian Journal of Statistics, 33*(3), 357–375.

Van der Vaart, A. W., & Wellner, J. A. (2007). Empirical processes indexed by estimated functions. In E. A. Cator (Ed.), *Asymptotics: Particles, processes and inverse problems* (pp. 234–252). Lithuania: Institute of Mathematical Statistics.

Chapter 5
Graphical Diagnostics, Tests, and Model Selection

The setting considered in this chapter is similar to that described in Chap. 4: We assume to have at hand a random sample $\boldsymbol{X}_1, \ldots, \boldsymbol{X}_n$ from a df H with continuous univariate margins $F_1, \ldots, F_d$. As we know by now, from Sklar's Theorem, see Theorem 2.3.1, H can be expressed as

$$H(\boldsymbol{x}) = C\big(F_1(x_1), \ldots, F_d(x_d)\big), \quad \boldsymbol{x} \in \mathbb{R}^d, \tag{5.1}$$

in terms of a unique d-dimensional copula C. In Sect. 4.1, we discussed the estimation of the copula under the realistic assumption that $F_1, \ldots, F_d$ are unknown and that C belongs to an absolutely continuous parametric family of copulas $\mathcal{C} = \{C_{\boldsymbol{\theta}} : \boldsymbol{\theta} \in \Theta\}$, where Θ, the parameter space, is a subset of $\mathbb{R}^p$ for some integer $p \geq 1$.

The aim of this chapter is to present some graphical and formal statistical procedures that can be used to guide the choice of the hypothesized parametric copula family $\mathcal{C}$.

Should the margins $F_1, \ldots, F_d$ be known, we could form the sample $\boldsymbol{U}_1, \ldots, \boldsymbol{U}_n$, where $\boldsymbol{U}_i$ is defined in (4.2), and carry out inference on C using standard statistical procedures for iid observations. As the sample $\boldsymbol{U}_1, \ldots, \boldsymbol{U}_n$ is unobservable, a natural path is to form parametric or nonparametric pseudo-observations instead, as defined by (4.6) and (4.9), respectively. To avoid biases arising from potential misspecifications of the margins, many authors advocate to base the inference on C on the latter, which amounts to forming the multivariate scaled ranks (4.10) as explained in Sect. 4.1.2. We adopt the same perspective in this chapter, which implies that all the presented procedures are margin-free.

We shall first briefly discuss some basic graphical diagnostics to guide the choice of $\mathcal{C}$. Several formal statistical tests will be presented next. The last section of the chapter will be devoted to the key issue of model selection, that is, copula family selection.

M. Hofert et al., *Elements of Copula Modeling with R*, Use R!,
https://doi.org/10.1007/978-3-319-89635-9_5

5.1 Basic Graphical Diagnostics

Let us assume for simplicity that $d = 2$; if $d > 2$ but is not too large, the suggested diagnostics can be performed on all $\binom{d}{2}$ bivariate margins; see also Sect. 5.2.6. One of the first steps when analyzing $\boldsymbol{X}_1, \ldots, \boldsymbol{X}_n$ consists of producing a scatter plot of the pseudo-observations defined in (4.9). Such a scatter plot could for instance confirm that the underlying copula C is indeed different from the independence copula Π, thereby justifying its modeling by a parametric family. Note in passing that one graphical tool (not presented here) dedicated to detecting departures from independence is the so-called *K-plot* of Genest and Boies (2003); see also Genest and Favre (2007).

The scatter plot of the pseudo-observations can for instance also suggest that the underlying copula is radially symmetric, that is, that $C = \bar{C}$, or that it is not exchangeable, that is, that $C(u_1, u_2) = C(u_2, u_1)$ does not hold for all $u_1, u_2 \in [0, 1]$; see Sect. 2.5.

To assess the presence of tail dependence, see Sect. 2.6.3, the pseudo-observations (4.9) can be transformed to *normal scores* first, by applying the quantile function of the standard normal to the component samples of the pseudo-observations. The rationale behind this approach is that normal scores are approximately marginally standard normal and that the resulting scatter plot could thus highlight departures from bivariate normality. Indeed, a scatter plot of bivariate normal observations should look "elliptically contoured," as explained in Sect. 3.1.1. Heavier tails (having more probability mass than the tails of normal distributions) in the scatter plot of normal scores suggest tail dependence; see, for example, Joe (2014, Section 1.4) for more details. In addition, the non-exchangeability or radial symmetry of C might be easier to detect than from the scatter plot of pseudo-observations.

Example 5.1.1 (Pseudo-Observations and Normal Scores) Figure 5.1 shows the pseudo-observations (left) and normal scores (right) of the danube data set presented in Example 4.1.6.

```
> data(danube, package = "lcopula")
> U <- as.matrix(danube)
> plot(U,         xlab = "Donau", ylab = "Inn")
> plot(qnorm(U), xlab = "Donau", ylab = "Inn")
```

Both scatter plots seem to show a slightly different distribution of the observations below and above the main diagonal, thereby suggesting a small departure from exchangeability. The lack of symmetry with respect to the point $(1/2, 1/2)$ also suggests that C is not radially symmetric. Finally, both the scatter plots of pseudo-observations and normal scores suggest the presence of (at least) upper tail dependence. □

The previous basic graphical diagnostics might suggest to mainly focus on certain parametric copula families for C. For each such family, C could be estimated

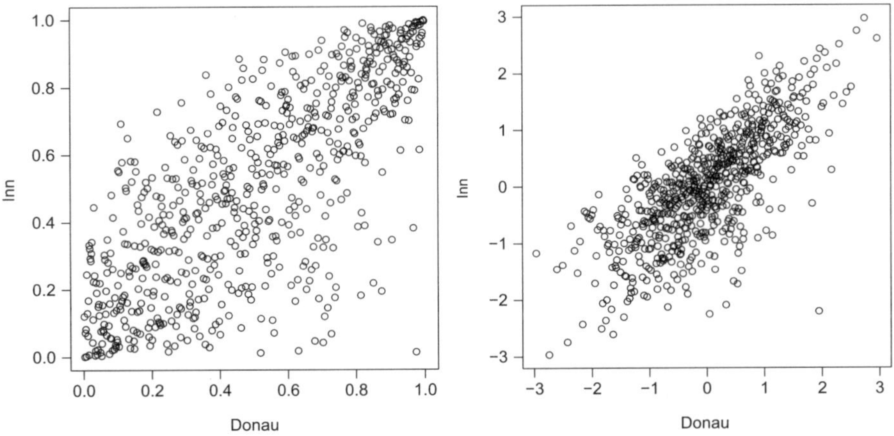

Fig. 5.1 Pseudo-observations (left) and normal scores (right) of the `danube` data set

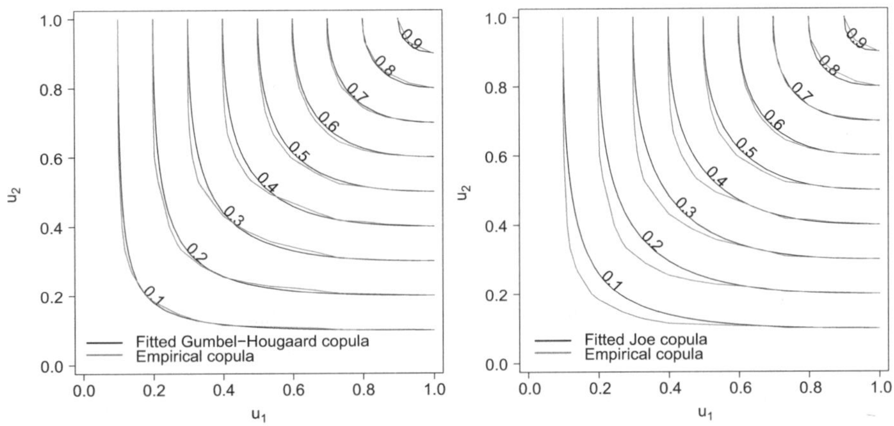

Fig. 5.2 Contour plots of the fitted Gumbel–Hougaard copula (left) and Joe copula (right) overlaid with the contours of the empirical copula for the `danube` data set

using one of the approaches described in Sect. 4.1. To assess which estimated copula fits best, some authors like to compare contour plots of the parametric estimates with those of the empirical copula of the data introduced in Sect. 4.2.1.

Example 5.1.2 (Comparing (Non)parametric Estimates of the Copula) The scatter plots produced in Example 5.1.1 suggest for instance to consider the Joe or the Gumbel–Hougaard family as possible choices for C, as both of these families exhibit upper tail dependence. Figure 5.2, produced with the following code (based on the package `latticeExtra` of Sarkar (2016) for overlaying plots), compares the corresponding contours with those of the empirical copula:

```
> ## Fit a Gumbel-Hougaard copula and compute the contours
> fg <- fitCopula(gumbelCopula(), data = U)
> cpG <- contourplot2(fg@copula, FUN = pCopula, region = FALSE,
                      key = list(corner = c(0.04, 0.04),
                                 lines = list(col = 1:2, lwd = 2),
                                 text = list(c("Fitted Gumbel-Hougaard copula",
                                               "Empirical copula"))))
> ## Fit a Joe copula and compute the contours
> fj <- fitCopula(joeCopula(), data = U)
> cpJ <- contourplot2(fj@copula, FUN = pCopula, region = FALSE,
                      key = list(corner = c(0.04, 0.04),
                                 lines = list(col = 1:2, lwd = 2),
                                 text = list(c("Fitted Joe copula",
                                               "Empirical copula"))))
> ## Compute the contours of the empirical copula
> u <- seq(0, 1, length.out = 16)
> grid <- as.matrix(expand.grid(u1 = u, u2 = u))
> val <- cbind(grid, z = C.n(grid, X = U))
> cpCn <- contourplot2(val, region = FALSE, labels = FALSE, col = 2)
> ## Plots (lattice objects)
> library(latticeExtra)
> cpG + cpCn
> cpJ + cpCn
```

The contour plots in Fig. 5.2 suggest that the Gumbel–Hougaard family fits better than the Joe family.

Because of the upper tail dependence and the possible slight non-exchangeability of C seeming to appear in Fig. 5.1, it is also natural to consider the Khoudraji–Gumbel–Hougaard family $\text{kho}_s(\Pi, C_\theta^{\text{GH}})$ constructed as explained in Sect. 3.4.2 as candidate family. The fitting of families constructed using Khoudraji's device can be numerically challenging as already discussed in Example 4.1.13, which is why the more robust Nelder–Mead optimization routine is used in the following code chunk. Also, Starting values for the parameters need to be provided.

```
> (fk <- fitCopula(khoudrajiCopula(copula2 = gumbelCopula()), data = U,
                   start = c(1.1, 0.5, 0.5), optim.method = "Nelder-Mead"))

Call: fitCopula(copula, data = data, start = ..1, optim.method = "Nelder-Mead")
Fit based on "maximum pseudo-likelihood" and 659 2-dimensional observations.
Copula: khoudrajiExplicitCopula
c2.alpha   shape1   shape2
  2.2661   0.9201   1.0000
The maximized loglikelihood is 281.9
Optimization converged
```

As we have already seen in Sect. 4.1.5, one way to make the optimization problem easier is to decrease the number of parameters by keeping some parameters fixed to certain predefined values. The latter can be done using the function `fixParam()`. Based on the previous fitting, we decide to fix the second shape parameter to one. The default (faster) BFGS optimization routine also happens to lead to convergence in this case:

```
> (fk2 <- fitCopula(khoudrajiCopula(copula2 = gumbelCopula(),
                                    shapes = fixParam(c(NA_real_, 1),
                                                      c(FALSE, TRUE))),
                    data = U, start = c(1.1, 0.5)))

Call: fitCopula(copula, data = data, start = ..1)
Fit based on "maximum pseudo-likelihood" and 659 2-dimensional observations.
Copula: khoudrajiExplicitCopula
c2.alpha   shape1
  2.2729   0.9218
The maximized loglikelihood is 281.9
Optimization converged
```

Comparing the last two outputs, we see that the difference between the two- and three-parameter models is very small in terms of parameter estimates. The contour plot of the fitted two-parameter Khoudraji–Gumbel–Hougaard copula overlaid by those of the empirical copula for the `danube` data set is represented in Fig. 5.3:

```
> cpK <- contourplot2(fk2@copula, FUN = pCopula, region = FALSE,
                      key = list(corner = c(0.04, 0.04),
                                 lines = list(col = 1:2, lty = 1),
                                 text = list(c("Fitted Khoudraji-Gumbel-
                                     Hougaard copula",
                                                "Empirical copula"))))
> cpK + cpCn
```

Whether the Khoudraji–Gumbel–Hougaard family fits better than the Gumbel–Hougaard family is unclear at this point and will be investigated in the next example, as well as in Examples 5.2.16 and 5.3.1 based on formal inference procedures.

□

Under the additional assumption that the unknown copula C in (5.1) is of extreme-value type (which might be strongly justified, typically, in certain envi-

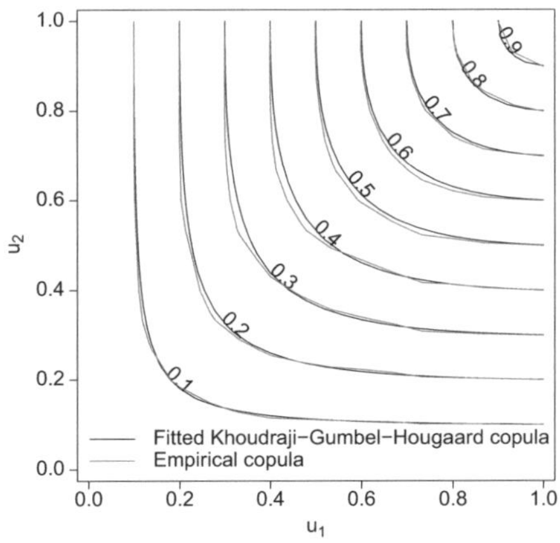

Fig. 5.3 Contour plot of the fitted Khoudraji–Gumbel–Hougaard copula overlaid by those of the empirical copula for the `danube` data set

ronmental applications), a meaningful alternative to contour plots consists of superimposing the graphs of the Pickands dependence function, see Sect. 3.3, of the estimated extreme-value copula with a nonparametric estimate of the Pickands dependence function such as one of those presented in Sect. 4.2.2.

Example 5.1.3 (Comparing (Non)parametric Estimates of the Pickands Dependence Function) As a continuation of Example 5.1.2, Fig. 5.4 displays the graphs of the Pickands dependence functions of the fitted Gumbel–Hougaard and two-parameter Khoudraji–Gumbel–Hougaard copulas (denoted by $A^{\mathrm{GH}}_{\theta_n}$ and $A^{\mathrm{KGH}}_{\boldsymbol{\theta}_n}$, respectively) as well as the corresponding (nonparametric) CFG estimate $A^{\mathrm{CFG}}_{n,c}$; see also Example 4.2.3. The code producing the plot is as follows:

```
> curve(A(fg@copula, x), from = 0, to = 1, ylim = c(0.5, 1),
        lwd = 2, xlab = "t", ylab = "A(t)", col = 1) # parametric
> curve(A(fk2@copula, x), 0,1, lwd = 2, add = TRUE, col = 2) # parametric
> curve(An.biv(U, x),      0,1, lwd = 2, add = TRUE, col = 3) # nonparametric
> lines(c(0, 0.5, 1), c(1, 0.5, 1), lty = 2)
> lines(c(0, 1),      c(1, 1),      lty = 2)
> legend("bottomright", bty = "n", lwd = 2, col=1:3,
         legend = expression({A[theta[n]]^{GH}}(t),
                             {A[bold(theta)[n]]^{KGH}}(t),
                             {A[list(n,c)]^{CFG}}(t)),
         inset = 0.02, y.intersp = 1.2)
```

The three curves are rather close. After a careful inspection, it appears that $A^{\mathrm{KGH}}_{\boldsymbol{\theta}_n}$ is slightly closer to the nonparametric estimate than $A^{\mathrm{GH}}_{\theta_n}$. Whether this reveals a good fit of one or both families, however, is subjective as the assessment of the distance between curves should take into account the sampling variability of both the parametric and nonparametric estimators of the unknown Pickands dependence function. □

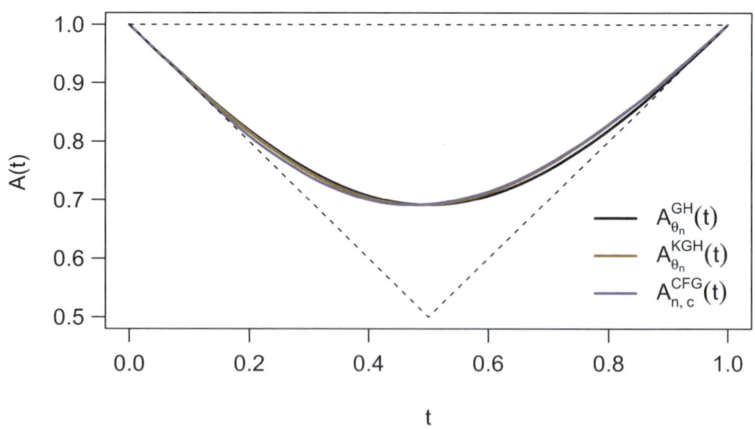

Fig. 5.4 Parametric ($A^{\mathrm{GH}}_{\theta_n}$ and $A^{\mathrm{KGH}}_{\boldsymbol{\theta}_n}$) and nonparametric ($A^{\mathrm{CFG}}_{n,c}$) estimates of the Pickands dependence function for the `danube` data set

5.2 Hypothesis Tests

Although plots and graphical diagnostics are crucial in any data analysis, they should be used with care if they do not incorporate in some way the influence of sampling variability. A prominent example is that of a Q-Q plot without confidence bands as there is no objective way to decide between accepting or rejecting a fitted univariate distribution in the case of a moderate departure from the line $y = x$; what a "small" or "large" departure from this line is depends in general on the type of problem at hand and in particular on the sample size.

In the rest of this section, we review selected *formal statistical tests*, that is, hypothesis tests which compute p-values. Such tests can help to guide the choice of the hypothesized copula family $\mathcal{C}$.

5.2.1 *Tests of Independence*

In the setting under consideration, a first natural question is whether C in (5.1) is actually different from the independence copula Π. If it is not, only marginal modeling is necessary, which can be carried out using classical statistical approaches for univariate iid observations. Formally, this amounts to testing

$$\mathcal{H}_0 : C = \Pi \quad \text{versus} \quad \mathcal{H}_1 : C \neq \Pi. \tag{5.2}$$

In dimension two, a classical alternative consists of testing

$$\mathcal{H}_0 : \tau = 0 \quad \text{versus} \quad \mathcal{H}_1 : \tau \neq 0,$$

or

$$\mathcal{H}_0 : \rho_s = 0 \quad \text{versus} \quad \mathcal{H}_1 : \rho_s \neq 0.$$

From Chap. 2, we know that $C = \Pi$ implies $\tau = 0$ and $\rho_s = 0$, but that the converse is false in general. This lack of equivalence, however, is usually not an issue in practice as copulas $C \neq \Pi$ such that $\tau = 0$ or $\rho_s = 0$ do not seem to arise often in applications. In any case, the latter configuration can be easily discarded by a scatter plot of the observations as illustrated in Example 5.2.2 below.

Example 5.2.1 (Test of Uncorrelatedness) We consider again the danube data set presented initially in Example 4.1.6 and formally test uncorrelatedness in terms of Kendall's tau between the two underlying variables:

```
> data(danube, package = "lcopula")
> cor.test(danube[,1], danube[,2], method = "kendall")
```

```
        Kendall's rank correlation tau

data:  danube[, 1] and danube[, 2]
z = 21.064, p-value < 2.2e-16
alternative hypothesis: true tau is not equal to 0
sample estimates:
      tau
0.5484731
```

As expected from the graphical diagnostics, the extremely small p-value provides strong evidence against the null hypothesis that $\tau = 0$, and therefore strong evidence in favor of the alternative hypothesis that $\tau \neq 0$, which in turn implies that $C \neq \Pi$. An analog test based on Spearman's rho can be carried out by setting `method = "spearman"` in `cor.test()`. □

Example 5.2.2 (A Fallacy of a Test of Uncorrelatedness) Let X be a standard normal random variable and let $Y = X^2$. Clearly, X and Y are not independent, so the copula C of (X, Y) is different from the independence copula and yet $\tau = 0$. The following code illustrates the inappropriateness of the test of uncorrelatedness based on Kendall's tau in this case:

```
> set.seed(1515)
> x <- rnorm(200)
> y <- x^2
> cor.test(x, y, method = "kendall")

        Kendall's rank correlation tau

data:  x and y
z = 0.6594, p-value = 0.5096
alternative hypothesis: true tau is not equal to 0
sample estimates:
       tau
0.03135678
```

The large p-value provides no evidence against the null hypothesis that $\tau = 0$, which is therefore a plausible hypothesis. A scatter plot, however, immediately reveals that $C \neq \Pi$. □

The tests of uncorrelatedness based on Kendall's tau or Spearman's rho could be adapted to $d > 2$ by considering the higher-dimensional extensions of these quantities studied in Joe (1990) or Schmid and Schmidt (2007). A more direct way of testing $\mathcal{H}_0$ in (5.2) when $d \geq 2$ consists of assessing the difference between a nonparametric estimate of C and an estimate of the latter under $\mathcal{H}_0$. From Sect. 4.2.1, a natural nonparametric estimator of C is the empirical copula, while an obvious estimator of C under $\mathcal{H}_0$ is Π, leading for instance to the test statistic

$$S_n^{\Pi} = \int_{[0,1]^d} n(C_n(\boldsymbol{u}) - \Pi(\boldsymbol{u}))^2 \, \mathrm{d}\boldsymbol{u}. \tag{5.3}$$

Such a test was initially suggested by Deheuvels (1981) and was thoroughly revisited in Genest and Rémillard (2004) and Genest et al. (2007). In particular, the asymptotics of S_n^Π under $\mathcal{H}_0$ in (5.2) follow immediately from those of the empirical copula process (4.13).

From a practical perspective, an approximate p-value for S_n^Π can be obtained by simulation. Given a large integer N, say $N = 1000$, independent realizations $S_n^{\Pi,(1)}, \dots, S_n^{\Pi,(N)}$ of S_n^Π under $\mathcal{H}_0$ are simulated as explained in Genest and Rémillard (2004), and an approximate p-value is computed as

$$\frac{1}{N+1}\left(\sum_{k=1}^{N} \mathbf{1}(S_n^{\Pi,(k)} \geq S_n^\Pi) + \frac{1}{2}\right).$$

The above slight modification of the classical formula $N^{-1}\sum_{k=1}^{N} \mathbf{1}(S_n^{\Pi,(k)} \geq S_n^\Pi)$ is used to ensure that the p-value is in the open interval (0, 1) so that transformations by quantile functions of continuous distributions are always well defined. This approach is adopted in the R package `copula` for all tests for which approximate p-values are computed using simulation or resampling, the default value for N being 1000. Consequently, for such tests, approximate p-values are numbers in the set $\{(0 + 1/2)/(N + 1), \dots, (N + 1/2)/(N + 1)\}$.

Example 5.2.3 (Test of Independence Based on S_n^Π) Assume that C is C_2^{F}, the Frank copula with parameter 2, and let us apply the test of independence studied in Genest and Rémillard (2004) to a sample of size n from any meta-Frank distribution constructed from C_2^{F}:

```
> n <- 100
> d <- 3
> set.seed(1969)
> U <- rCopula(n, frankCopula(2, dim = d))
```

To carry out the test, we first generate 1000 realizations of the test statistic under independence for sample size n and dimension d using the function `indepTestSim()` and then compute the approximate p-value using `indepTest()`:

```
> dist <- indepTestSim(n, p = d, verbose = FALSE)
> indepTest(U, d = dist)

Global Cramer-von Mises statistic: 0.1295138 with p-value 0.0004995005
Combined p-values from the Mobius decomposition:
   0.0004995005  from Fisher's rule,
   0.008491508  from Tippett's rule.
```

The first p-value is the p-value for the asymptotically equivalent version of S_n^Π considered in Genest and Rémillard (2004). The definitions of the statistics corresponding to the remaining two p-values can be found in the latter reference and in Kojadinovic and Holmes (2009, Section 3.5).

Hence, as expected, the tests provide strong evidence against (respectively, in favor of) the null hypothesis that $C = \Pi$ (respectively, the alternative that $C \neq \Pi$). □

5.2.2 *Tests of Exchangeability*

Many popular parametric copula families are exchangeable, see Sect. 2.5, and it is therefore of interest to assess whether such a hypothesis is plausible for the data at hand. For $d = 2$, a natural test statistic is

$$S_n^{\text{ex}_C} = \int_{[0,1]^2} n(C_n(u_1, u_2) - C_n(u_2, u_1))^2 \, \mathrm{d}C_n(\boldsymbol{u}), \tag{5.4}$$

where C_n is the empirical copula defined in (4.12). The corresponding test is studied in Genest et al. (2012) as a particular case of the one studied in Rémillard and Scaillet (2009); see also Kojadinovic and Yan (2012, Section 5). An extension to arbitrary dimensions $d > 2$ is considered in Harder and Stadtmüller (2017).

Example 5.2.4 (Test of Exchangeability Based on $S_n^{\text{ex}_C}$) The test studied in Genest et al. (2012) is implemented in the R function `exchTest()`. We can easily apply it to the `danube` data set:

```
> set.seed(1453)
> exchTest(as.matrix(danube))

        Test of exchangeability for bivariate copulas with argument
        'm' set to 0

data:  as.matrix(danube)
statistic = 0.051969, p-value = 0.001499
```

Setting the seed prior to the function call is done to guarantee reproducibility since the p-value for $S_n^{\text{ex}_C}$ is computed using resampling, which involves pseudo-random number generation. Looking at the p-value, we conclude that there is strong evidence against exchangeability in the `danube` data set. □

Remark 5.2.5

1) As with any statistical test, it is important to distinguish between the so-called "presence of the effect" and the "practical importance of the effect"; see, for example, Wild and Seber (1999, Chapter 9). The result of the test indicates that it is extremely likely that the underlying unknown copula is non-exchangeable. However, it does not say anything about the strength of the non-exchangeability.
2) We have already carried out two tests on the `danube` data set, and the results of more tests on the same data set will be presented in the remaining sections of this chapter. Consequently, with the issue of multiple testing in mind, rejection or non-rejection of a given null hypothesis on the underlying unknown copula

C should not be decided by comparing its p-value with, say, the classical $\alpha = 5\%$ significance level. In general, for a study involving m tests, a Bonferroni correction could be used. It consists of using α/m as significance level for each test. Such a correction is, however, known to lead to a too conservative control of the so-called family-wise error rate when m is large or the test statistics are positively dependent. In the case of a study on the `danube` data set, for instance, carrying out each test, say, at the 1% level might be an acceptable compromise.

□

Under the additional assumption that the unknown bivariate copula is of the extreme-value type, a potentially more powerful approach uses the fact that the exchangeability of C is equivalent to the symmetry of its associated Pickands dependence function A with respect to the vertical line $t = 1/2$. A natural alternative test statistic is therefore

$$S_n^{\mathrm{ex}_A} = \int_{[0,1]} n(A_n(t) - A_n(1-t))^2 \,\mathrm{d}t,$$

where A_n is a nonparametric estimator of A such as the Pickands estimator or the CFG estimator presented in Sect. 4.2.2; see Kojadinovic and Yan (2012) for more details.

Example 5.2.6 (Test of Exchangeability Based on $S_n^{\mathrm{ex}_A}$) The aforementioned test is implemented in the R function `exchEVTest()`. It relies on resampling to compute an approximate p-value for $S_n^{\mathrm{ex}_A}$. To illustrate its computational cost, we shall record and report the run time of one execution of `exchEVTest()` on the `danube` data set using the following wrapper `withTime()` of R's `system.time()`:

```
> withTime <- function(expr, ...)
  {
      st <- system.time(r <- expr, ...)
      list(value = r,  sys.time = st)
  }
```

Combined with `exchEVTest()`, this leads to the following code:

```
> set.seed(1492)
> withTime(exchEVTest(as.matrix(danube)))

        Test of exchangeability for bivariate extreme-value copulas
        with argument 'estimator' set to 'CFG' and argument 'm' set
        to 100

data:  as.matrix(danube)
statistic = 0.14123, p-value = 0.003497

User time:  2.9 sec
```

Under the additional assumption of bivariate extreme-value dependence, the same conclusion as for the test implemented in the R function `exchTest()` is reached: there is strong evidence against exchangeability. □

5.2.3 A Test of Radial Symmetry

Similar to exchangeability, we can also test radial symmetry. Starting from the definition of radial symmetry of a copula, see Sect. 2.5, a natural test statistic is

$$S_n^{\mathrm{sym}} = \int_{[0,1]^d} n(C_n(\boldsymbol{u}) - \bar{C}_n(\boldsymbol{u}))^2 \,\mathrm{d}C_n(\boldsymbol{u}), \tag{5.5}$$

where C_n is the empirical copula of $\boldsymbol{X}_1, \ldots, \boldsymbol{X}_n$ as defined in (4.12) and $\bar{C}_n$ is the empirical copula of $-\boldsymbol{X}_1, \ldots, -\boldsymbol{X}_n$. Recall that C_n is merely the empirical df of the sample of pseudo-observations $\boldsymbol{U}_{1,n}, \ldots, \boldsymbol{U}_{n,n}$ defined in (4.9) while some thought reveals that $\bar{C}_n$ is the empirical df of the sample $\mathbf{1} - \boldsymbol{U}_{1,n}, \ldots, \mathbf{1} - \boldsymbol{U}_{n,n}$.

The test based on S_n^{sym} is studied in the bivariate case by Genest and Nešlehová (2014). An alternative multivariate version of it is considered in Kojadinovic (2017).

Example 5.2.7 (Test of Radial Symmetry Based on S_n^{sym}) The test based on S_n^{sym} is implemented in the R function `radSymTest()`. It also relies on resampling for computing an approximate p-value. Applying it to the `danube` data set, we obtain:

```
> set.seed(1453)
> withTime(radSymTest(as.matrix(danube)))

        Test of radial symmetry based on the empirical copula

data:  as.matrix(danube)
statistic = 0.22975, p-value = 0.0004995

User time:  1.8 sec
```

We conclude from the p-value that there is strong evidence against radial symmetry in the `danube` data set, which is in accordance with the scatter plot of the pseudo-observations displayed on the left-hand side of Fig. 5.1.

For the `rdj` data set introduced in Example 4.1.3, we obtain:

```
> data(rdj)
> Xrdj <- as.matrix(rdj[,-1]) # omitting component 'Date'
> set.seed(1389)
> withTime(radSymTest(Xrdj))
```

```
        Test of radial symmetry based on the empirical copula

data:  Xrdj
statistic = 0.051834, p-value = 0.521

User time:  7.9 sec
```

There is thus no evidence against radial symmetry in the `rdj` data set. □

5.2.4 Tests of Extreme-Value Dependence

Maybe somewhat surprisingly, extreme-value copulas do not only arise in applications when the available data $\boldsymbol{X}_1, \dots, \boldsymbol{X}_n$ are componentwise maxima as is sometimes the case in hydrology, climatology, or meteorology for instance. It is thus of practical interest to have tests to assess whether the assumption of extreme-value dependence for C in (5.1) is plausible. More formally, the underlying null and alternative hypotheses are

$$\mathcal{H}_0 : C \in \mathcal{E} \quad \text{against} \quad \mathcal{H}_1 : C \notin \mathcal{E},$$

where $\mathcal{E}$ denotes the set of extreme-value copulas.

In this section, we focus on the three tests implemented in the R package `copula`. Note that a recent review of tests of extreme-value dependence can be found in Bücher and Kojadinovic (2015).

In the bivariate case, a first powerful test exploits the seminal ideas of Ghoudi et al. (1998). It is based on the moments of the random variable $W = H(X_1, X_2) = C\big(F_1(X_1), F_2(X_2)\big)$, the df of which, denoted by K, is called *Kendall distribution function* (the latter is implemented in the function `pK()` in the R package `copula`). Under $\mathcal{H}_0$, Ghoudi et al. (1998) showed that

$$K(t) = \mathbb{P}(W \le t) = t - (1 - \tau)t \log t, \quad t \in (0, 1],$$

where τ denotes Kendall's tau, and that, for $k \in \mathbb{N}$, $\mu_k = \mathbb{E}(W^k) = (k\tau + 1)/(k + 1)^2$, which, for instance, implies that

$$-1 + 8\mu_1 - 9\mu_2 = 0. \tag{5.6}$$

In order to test $\mathcal{H}_0$, Ghoudi et al. (1998) suggest to assess whether a sample version of the left-hand side of (5.6) is significantly different from zero, leading to the statistic

$$S_n^{\text{ev}K} = -1 + \frac{8}{n(n-1)} \sum_{i \ne j} I_{ij} - \frac{9}{n(n-1)(n-2)} \sum_{i \ne j \ne k} I_{ij} I_{kj}, \tag{5.7}$$

where $I_{ij} = 1(X_{i1} \le X_{j1}, X_{i2} \le X_{j2})$. This test is revisited in Ben Ghorbal et al. (2009) where two alternative strategies for computing approximate p-values are proposed.

Example 5.2.8 (Test of Extreme-Value Dependence Based on $S_n^{\mathrm{ev}_K}$) The test based on (5.7) was implemented by Ben Ghorbal et al. (2009) and its code was generously donated to the R package `copula`. The corresponding R function is called `evTestK()`. Applying it to the `danube` data set shows that there is no evidence against $\mathcal{H}_0$, implying that the hypothesis of extreme-value dependence for C is plausible.

```
> set.seed(1805)
> evTestK(as.matrix(danube))

        Test of bivariate extreme-value dependence based on Kendall's
        distribution with argument 'method' set to "fsample"

data:  as.matrix(danube)
statistic = -0.039908, p-value = 0.9682
```

□

A second bivariate test implemented in the R package `copula` exploits the fact that, under $\mathcal{H}_0$, C can be represented as in (3.12). Given a nonparametric estimator A_n of A and using the empirical copula C_n defined in (4.12), the previous representation can be tested empirically, leading to the test statistic

$$S_n^{\mathrm{ev}_A} = \int_{[0,1]^2} n\left(C_n(u_1,u_2) - \exp\left(\log(u_1u_2)A_n\left(\frac{\log(u_2)}{\log(u_1u_2)}\right)\right)\right)^2 \mathrm{d}C_n(u_1,u_2).$$

More details can be found in Kojadinovic and Yan (2012).

Example 5.2.9 (Test of Extreme-Value Dependence Based on $S_n^{\mathrm{ev}_A}$) The test based on $S_n^{\mathrm{ev}_A}$ is implemented in the R function `evTestA()`. It uses resampling for the computation of an approximate p-value for $S_n^{\mathrm{ev}_A}$ and is particularly computationally demanding. Applying it to the `danube` data set, we obtain:

```
> set.seed(1815)
> withTime(evTestA(as.matrix(danube)))

        Test of bivariate extreme-value dependence based on the CFG
        estimator with argument 'derivatives' set to 'An'

data:  as.matrix(danube)
statistic = 0.026716, p-value = 0.09041

User time:  192.7 sec
```

Hence, the same conclusion as for the test implemented in `evTestK()` is reached: $\mathcal{H}_0$ is plausible for the `danube` data set. □

The last test that we mention can also be used when $d > 2$. It consists of assessing empirically whether (3.10) holds using the empirical copula C_n in (4.12). The recommended test statistic is

$$S_n^{\mathrm{ev}_C} = T_{3,n} + T_{4,n} + T_{5,n}, \tag{5.8}$$

where

$$T_{r,n} = \int_{[0,1]^d} n\left(\left(C_n(u_1^{1/r}, \ldots, u_d^{1/r})\right)^r - C_n(\boldsymbol{u})\right)^2 \mathrm{d}C_n(\boldsymbol{u}).$$

More details can be found in Kojadinovic et al. (2011).

Example 5.2.10 (Test of Extreme-Value Dependence Based on $S_n^{\mathrm{ev}_C}$) The previous test also involves resampling. It is implemented in the R function `evTestC()`. Applying it to the `rdj` data set, we obtain:

```
> set.seed(1905)
> withTime(evTestC(Xrdj))

        Max-stability based test of extreme-value dependence for
        multivariate copulas

data:  Xrdj
statistic = 1.7305, p-value = 0.0004995

User time:  5.3 sec
```

As we could have expected given the financial nature of the observations, there is strong evidence against extreme-value dependence in the `rdj` data set, which, for instance, discards the Gumbel–Hougaard family as a plausible model for C. □

As a final remark, let us mention that, in the same way that extreme-value dependence can be tested, so can *Archimedeanity*, that is, that C in (5.1) is an Archimedean copula. In the bivariate case, such tests are developed in Bücher et al. (2012).

5.2.5 *Goodness-of-Fit Tests*

The classes of tests described in the preceding sections can be used to narrow down the search for an adequate parametric copula family. For each family $\mathcal{C}$ that was not previously rejected, it is then typically of interest to assess whether the unknown copula C in (5.1) actually belongs to $\mathcal{C}$. This *goodness-of-fit issue* amounts formally to testing

$$\mathcal{H}_0 : C \in \mathcal{C} \quad \text{versus} \quad \mathcal{H}_1 : C \notin \mathcal{C}.$$

A large number of goodness-of-fit tests were proposed in the literature as can be concluded for instance from the recent reviews of Genest et al. (2009) and Fermanian (2013). Among the existing procedures, so-called *blanket tests* require neither an arbitrary categorization of the data, nor any choice of smoothing parameter, weight function, kernel, window, etc. Among blanket tests, one approach that appears to perform particularly well according to the large scale simulations carried out in Genest et al. (2009) and Berg (2009) is based on the empirical copula C_n defined in (4.12). The empirical copula is a consistent estimator of the unknown copula C whether $\mathcal{H}_0$ is true or not. Hence, as suggested in Fermanian (2005), Quessy (2005), and Genest and Rémillard (2008), a natural goodness-of-fit test consists of comparing C_n with an estimate $C_{\boldsymbol{\theta}_n}$ of C obtained under the assumption that $C \in \mathcal{C}$ holds. In the previous statement, $\boldsymbol{\theta}_n$ is an estimator of $\boldsymbol{\theta}$ computed from the pseudo-observations $\boldsymbol{U}_{1,n}, \ldots, \boldsymbol{U}_{n,n}$ defined in (4.9) such as the maximum pseudo-likelihood estimator presented in Sect. 4.1.2. According to the large scale simulations carried out in Genest et al. (2009), the most powerful version of this procedure is based on the Cramér-von Mises statistic

$$S_n^{\text{gof}} = \int_{[0,1]^d} n(C_n(\boldsymbol{u}) - C_{\boldsymbol{\theta}_n}(\boldsymbol{u}))^2 \, \mathrm{d}C_n(\boldsymbol{u}) = \sum_{i=1}^{n} (C_n(\boldsymbol{U}_{i,n}) - C_{\boldsymbol{\theta}_n}(\boldsymbol{U}_{i,n}))^2. \tag{5.9}$$

An approximate p-value for the test based on S_n^{gof} can be obtained by means of a *parametric bootstrap* whose asymptotic validity is investigated in Genest and Rémillard (2008).

Algorithm 5.2.11 (Parametric Bootstrap)

1) Compute the pseudo-observations $\boldsymbol{U}_{1,n}, \ldots, \boldsymbol{U}_{n,n}$ defined in (4.9).
2) Compute an estimate $\boldsymbol{\theta}_n$ of $\boldsymbol{\theta}$ from the pseudo-observations $\boldsymbol{U}_{1,n}, \ldots, \boldsymbol{U}_{n,n}$.
3) Compute the test statistic S_n^{gof}.
4) For some large integer N, repeat the following steps for every $k \in \{1, \ldots, N\}$:

   4.1) Generate a pseudo-random sample $\boldsymbol{U}_1^{(k)}, \ldots, \boldsymbol{U}_n^{(k)}$ from the fitted copula $C_{\boldsymbol{\theta}_n}$ and compute the corresponding pseudo-observations $\boldsymbol{U}_{1,n}^{(k)}, \ldots, \boldsymbol{U}_{n,n}^{(k)}$.
   4.2) Compute an estimate $\boldsymbol{\theta}_n^{(k)}$ of $\boldsymbol{\theta}$ from the pseudo-observations $\boldsymbol{U}_{1,n}^{(k)}, \ldots, \boldsymbol{U}_{n,n}^{(k)}$ using the same (rank-based) estimator as in Step 2).
   4.3) Compute the corresponding version $S_n^{\text{gof},(k)}$ of S_n^{gof} as:

$$S_n^{\text{gof},(k)} = \sum_{i=1}^{n} \left(C_n^{(k)}(\boldsymbol{U}_{i,n}^{(k)}) - C_{\boldsymbol{\theta}_n^{(k)}}(\boldsymbol{U}_{i,n}^{(k)}) \right)^2,$$

where

$$C_n^{(k)}(\boldsymbol{u}) = \frac{1}{n}\sum_{i=1}^{n} \mathbf{1}\big(\boldsymbol{U}_{i,n}^{(k)} \le \boldsymbol{u}\big), \quad \boldsymbol{u} \in [0,1]^d.$$

Under $\mathcal{H}_0$, $S_n^{\mathrm{gof},(k)}$ can be thought of as an approximately independent copy of S_n^{gof}.

5) An approximate p-value for the test is given by $(1/2 + \sum_{k=1}^{N} \mathbf{1}(S_n^{\mathrm{gof},(k)} \ge S_n^{\mathrm{gof}}))/(N+1)$. □

The main advantage of the above procedure is its conceptual simplicity. The goodness of fit of a parametric family $\mathcal{C}$ can be assessed as long as one knows how to estimate the unknown parameter vector $\boldsymbol{\theta}$ and how to generate pseudo-random samples from $C_{\boldsymbol{\theta}}$. If $C_{\boldsymbol{\theta}}$ cannot be evaluated, the procedure can be adapted by adding a "second-level" bootstrap; see Genest and Rémillard (2008).

Similar parametric bootstrap-based procedures can be implemented for many other goodness-of-fit statistics; for reviews, see, for example, Genest et al. (2009) and Berg (2009).

Example 5.2.12 (Parametric Bootstrap-Based Tests) The key R function for carrying out goodness-of-fit tests in the package `copula` is called `gofCopula()`. By default, it is based on the statistic S_n^{gof} and implements the above parametric bootstrap. We can then assess, for instance, whether the Clayton copula family is a good choice for the `rdj` data set introduced in Example 4.1.3:

```
> set.seed(1598)
> withTime(gofCopula(claytonCopula(dim = 3), x = Xrdj, optim.method = "BFGS"))

        Parametric bootstrap-based goodness-of-fit test of Clayton
        copula, dim. d = 3, with 'method'="Sn", 'estim.method'="mpl":

data:  x
statistic = 0.87338, parameter = 0.58565, p-value = 0.0004995

User time:  76.6 sec
```

Estimation of the unknown parameter of the Clayton copula is carried out by maximum pseudo-likelihood estimation, which is the default estimation method. The underlying numerical optimization routine can be changed via the argument `optim.method` passed to `fitCopula()`, which is set to `"BFGS"` in the code above to avoid a few warnings related to possible convergence problems in this case. As one can see from the output, the parameter estimate is $\theta_n = 0.58565$, the test statistic is $S_n^{\mathrm{gof}} = 0.87338$ and the corresponding p-value (computed as in Algorithm 5.2.11 with $N = 1000$) is 0.0004995, which corresponds to 0.5/1001. As could have been expected given the fact that the observations under consideration are trivariate financial log-returns, the Clayton family is strongly rejected. □

The main inconvenience of the above goodness-of-fit test based on the parametric bootstrap is its high computational cost, as each iteration requires both random number generation from the fitted copula and estimation of the copula parameters. As the sample size increases, its application becomes prohibitive. In order to circumvent this high computational cost, a faster, large-sample testing procedure based on multiplier central limit theorems was proposed in Kojadinovic et al. (2011). As n reaches 300, the Monte Carlo experiments carried out for $d = 2, 3$, and 4 in Kojadinovic and Yan (2011) indicate that one can safely use the multiplier approach as an alternative to the parametric bootstrap. The previous statement means that under $\mathcal{H}_0$ both resampling procedures are asymptotically equivalent and thus asymptotically valid. Under the alternative, however, they might not necessarily return similar p-values.

The price to pay for the higher computational efficiency of the multiplier resampling approach is more programming work as certain partial derivatives need to be computed for each hypothesized parametric copula family $\mathcal{C}$. For those families in `copula` for which these partial derivatives are not implemented, numerical differentiation based on the R package `numDeriv` of Gilbert and Varadhan (2016) is used.

Example 5.2.13 (Multiplier Goodness-of-Fit Tests) For the `rdj` data set, given the large sample size, the multiplier resampling approach can be safely used as an alternative to the default parametric bootstrap by utilizing `gofCopula(, simulation = "mult")`:

```
> set.seed(1610)
> withTime(gofCopula(claytonCopula(dim = 3), x = Xrdj, simulation = "mult"))

        Multiplier bootstrap-based goodness-of-fit test of Clayton
        copula, dim. d = 3, with 'method'="Sn", 'estim.method'="mpl":

data:  x
statistic = 0.87338, parameter = 0.58565, p-value = 0.0004995

User time:  4.4 sec
```

As previously, the Clayton family is strongly rejected. We can also test whether the normal family fits:

```
> set.seed(1685)
> gofCopula(normalCopula(dim = 3, dispstr = "un"), x = Xrdj,
            simulation = "mult")

        Multiplier bootstrap-based goodness-of-fit test of Normal
        copula, dim. d = 3, with 'method'="Sn", 'estim.method'="mpl":

data:  x
statistic = 0.04101, parameter1 = 0.57812, parameter2 =
0.34002, parameter3 = 0.40169, p-value = 0.09041
```

A similar hypothesis for the t family in which the degrees of freedom are kept fixed at 10 can be tested as follows:

```
> set.seed(1792)
> gofCopula(tCopula(dim = 3, dispstr = "un", df.fixed = TRUE, df = 10),
            x = Xrdj, simulation = "mult")

        Multiplier bootstrap-based goodness-of-fit test of t-copula,
        dim. d = 3, with 'method'="Sn", 'estim.method'="mpl":

data:  x
statistic = 0.022475, parameter1 = 0.59238, parameter2 =
0.36251, parameter3 = 0.42644, p-value = 0.6538
```

Hence, neither the normal family nor the t family with 10 degrees of freedom is rejected. Both assumptions on C are thus plausible. The fact that the p-value for the t family with 10 degrees of freedom is higher should not in principle be used as an argument to select it. □

Example 5.2.14 (Empirical Levels of the Multiplier Goodness-of-Fit Test for the Joe Family) The multiplier goodness-of-fit test is expected to be a valid large-sample alternative to the parametric bootstrap. The following code shows how the user can empirically assess whether such a test holds its level in the case of small samples, for instance, when assessing the fit of the Joe family:

```
> theta <- iTau(joeCopula(), tau = 0.5) # Joe copula parameter
> ##' @title P-value of multiplier goodness-of-fit test on data
> ##'        generated under the null hypothesis
> ##' @param n sample size
> ##' @param theta Joe copula parameter
> ##' @return p-value of the multiplier goodness-of-fit test
> pvalMult <- function(n, theta)
  {
      U <- rCopula(n, copula = joeCopula(theta))
      gofCopula(joeCopula(), x = pobs(U), simulation = "mult",
                optim.method = "BFGS")$p.value
  }
```

Let us then compute, say, 1000 p-values under $\mathcal{H}_0$:

```
> set.seed(1940)
> pv <- withTime(replicate(1000, pvalMult(n = 100, theta = theta)))
```

Recall that, because of the use of `withTime()`, which is defined in Example 5.2.6, the output of `replicate()` is stored in `pv$value`, while the corresponding run time is stored in `pv$sys.time`:

```
> pv$sys.time # the run time

User time:  34.1 sec
```

If the test holds its level, the p-values should be approximately uniformly distributed. The lower tail is of particular interest in practice:

```
> alpha <- c(0.01, 0.05, 0.1) # nominal levels
> rbind(nom.level = alpha, emp.level = ecdf(pv$value)(alpha)) # levels

           [,1]  [,2]  [,3]
nom.level 0.010 0.050 0.100
emp.level 0.008 0.044 0.096
```

It thus seems that the multiplier goodness-of-fit test based on maximum pseudo-likelihood estimation for the Joe family holds its level reasonably well when applied to small samples with moderate dependence. Before drawing a final conclusion, other values of Kendall's tau should be considered as well.

The previous code can be adapted to assess whether any test of interest available in R holds its level. □

When $d = 2$ and C is a one-parameter copula family, method-of-moment estimation based on Kendall's tau or Spearman's rho can be used as an alternative to maximum pseudo-likelihood estimation and often results in faster goodness-of-fit tests.

Example 5.2.15 (Goodness-of-Fit Tests Based on Method-of-Moments Estimation) Let us again consider the `danube` data set presented initially in Example 4.1.6. As we have seen in Examples 5.2.8 and 5.2.9, there is no evidence against extreme-value dependence for these observations, which suggests to consider a copula family of the extreme-value type as a possible model. Although, from Examples 5.2.4 and 5.2.6, there is strong evidence against exchangeability, for pedagogical purposes, we shall still assess the fit of the Gumbel–Hougaard family. To this end, we first apply the parametric bootstrap based on the statistic S_n^{gof} in (5.9) and then do the same with the multiplier bootstrap. Instead of using maximum pseudo-likelihood estimation, we use method-of-moments estimation based on Kendall's tau via `gofCopula(, estim.method = "itau")`:

```
> Xd <- as.matrix(danube)
> set.seed(1613)
> gofCopula(gumbelCopula(), x = Xd, estim.method = "itau")

        Parametric bootstrap-based goodness-of-fit test of Gumbel
        copula, dim. d = 2, with 'method'="Sn",
        'estim.method'="itau":

data:  x
statistic = 0.024399, parameter = 2.2147, p-value = 0.05145

> set.seed(1914)
> gofCopula(gumbelCopula(), x = Xd, estim.method = "itau",
            simulation = "mult")
```

```
	Multiplier bootstrap-based goodness-of-fit test of Gumbel
	copula, dim. d = 2, with 'method'="Sn",
	'estim.method'="itau":

data:  x
statistic = 0.024399, parameter = 2.2147, p-value = 0.07343
```

As one can see, both versions of the test based on S_n^{gof} give similar p-values providing only weak evidence against the Gumbel–Hougaard family.

The Joe family, however, is strongly rejected:

```
> set.seed(1848)
> gofCopula(joeCopula(), x = Xd, estim.method = "itau")

	Parametric bootstrap-based goodness-of-fit test of Joe
	copula, dim. d = 2, with 'method'="Sn",
	'estim.method'="itau":

data:  x
statistic = 0.064507, parameter = 3.2713, p-value = 0.0004995
```

□

Example 5.2.16 (Goodness of Fit of the Khoudraji–Gumbel–Hougaard Family) Because of the possible slight non-exchangeability in the `danube` data set (see Fig. 5.1) and given that it is plausible that the underlying copula is of the extreme-value type following Examples 5.2.8 and 5.2.9, it is meaningful to consider a non-exchangeable extreme-value copula family as candidate family for the `danube` data set. The Khoudraji–Gumbel–Hougaard copula $\text{kho}_{\boldsymbol{s}}(\Pi, C_\theta^{\text{GH}})$ given by (3.14) is such a family and was already considered in Example 5.1.2 in its two-parameter version. For such a multiparameter model, method-of-moments estimation based on Kendall's tau or Spearman's rho cannot be applied, which is why default maximum pseudo-likelihood estimation is used:

```
> set.seed(1969) # Parametric bootstrap-based test
> withTime(
      gofCopula(khoudrajiCopula(copula2 = gumbelCopula(),
                                shapes = fixParam(c(NA_real_, 1), c(FALSE, TRUE))),
                start = c(1.1, 0.5), x = Xd, optim.method = "Nelder-Mead")
  )

	Parametric bootstrap-based goodness-of-fit test of Khoudraji
	copula, dim. d = 2, constructed from Independence copula
	Gumbel copula, with 'method'="Sn", 'estim.method'="mpl":

data:  x
statistic = 0.029712, parameter1 = 2.2729, parameter2 =
0.9218, p-value = 0.04745

User time:  86.1 sec
```

```
> set.seed(1969) # Multiplier-based test
> withTime(
      gofCopula(khoudrajiCopula(copula2 = gumbelCopula(),
                                shapes = fixParam(c(NA_real_, 1), c(FALSE, TRUE))),
                start = c(1.1, 0.5), x = Xd, optim.method = "Nelder-Mead",
                simulation = "mult")
  )

        Multiplier bootstrap-based goodness-of-fit test of Khoudraji
        copula, dim. d = 2, constructed from Independence copula
        Gumbel copula, with 'method'="Sn", 'estim.method'="mpl":

data:  x
statistic = 0.029712, parameter1 = 2.2729, parameter2 =
0.9218, p-value = 0.04046

User time:  1.2 sec
```

Both p-values provide only weak evidence against the two-parameter Khoudraji–Gumbel–Hougaard family, which could thus be regarded as a plausible model along with the simpler Gumbel–Hougaard family. Model selection (ranking) will be discussed in Sect. 5.3, and in Example 5.3.1 in particular. □

Notice finally that alternatives to the aforementioned goodness-of-fit statistics are available in the R package `copula`. Most rely on the parametric bootstrap for the computation of approximate p-values. More details can be found on the help pages of the functions `gofCopula()` and `gofTstat()`.

5.2.6 A Mixture of Graphical and Formal Goodness-of-Fit Tests

As the dimension d of the data increases, the application of the goodness-of-fit tests presented in the previous section becomes prohibitive, especially if the number of parameters of the hypothesized copula family increases with d. A prominent example of such a situation is when assessing the fit of elliptical copula families such as the normal or the t, as their number of parameters is of the order d^2. In such cases, one could apply the formal goodness-of-fit tests described in the previous section on lower-dimensional margins and adjust the obtained p-values, for instance, to take into account the number of tests. However, the resulting approach might turn out to be too computationally demanding and its results not clearly interpretable. With such constraints in mind, Hofert and Mächler (2014) proposed a mixture of graphical and formal goodness-of-fit tests based on pairwise Rosenblatt transforms; see Sect. 2.7. A realistic application is presented in the next example; see also the demo `gof_graph` for additional visualizations.

Example 5.2.17 (A Mixture of Graphical and Formal Goodness-of-Fit Tests) We consider the log-returns of all $d = 20$ constituents of the Swiss Market Index (SMI) from 2011-09-09 to 2012-03-28 already used in Example 4.1.9:

```
> ## Load the data, compute the log-returns and the pseudo-observations
> data(SMI.12) # load the SMI constituent data
> library(qrmtools)
> X <- returns(SMI.12) # compute log-returns
> U <- pobs(X) # compute pseudo-observations
> d <- ncol(U) # 20 dimensions
```

We then fit a t copula to the pseudo-observations using the method of Mashal and Zeevi (2002) briefly presented in Sect. 4.1.3:

```
> fit <- fitCopula(tCopula(dim = d, dispstr = "un"), data = U,
                   method = "itau.mpl")
> ## Extract parameter estimates
> len <- length(coef(fit))
> stopifnot(len == d*(d-1)/2 + 1) # sanity check
> p <- coef(fit)[seq_len(len-1)] # correlations
> ## Note: The estimated correlation matrix can be obtained via p2P(p, d = d)
> nu <- coef(fit)[len] # degrees of freedom nu
```

Next, we apply the graphical test of Hofert and Mächler (2014) (based on the functions `pairwiseCcop()`, `pairwiseIndepTest()`, `pviTest()` and `pairsRosenblatt()`; see the respective help pages for details) to the pseudo-observations to assess the fit of the estimated t copula:

```
> ## Define the H_0 copula
> cop.t <- ellipCopula("t", df = nu, param = p, dim = d, dispstr = "un")
> ## Build the array of pairwise H_0-transformed data columns
> cu.u.t <- pairwiseCcop(U, cop.t, df = nu)
> ## Compute pairwise matrix (d by d) of p-values and corresponding colors
> set.seed(1389) # for reproducibility
> pw.indep.t <- pairwiseIndepTest(cu.u.t, N = 256, verbose = FALSE)
> p.val.t <- pviTest(pw.indep.t) # matrix of p-values
> ## Pairwise Rosenblatt plot (test for the fitted t copula)
> title <- list("Pairwise Rosenblatt transformed pseudo-observations",
                quote(bold("to test")~~italic(H[0]:C)~~"is"~~italic(t)))
> cols <- gray(seq(0.1, 1, length.out = 6))
> pairsRosenblatt(cu.u.t, pvalueMat = p.val.t, method = "QQchisq", pch=".",
                  xaxt = "n", yaxt = "n",
                  colList = pairsColList(p.val.t, bucketCols = cols,
                                         BWcutoff = 0),
                  main.centered = TRUE, main = title, line.main = c(2, -0.8))
```

The resulting plot is shown in Fig. 5.5. As already explained, for pedagogical purposes, we ignore in this chapter the possible serial dependence in the financial log-returns (see Chap. 6 for a realistic preprocessing). The main idea behind this test is to apply, to all bivariate margins of the pseudo-observations, the Rosenblatt transformation with the estimated hypothesized copula. Under $\mathcal{H}_0$, the resulting bivariate samples should thus approximately follow a uniform distribution on $[0, 1]^2$, which can be tested with the formal test based on (5.3). Note that, when applying the latter, the fact that the parameters of the hypothesized copula are estimated is not taken into account. The corresponding (distorted) p-values, referred to as pp-values ("pseudo p-values") in Hofert and Mächler (2014), are used to determine the gray-scale background in Fig. 5.5. There are various options available in `pairsRosenblatt()`; see the help page for more details. Here, we opt to overlay the plot with pairwise Q-Q plots against a χ_2^2 distribution after mapping the Rosenblatt-transformed pseudo-observations to univariate observations as explained in Hofert and Mächler (2014). Note that all plots in the jth column of the plot matrix in Fig. 5.5 are obtained by conditioning on the jth variable in the pairwise Rosenblatt transform, which implies that the plot matrix is typically not symmetric. As this figure reveals, there is no strong evidence against the hypothesized t copula family here.

Graphical goodness-of-fit tests of this kind can be helpful to detect bivariate (or higher-dimensional) marginal departures from the hypothesized copula under $\mathcal{H}_0$. This can be useful especially in higher dimensions where it is likely that no model

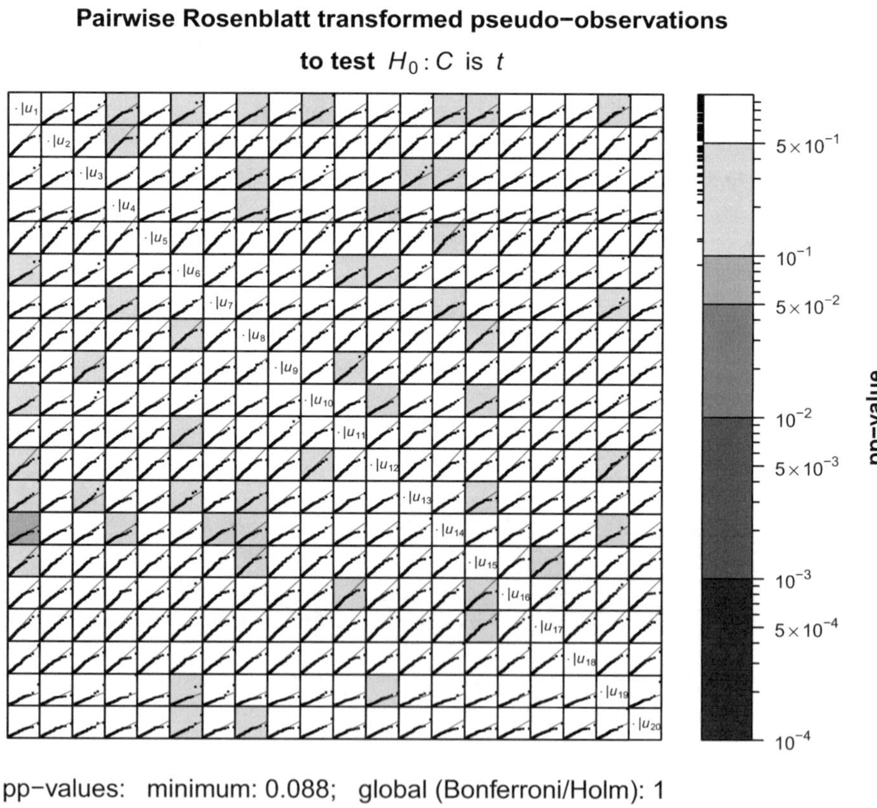

Fig. 5.5 Graphical output of one of the tests of Hofert and Mächler (2014) assessing the fit of a t copula in a pairwise manner based on pseudo-observations computed from log-returns of the 20 constituents of the Swiss Market Index (SMI) from 2011-09-09 to 2012-03-28

fits perfectly and thus formal goodness-of-fit tests often result in rejection of $\mathcal{H}_0$ without leading to further insight in why the hypothesized model was rejected.

The graphical display underlying Fig. 5.5 is a scatter-plot matrix, which (even when the individual cells are shrunk down significantly and only display colors) can be difficult to visually examine when the dimension is large. To address this (and other) limitation(s) of scatter-plot matrices, so-called *zenplots (zigzag expanded navigation plots)* have recently been developed; see the R package `zenplots` of Hofert and Oldford (2016, 2017) for an application to constituents of the S&P 500. □

5.3 Model Selection

Goodness-of-fit testing is a key step when searching for an acceptable parametric copula family for the data at hand. Its results can, however, be disappointing in practice. When the sample size is small, it is not uncommon that none of the candidate parametric copula families are rejected, meaning that all candidate models are plausible. On the contrary, when the dimension or the sample size is large, it can easily happen that all the candidate families are rejected.

While a goodness-of-fit test's aim is to assess whether a given parametric copula family fits the data at hand, the role of model selection in the setting considered in this chapter is to rank the candidate copula families, resulting, in practice, in the selection of the best ranked family. Ideally, one would not want the selected family to be rejected by a goodness-of-fit test but this can happen in applications.

In the classical likelihood-based setting of parametric statistics, a famous criteria for model selection is the *Akaike information criterion (AIC)*. In the "full" likelihood estimation context of Sect. 4.1.1 in which the marginal and copula parameters are estimated simultaneously, its use is thus fully justified. In that case, the formula is

$$\mathrm{AIC} = 2(\ell_{n,\max} - p), \tag{5.10}$$

where $\ell_{n,\max}$ is the maximized likelihood and p is the total number of marginal and copula parameters.

Many authors such as Chen and Fan (2005), McNeil et al. (2015), or Joe (2014) also apply the above AIC formula in the two-stage estimation contexts of the inference functions for margins method described in Sect. 4.1.1 or the maximum pseudo-likelihood approach presented in Sect. 4.1.2. In the latter case for instance, $\ell_{n,\max}$ is the maximized pseudo-likelihood and p is the number of copula parameters.

The earliest theoretical investigation of model selection for copula families seems to be due to Grønneberg and Hjort (2014). The authors conclude that the use of the classical AIC formula is not justified in the context of the previously mentioned two-stage estimation methods, except under restrictive conditions not satisfied by most parametric copula families of interest. They propose several alternative selection criteria which do not suffer from the theoretical deficiencies of AIC in the context under consideration. One of these criteria, namely the cross-validation copula information criterion defined in Grønneberg and Hjort (2014, Equation (44)), is compared with the naive use of AIC in the context of maximum pseudo-likelihood estimation in a bivariate numerical experiment involving five common one-parameter copula families in Jordanger and Tjøstheim (2014). The conclusion of the study is that the two selection criteria do not seem to differ significantly overall. Given the computational cost of the cross-validation copula information criterion, Jordanger and Tjøstheim (2014) suggest to keep using the significantly simpler AIC formula in the context of maximum pseudo-likelihood estimation. One should, however, remain cautious because the aforementioned numerical study, despite its quality, considers neither some of the pathological examples put forward in Grønneberg and Hjort (2014) nor higher dimensional multiparameter copula families.

As explained in Grønneberg and Hjort (2014, Section 4), the aforementioned cross-validation copula information criterion is, up to a multiplicative constant, the first-order equivalent of the cross-validation criterion

$$\widehat{\mathrm{xv}}_n = \frac{1}{n} \sum_{i=1}^{n} \log c_{\boldsymbol{\theta}_{n,-i}}\big(\boldsymbol{F}_{n,-i}(\boldsymbol{X}_i)\big), \tag{5.11}$$

where $\boldsymbol{\theta}_{n,-i}$ is the maximum pseudo-likelihood estimate computed from the sample $\boldsymbol{X}_1, \dots, \boldsymbol{X}_{i-1}, \boldsymbol{X}_{i+1}, \dots, \boldsymbol{X}_n$ and

$$\boldsymbol{F}_{n,-i}(\boldsymbol{x}) = \big(F_{n,1,-i}(x_1), \dots, F_{n,d,-i}(x_d)\big), \quad \boldsymbol{x} \in \mathbb{R},$$

with

$$F_{n,j,-i}(x) = \begin{cases} \dfrac{1}{n} \displaystyle\sum_{\substack{k=1 \\ k \neq i}}^{n} \mathbf{1}(X_{kj} \le x), & \text{if } x \ge \displaystyle\min_{k \in \{1,\dots,n\} \setminus \{i\}} X_{kj}, \\ 1/n, & \text{otherwise}. \end{cases} \tag{5.12}$$

The above formula is a natural adaptation of classical *leave-one-out cross validation*; see, for example, Claeskens et al. (2008, Section 2.9) as well as Grønneberg and Hjort (2014, Remark 1) for a discussion of (5.12) and possible alternatives. By construction, it will penalize copula families with too many parameters that tend to overfit. It has the advantage of being easy to implement but the inconvenience of being computationally expensive.

Example 5.3.1 (Cross-Validation for the `danube` *Data Set)* The R function implementing (5.11) is `xvCopula()`. Note that it actually returns $n\widehat{\mathrm{xv}}_n$ to be of the same order of magnitude as the maximized likelihood. For the `danube` data set, we use it to compare the Joe family, the Gumbel–Hougaard family, and the two-parameter Khoudraji–Gumbel–Hougaard family considered in Example 5.2.16:

```
> Xdan <- as.matrix(danube)
> withTime(xvCopula(joeCopula(), x = Xdan))

[1] 246.112
User time:  43.4 sec

> withTime(xvCopula(gumbelCopula(), x = Xdan))

[1] 276.3542
User time:  62.6 sec

> withTime(
      xvCopula(khoudrajiCopula(copula2 = gumbelCopula(),
                               shapes = fixParam(c(NA_real_, 1),
                                                 c(FALSE, TRUE))),
               x = Xdan, start = c(1.1, 0.5), optim.method = "Nelder-Mead")
  )

[1] 278.7986
User time:  68 sec
```

In terms of leave-one-out cross-validation, the Khoudraji–Gumbel–Hougaard family should thus be slightly preferred to the Gumbel–Hougaard family, which in turn should be preferred to the Joe family.

To lower the computational cost, a k-fold cross-validation analog can be used. It consists of dividing the data into $k < n$ blocks of rows of approximately the same size instead of n as in the leave-one-out approach. In practice, before forming the blocks, the rows are randomly shuffled, making k-fold cross-validation dependent on the seed.

```
> k <- 50
> set.seed(7)
> withTime(xvCopula(joeCopula(), x = Xdan, k = k))

[1] 246.3862
User time:  3.2 sec

> set.seed(13)
> withTime(xvCopula(gumbelCopula(), x = Xdan, k = k))

[1] 276.5334
User time:  4.7 sec

> set.seed(14)
> withTime(
      xvCopula(khoudrajiCopula(copula2 = gumbelCopula(),
                               shapes = fixParam(c(NA_real_, 1),
                                                 c(FALSE, TRUE))),
               x = Xdan, k = k, start = c(1.1, 0.5),
               optim.method = "Nelder-Mead")
  )
```

```
[1] 278.4246
User time:  5 sec
```

Hence, in the above setting, leave-one-out cross-validation and its k-fold analog induce the same order on the candidate copula families, the latter being, however, substantially faster. □

Example 5.3.2 (Cross-Validation for the `rdj` *Data Set)* We also compute $n\widehat{\mathrm{xv}}_n$ for the `rdj` data set for the normal family and the t family with 10 degrees of freedom:

```
> withTime(xvCopula(normalCopula(dim = 3, dispstr = "un"), x = Xrdj))
```

```
[1] 370.2897
User time:  274.7 sec
```

```
> withTime(xvCopula(tCopula(dim = 3, dispstr = "un", df = 10, df.fixed = TRUE),
                    x = Xrdj))
```

```
[1] 411.3878
User time:  541.3 sec
```

The k-fold cross-validation analog leads to the same ranking:

```
> set.seed(22)
> withTime(xvCopula(normalCopula(dim = 3, dispstr = "un"), x = Xrdj, k = k))
```

```
[1] 370.1114
User time:  10.8 sec
```

```
> set.seed(4)
> withTime(xvCopula(tCopula(dim = 3, dispstr = "un", df = 10, df.fixed = TRUE),
                    x = Xrdj, k = k))
```

```
[1] 411.5618
User time:  21 sec
```

We can even compute these selection criteria for the t copula when the degrees of freedom are treated as a parameter to be estimated. For instance:

```
> set.seed(1980)
> withTime(xvCopula(tCopula(dim = 3, dispstr = "un"), x = Xrdj, k = k))
```

```
[1] 414.1781
User time:  37.1 sec
```

The previous results suggest to consider the following estimate of the t copula for the `rdj` data set:

```
> summary(fitCopula(tCopula(dim = 3, dispstr = "un"), data = pobs(Xrdj)))
```

```
Call: fitCopula(copula, data = data)
Fit based on "maximum pseudo-likelihood" and 1262 3-dimensional observations.
t-copula, dim. d = 3
      Estimate Std. Error
rho.1   0.5877      0.020
rho.2   0.3593      0.026
rho.3   0.4225      0.025
df      6.5025         NA
The maximized loglikelihood is 419.3
Optimization converged
Number of loglikelihood evaluations:
function gradient
      31       10
```

□

References

Ben Ghorbal, M., Genest, C., & Nešlehová, J. G. (2009). On the test of Ghoudi, Khoudraji, and Rivest for extreme-value dependence. *The Canadian Journal of Statistics, 37*(4), 534–552.

Berg, D. (2009). Copula goodness-of-fit testing: An overview and power comparison. *The European Journal of Finance, 15*, 675–701.

Bücher, A., Dette, H., & Volgushev, S. (2012). A test for Archimedeanity in bivariate copula models. *Journal of Multivariate Analysis, 110*, 121–132.

Bücher, A., & Kojadinovic, I. (2015). An overview of nonparametric tests of extreme-value dependence and of some related statistical procedures. In D. Dey & J. Yan (Eds.) *Extreme value modeling and risk analysis: Methods and applications* (pp. 377–398). Boca Baton, FL: Chapman & Hall/CRC.

Chen, X., & Fan, Y. (2005). Pseudo-likelihood ratio tests for semiparametric multivariate copula model selection. *The Canadian Journal of Statistics, 33*, 389–414.

Claeskens, G., & Hjort, N. L. (2008). *Model selection and model averaging*. London: Cambridge University Press.

Deheuvels, P. (1981). A non parametric test for independence. *Publications de l'Institut de Statistique de l'Université de Paris, 26*, 29–50.

Fermanian, J.-D. (2005). Goodness-of-fit tests for copulas. *Journal of Multivariate Analysis, 95*(1), 119–152.

Fermanian, J.-D. (2013). An overview of the goodness-of-fit test problem for copulas. In P. Jaworski, F. Durante, & W. K. Härdle (Eds.), *Copulae in mathematical and quantitative finance. Lectures notes in statistics* (pp. 61–89). Berlin: Springer.

Genest, C., & Boies, J.-C. (2003). Detecting dependence with Kendall plots. *The American Statistician, 57*, 275–284.

Genest, C., & Favre, A.-C. (2007). Everything you always wanted to know about copula modeling but were afraid to ask. *Journal of Hydrological Engineering, 12*, 347–368.

Genest, C., Nešlehová, J., & Quessy, J.-F. (2012). Tests of symmetry for bivariate copulas. *Annals of the Institute of Statistical Mathematics, 64*, 811–834.

Genest, C., & Nešlehová, J. G. (2014). On tests of radial symmetry for bivariate copulas. *Statistical Papers, 55*, 1107–1119.

Genest, C., Quessy, J.-F., & Rémillard, B. (2007). Asymptotic local efficiency of Cramér–von Mises tests for multivariate independence. *The Annals of Statistics, 35*, 166–191.

Genest, C., & Rémillard, B. (2004). Tests of independence and randomness based on the empirical copula process. *Test, 13*(2), 335–369.

Genest, C., & Rémillard, B. (2008). Validity of the parametric bootstrap for goodness-of-fit testing in semiparametric models. *Annales de l'Institut Henri Poincaré: Probabilités et Statistiques, 44*, 1096–1127.

Genest, C., Rémillard, B., & Beaudoin, D. (2009). Goodness-of-fit tests for copulas: A review and a power study. *Insurance: Mathematics and Economics, 44*, 199–213.

Ghoudi, K., Khoudraji, A., & Rivest, L.-P. (1998). Propriétés statistiques des copules de valeurs extrêmes bidimensionnelles. *The Canadian Journal of Statistics, 26*(1), 187–197.

Gilbert, P., & Varadhan, R. (2016). numDeriv: Accurate numerical derivatives, R package version 2016.8-27. https://CRAN.R-project.org/package=numDeriv

Grønneberg, S., & Hjort, N. L. (2014). The copula information criteria. *Scandinavian Journal of Statistics, 41*, 436–459.

Harder, M., & Stadtmüller, U. (2017). Testing exchangeability of copulas in arbitrary dimension. *Journal of Nonparametric Statistics, 29*(1), 40–60.

Hofert, M., & Mächler, M. (2014). A graphical goodness-of-fit test for dependence models in higher dimensions. *Journal of Computational and Graphical Statistics, 23*(3), 700–716.

Hofert, M., & Oldford, R. W. (2016). zenplots: Zigzag expanded navigation plots, R package version 0.0-1. https://CRAN.R-project.org/package=zenplots

Hofert, M., & Oldford, R. W. (2017). Visualizing dependence in high-dimensional data: An application to S&P 500 constituent data. *Econometrics and Statistics, 8*, 161–183.

Joe, H. (1990). Multivariate concordance. *Journal of Multivariate Analysis, 35*(1), 12–30.

Joe, H. (2014). Dependence modeling with copulas. Boca Raton, FL: Chapman & Hall/CRC.

Jordanger, L. A., & Tjøstheim, D. (2014). Model selection of copulas: AIC versus a cross validation copula information criterion. *Statistics & Probability Letters, 92*, 249–255.

Kojadinovic, I. (2017). Some copula inference procedures adapted to the presence of ties. *Computational Statistics and Data Analysis, 142*, 24–41.

Kojadinovic, I., & Holmes, M. (2009). Tests of independence among continuous random vectors based on Cramér–von Mises functionals of the empirical copula process. *Journal of Multivariate Analysis, 100*(6), 1137–1154.

Kojadinovic, I., Segers, J., & Yan, J. (2011). Large-sample tests of extreme-value dependence for multivariate copulas. *The Canadian Journal of Statistics, 39*(4), 703–720.

Kojadinovic, I., & Yan, J. (2011). A goodness-of-fit test for multivariate multiparameter copulas based on multiplier central limit theorems. *Statistics and Computing, 21*(1), 17–30.

Kojadinovic, I., & Yan, J. (2012). A nonparametric test of exchangeability for extreme-value and left-tail decreasing bivariate copulas. *Scandinavian Journal of Statistics, 39*(3), 480–496.

Kojadinovic, I., Yan, J., & Holmes, M. (2011). Fast large-sample goodness-of-fit for copulas. *Statistica Sinica, 21*(2), 841–871.

Mashal, R., & Zeevi, A. (2002). *Beyond correlation: Extreme co-movements between financial assets*. Technical report, Columbia School of Business. https://www0.gsb.columbia.edu/faculty/azeevi/PAPERS/BeyondCorrelation.pdf

McNeil, A. J., Frey, R., & Embrechts, P. (2015). *Quantitative risk management: Concepts, techniques and tools* (2nd ed.). Princeton, NJ: Princeton University Press.

Quessy, J.-F. (2005), *Méthodologie et application des copules: tests d'adéquation, tests d'indépendance, et bornes sur la valeur-à-risque*. PhD thesis, Université Laval, Québec, Canada.

Rémillard, B., & Scaillet, O. (2009), Testing for equality between two copulas. *Journal of Multivariate Analysis, 100*(3), 377–386.

Sarkar, D. (2016). latticeExtra: Extra graphical utilities based on lattice, R package version 0.6-28. http://CRAN.R-project.org/package=latticeExtra

Schmid, F., & Schmidt, R. (2007). Multivariate extensions of Spearman's rho and related statistics. *Statistics & Probability Letters, 77*(4), 407–416.

Wild, C. J., & Seber, G. A. F. (1999). *Chance encounters: A first course in data analysis and inference*. New York: Wiley.

Chapter 6
Ties, Time Series, and Regression

Chapters 4 and 5 described statistical procedures for modeling the copula of a continuous d-dimensional random vector $\boldsymbol{X}$ with df H from a random sample $\boldsymbol{X}_1, \ldots, \boldsymbol{X}_n$ from H. Since H has continuous univariate margins, ties in the d component samples of $\boldsymbol{X}_1, \ldots, \boldsymbol{X}_n$ cannot occur. If the iid assumption holds but ties occur, this is an indication of the fact that either the random phenomenon of interest is not continuous (a setting not covered in this book; see, for instance, Genest et al. (2014), Joe (2014), and the references therein), or of rounding or lack of measurement precision issues. Section 6.1 of this chapter is concerned with the latter situation. As we shall see, ignoring the presence of ties can strongly bias the results of many of the inference procedures presented in Chap. 5. Fortunately, a few of the statistical techniques under consideration still seem to provide meaningful results and certain others can be adapted for ties.

The second section of this chapter focuses on the frequently encountered situation in applications where the random vector of interest $\boldsymbol{X}$ is observed at successive points in time (that is, as a time series), which naturally leads to the question of whether the collected observations $\boldsymbol{X}_1, \ldots, \boldsymbol{X}_n$ are iid. Hence, prior to any modeling, a first crucial step is to assess whether $\boldsymbol{X}_1, \ldots, \boldsymbol{X}_n$ are observations of the same random phenomenon and, more generally, whether the statistical properties of the underlying time series remain constant over the observation period. The latter amounts to testing whether $\boldsymbol{X}_1, \ldots, \boldsymbol{X}_n$ can be regarded as a stretch from a stationary multivariate time series. If there is no evidence against the stationarity assumption, a natural next step is to test for serial independence. In case of evidence against serial independence, $\boldsymbol{X}_1, \ldots, \boldsymbol{X}_n$ cannot be considered as iid and most of the modeling steps described in Chaps. 4 and 5 should not be directly applied. Section 6.2 will describe statistical procedures for testing stationarity and serial independence along with a frequently encountered copula-based model for multivariate time series.

The third section of this chapter is concerned with yet another setting that frequently occurs in practice: A random vector $\boldsymbol{Z}$ of exogenous variables is

M. Hofert et al., *Elements of Copula Modeling with R*, Use R!,
https://doi.org/10.1007/978-3-319-89635-9_6

observed in addition to the random vector $\boldsymbol{X}$ of interest. Given a random sample $(\boldsymbol{X}_1, \boldsymbol{Z}_1), \dots, (\boldsymbol{X}_n, \boldsymbol{Z}_n)$ from $(\boldsymbol{X}, \boldsymbol{Z})$, the goal is then to model the conditional distribution of $\boldsymbol{X}$ given $\boldsymbol{Z}$ (as opposed to the unconditional distribution of $\boldsymbol{X}$). Under additional continuity assumptions, such a regression-like problem can be tackled using the notion of conditional copula. Section 6.3 will provide an overview of the main copula-based models for this task and the related statistical challenges.

6.1 Ties

Let $X_1, \dots, X_n$ be a univariate random sample from a df F. If $X_i = X_j$ for some $i \neq j$, the sample contains *ties*. It is well known that if F is continuous, ties occur with probability zero. By extension, we shall say that a d-dimensional data set $\boldsymbol{X}_1, \dots, \boldsymbol{X}_n$ contains *ties* if at least one component sample $X_{1j}, \dots, X_{nj}$, $j \in \{1, \dots, d\}$, contains ties.

If we assume, as in Chaps. 4 and 5, that the available data $\boldsymbol{X}_1, \dots, \boldsymbol{X}_n$ are iid from a df H with continuous univariate margins, ties should therefore not occur. Yet, in practice, because of rounding and lack of measurement precision, it is not uncommon to observe ties in a data set arising from the observation of a truly continuous random phenomenon.

Most inference procedures on the underlying unknown copula C described in Chaps. 4 and 5 are based on pseudo-observations from C (computed from the data using (4.9)) which are equivalent to the multivariate scaled ranks (4.10) in the absence of ties. Because ranks can be defined in several ways when ties are present, this equivalence does not necessarily hold anymore if ties are present.

The following example briefly presents the most frequently encountered definitions of ranks in the presence of ties.

Example 6.1.1 (Computing Ranks in the Presence of Ties) The R function `rank()` can be used to compute ranks for a sequence of numbers. The method for computing ranks is specified via the argument `ties.method` which is `"average"` by default. Reassuringly, when there are no ties, all methods lead to the same result. For instance:

```
> set.seed(1979)
> (U <- runif(8))

[1] 0.9088996 0.4820053 0.2529599 0.7284594 0.1655928 0.2641590
[7] 0.6108628 0.3877454

> R.avg    <- rank(U) # ties.method = "average"
> R.random <- rank(U, ties.method = "random")
> R.max    <- rank(U, ties.method = "max")
> R.min    <- rank(U, ties.method = "min")
> stopifnot(R.random == R.avg, R.max == R.avg, R.min == R.avg)
```

To simulate lack of measurement precision, we divide the unit interval into b bins of equal length and replace each original observation by the center of the bin to which it belongs:

```
> b <- 10 # number of bins
> (U.ties <- cut(U, breaks = 0:b/b, labels = 0.5:(b - 0.5)/b)) # a factor
```

```
[1] 0.95 0.45 0.25 0.75 0.15 0.25 0.65 0.35
Levels: 0.05 0.15 0.25 0.35 0.45 0.55 0.65 0.75 0.85 0.95
```

As one can see, the third and the sixth observations are now both equal to 0.25. The default setting `ties.method = "average"` then assigns to each of the two observations the average of the ranks they would obtain if there were no ties. In other words, the value 0.25 being the second smallest observation and appearing twice, the ranks assigned to the third and the sixth observations are $(2+3)/2 = 2.5$:

```
> rank(U.ties) # ties.method = "average"
```

```
[1] 8.0 5.0 2.5 7.0 1.0 2.5 6.0 4.0
```

Computing average ranks (also known as *mid-ranks*) is the default approach in nonparametric statistics to deal with ties. As seen above, two other solutions for assigning ranks to tied observations consist of computing maximum or minimum ranks instead of average ranks:

```
> rank(U.ties, ties.method = "max") # maximum rank for each tied observation
```

```
[1] 8 5 3 7 1 3 6 4
```

```
> rank(U.ties, ties.method = "min") # minimum rank for each tied observation
```

```
[1] 8 5 2 7 1 2 6 4
```

Notice the difference between the two previous outputs in the third and sixth components.

Yet another solution consists of assigning the ranks of tied observations pseudo-randomly, which thus depends on the seed of the underlying pseudo-random number generator:

```
> set.seed(8)
> rank(U.ties, ties.method = "random")
```

```
[1] 8 5 3 7 1 2 6 4
```

```
> set.seed(46)
> rank(U.ties, ties.method = "random")
```

```
[1] 8 5 2 7 1 3 6 4
```

As one can see, the ranks of the two tied observations (the third and the sixth), which, if there were no ties, should have been either 2 and 3 or 3 and 2, are assigned pseudo-randomly. □

When applying the same method to the component series of the multivariate random sample $\boldsymbol{X}_1, \ldots, \boldsymbol{X}_n$, the four rank-computing methods discussed in the previous example can lead to four different sets of multivariate scaled ranks of the form (4.10). Notice that only the use of maximum ranks preserves the equivalence between (4.9) and (4.10). Furthermore, the minimum, average, or maximum rank method can be regarded as being of the same nature in the sense that, unlike the random method, all three will result in ties in the multivariate scaled ranks.

Example 6.1.2 (Effect of Ties on Multivariate Scaled Ranks) To illustrate the effect of ties on multivariate scaled ranks, we first generate independent realizations from a bivariate Clayton copula with parameter $\theta = 5$ and then produce a discretized version of the generated data by proceeding as in Example 6.1.1 for each component sample:

```
> n <- 200
> set.seed(4809)
> U <- rCopula(n, claytonCopula(5))
> b <- 10 # number of bins
> U.ties <- (apply(U, 2, cut, breaks = 0:b/b, labels = FALSE) - 0.5) / b
```

Multivariate scaled ranks can then be computed using the function `pobs()` whose argument `ties.method` is passed on to the underlying call of `rank()` and thus controls the rank computation method. As illustrated by the following code, the function `pobs()` is indeed merely a convenient wrapper of `rank()`:

```
> ties.method <- "max" # can be changed to "min", "average" or "random"
> stopifnot(all.equal(pobs(U.ties, ties.method = ties.method),
                      apply(U.ties, 2, rank, ties.method = ties.method) /
                      (nrow(U.ties) + 1))) # check
```

Figure 6.1, produced with the following code, compares the multivariate scaled ranks computed from the original continuous sample, the multivariate scaled ranks computed from the discretized sample using the default `ties.method = "average"` and the multivariate scaled ranks computed from the discretized sample using `ties.method = "random"`:

```
> plot(pobs(U), xlab = "", ylab = "")
> plot(pobs(U.ties), xlim = 0:1, ylim = 0:1, xlab = "", ylab = "")
> set.seed(732)
> plot(pobs(U.ties, ties.method = "random"), xlab = "", ylab = "")
```

As one can see, the discretization step (here simulating lack of measurement precision, for instance) results in a loss of information. While, at first glance, the right plot seems to retain some of the characteristic features expected from a Clayton copula, it is not difficult to convince oneself that the middle plot could equally well result from the use of other copulas with a similar Kendall's tau as C_5^{C}. One can also

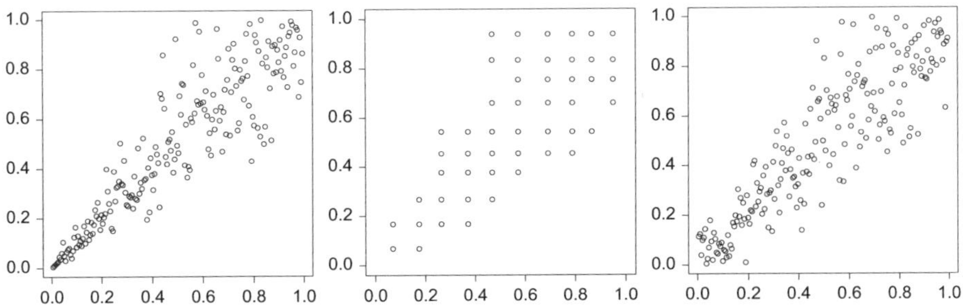

Fig. 6.1 Multivariate scaled ranks computed from the original continuous sample (left), from the discretized sample using the default `ties.method = "average"` (middle) and from the discretized sample using `ties.method = "random"` (right)

verify that the use of `ties.method = "min"` or `ties.method = "max"` leads to plots of the same nature as the middle one.

In the context of the current example, note that computing multivariate scaled ranks using `ties.method = "random"` is equivalent to the following: First, add to each discretized bivariate observation a realization of a bivariate random vector whose components are independent and whose margins are (for instance) $U(-a, a)$ with $0 < a \leq 1/(2b)$, and then compute bivariate scaled ranks from the resulting sample (which, by construction, does not contain ties anymore). This equivalence is illustrated by the following code:

```
> a <- runif(1, min = 0, max = 1/(2 * b)) # or a fixed number in (0, 1/(2b))
> set.seed(732)
> V <- pobs(U.ties + runif(2 * n, min = -a, max = a))
> set.seed(732)
> stopifnot(all.equal(V, pobs(U.ties, ties.method = "random"))) # check
```

As a consequence of the previous equivalence, the use of `ties.method = "random"` when computing multivariate scaled ranks does not actually do a better job than the use of `ties.method = "average"`. Roughly speaking, absence of information on local dependence due to discreteness is replaced by local independence. The latter is best seen by comparing the lower-left corners of the left-most and right-most plots of Fig. 6.1: The lower tail dependence characteristic of Clayton copulas present in the left panel disappeared completely in the right panel. □

The previous example illustrates well the effect of ties on dependence: The presence of ties can hide crucial features distinguishing one copula family from another. The effect increases with the amount of ties. As a consequence, inference on the unknown copula C is expected to become more difficult as the proportion of tied observations increases.

With inference under a parametric assumption on C as in Sect. 4.1 in mind, a natural question is how copula estimation is affected by the presence of ties. The following example attempts to empirically provide a first answer.

Example 6.1.3 (Estimation of Copula Parameters in the Presence of Ties) We consider a small simulation study involving only one-parameter copula families. We focus on maximum pseudo-likelihood estimation and on method-of-moments estimation based on Kendall's tau; see Sect. 4.1.2. The first family we consider is the Frank family. The precise setup is as follows:

```
> theta <- iTau(frankCopula(), tau = 0.25) # copula parameter
> fc25 <- frankCopula(theta) # corresponding copula object
```

Samples generated from the specified Frank copula will be discretized using the following function:

```
> ## Discretizes all components of U; returns a matrix of real numbers
> discrAll <- function(U, b)
      (apply(U, 2L, cut, breaks = 0:b/b, labels = FALSE) - 1/2) / b
```

Before continuing, note that, in the R package `copula`, when performing method-of-moments estimation based on Kendall's tau from data containing ties, Kendall's tau is estimated by means of its sample version adjusted for ties known as *tau b*; see Kendall (1945) and, for instance, Agresti (2010). Furthermore, whether `ties.method = "average"` or `ties.method = "random"` is used in `pobs()` to form the multivariate scaled ranks (4.10), the maximum pseudo-likelihood estimate for the hypothesized copula family is obtained through (4.11). In our experiments, numerical maximization will be performed with the BFGS optimization method. Whatever the estimation method, if `ties.method = "random"` is used, it is natural to carry out the estimation from the resulting tie-free pseudo-observations for different random seeds and to average the obtained estimates. To do so, we use the following wrapper of the function `fitCopula()`:

```
> ##' @title Breaking the ties m times and averaging the estimates
> ##' @param m number of replications
> ##' @param U.ties discretized sample
> ##' @param cop copula to be fitted
> ##' @param method fitting method of fitCopula()
> ##' @param optim.method optimization method of fitCopula()
> ##' @return average estimate over randomly broken ties when computing pobs()
> fitCopulaRand <- function(m, U.ties, cop, method, optim.method)
  {
      fit1 <- function() {
          V <- pobs(U.ties, ties.method = "random") # break ties at random
          fitCopula(cop, data = V, method = method,
                    optim.method = optim.method)@estimate # return param. est.
      }
      mean(replicate(m, fit1())) # average of estimates over m randomizations
  }
```

We then consider a function that generates one sample from a specified one-parameter copula and estimates its parameter from multivariate scaled ranks computed either from the initial continuous sample or from the discretized sample:

```
> ##' @title Parameter estimation from discrete data with various methods
> ##' @param n sample size
> ##' @param cop one-parameter copula object
> ##' @param discretize discretizing function
> ##' @param b number of bins
> ##' @param m number of randomizations
> ##' @param optim.method MPLE optimization method
> ##' @return parameter estimates for the different methods
> oneFit <- function(n, cop, discretize, b = 10, m = 30,
                     optim.method = "BFGS")
  {
      U <- rCopula(n, copula = cop) # a sample of size n from cop
      ## Rank-based fitting from the continuous sample
      V <- pobs(U) # pseudo-observations (no ties)
      mpl <- fitCopula(cop, data = V,
                       optim.method = optim.method)@estimate # MPLE
      itau <- fitCopula(cop, data = V,
                        method = "itau")@estimate # tau inversion
      ## Fitting from the discretized sample based on average ranks
      U.ties <- discretize(U, b = b) # discretize the sample
      W <- pobs(U.ties) # corresponding multivariate scaled average ranks
      mpl.ave  <- fitCopula(cop, data = W,
                            optim.method = optim.method)@estimate # MPLE
      itau.ave <- fitCopula(cop, data = W,
                            method = "itau")@estimate # tau inversion
      ## Average fits over randomized ranks based on the discretized sample
      mpl.rand  <- fitCopulaRand(m, U.ties, cop = cop, method = "mpl",
                                 optim.method = optim.method) # MPLE
      itau.rand <- fitCopulaRand(m, U.ties, cop = cop, method = "itau",
                                 optim.method = optim.method) # tau inversion
      ## Return the different estimates of theta
      c(mpl = mpl, mpl.ave = mpl.ave, mpl.rand = mpl.rand,
        itau = itau, itau.ave = itau.ave, itau.rand = itau.rand)
  }
```

The previous function is then executed 100 times with sample size `n = 100` and its run time is measured using `withTime()`:

```
> set.seed(2010)
> est.fc25 <- withTime(replicate(100, oneFit(n = 100, cop = fc25,
                                             discretize = discrAll)))
```

Recall that, because of `withTime()`, the output of `replicate()` is stored in `est.fc25$value` while the corresponding run time is stored in `est.fc25$sys.time`.

The simulations can be summarized using the following function which displays the copula family name, the population version of Kendall's tau, the run time, as well as the estimated relative biases, standard deviations and root mean squared errors (in percent) of the underlying estimators on the scale of Kendall's tau:

```
> ##' @title Relative bias, standard deviation and root mean squared error
> ##' @param est simulation object
> ##' @param cop underlying copula object
> ##' @return rel. bias, standard dev. and root mean squared error (in %)
> sumSims <- function(est, cop)
  {
      tau <- tau(cop) # population version of Kendall's tau
      cat(describeCop(cop, kind = "very short"),
          "with a Kendall's tau of", tau, "\n")
      print(est$sys.time) # print user time (first component of system.time())
      ## Estimates on Kendall's tau scale
      tau.n.s <- apply(est$value, 1:2, function(x) tau(setTheta(cop, x)))
      ## Relative biases on Kendall's tau scale
      bias <- (rowMeans(tau.n.s) - tau) / tau
      ## Standard deviations of estimates on Kendall's tau scale
      std <- apply(tau.n.s, 1, sd)
      ## Root mean squared errors on Kendall's tau scale
      rmse <- sqrt(rowMeans((tau.n.s - tau)^2))
      round(rbind(bias = bias, std = std, rmse = rmse) * 100, 2)
  }
> sumSims(est = est.fc25, cop = fc25)

Frank copula with a Kendall's tau of 0.25
User time:  54.8 sec
      mpl mpl.ave mpl.rand  itau itau.ave itau.rand
bias 0.28    0.39    -2.08 -0.52     8.09     -2.72
std  6.82    6.87     6.70  6.84     7.53      6.77
rmse 6.78    6.84     6.69  6.80     7.76      6.77
```

The results for a Frank copula with moderate dependence can be obtained as follows:

```
> fc5 <- frankCopula(iTau(frankCopula(), tau = 0.5))
> set.seed(2010)
> est.fc5 <- withTime(replicate(100, oneFit(n = 100, cop = fc5,
                                            discretize = discrAll)))
> sumSims(est = est.fc5, cop = fc5)

Frank copula with a Kendall's tau of 0.5
User time:  54.5 sec
       mpl mpl.ave mpl.rand  itau itau.ave itau.rand
bias -0.47   -0.48    -2.78 -0.62     7.93     -2.86
std   5.27    5.35     5.21  5.41     5.90      5.30
rmse  5.25    5.33     5.36  5.39     7.08      5.46
```

For a Frank copula with strong dependence, we obtain:

```
> fc75 <- frankCopula(iTau(frankCopula(), tau = 0.75))
> set.seed(2010)
> est.fc75 <- withTime(replicate(100, oneFit(n = 100, cop = fc75,
                                             discretize = discrAll)))
> sumSims(est = est.fc75, cop = fc75)
```

```
Frank copula with a Kendall's tau of 0.75
User time:  58.7 sec
       mpl mpl.ave mpl.rand  itau itau.ave itau.rand
bias -0.76   -1.90    -4.65 -0.20     6.44     -4.15
std   2.74    2.78     2.60  2.79     2.87      2.65
rmse  2.78    3.11     4.34  2.78     5.61      4.08
```

Let us also consider the Clayton copula family:

```
Clayton copula with a Kendall's tau of 0.25
User time:  46.7 sec
      mpl mpl.ave mpl.rand  itau itau.ave itau.rand
bias 3.29    9.26    -7.75 -0.99     5.92     -4.62
std  6.15    6.06     5.18  6.09     6.69      6.02
rmse 6.18    6.46     5.50  6.07     6.81      6.10

Clayton copula with a Kendall's tau of 0.5
User time:  54.7 sec
       mpl mpl.ave mpl.rand  itau itau.ave itau.rand
bias -0.16    0.41   -12.15 -0.65     6.66     -3.93
std   4.24    4.53     3.40  4.83     5.16      4.62
rmse  4.22    4.51     6.95  4.82     6.12      5.00

Clayton copula with a Kendall's tau of 0.75
User time:  49 sec
       mpl mpl.ave mpl.rand itau itau.ave itau.rand
bias -1.07   -2.64   -15.94 0.13     5.88     -4.62
std   2.58    3.16     2.18 2.88     2.85      2.54
rmse  2.69    3.71    12.15 2.87     5.24      4.29
```

Similarly, for the Gumbel–Hougaard family, we obtain:

```
Gumbel copula with a Kendall's tau of 0.25
User time:  92.4 sec
      mpl mpl.ave mpl.rand  itau itau.ave itau.rand
bias 1.26    7.61    -6.23 -1.97     5.34     -5.11
std  6.52    6.33     5.85  6.56     6.91      6.22
rmse 6.49    6.58     6.02  6.55     7.01      6.32

Gumbel copula with a Kendall's tau of 0.5
User time:  87.6 sec
       mpl mpl.ave mpl.rand  itau itau.ave itau.rand
bias -0.70    1.03    -8.40 -1.80     5.35     -5.12
std   5.36    5.09     4.52  5.45     5.62      5.05
rmse  5.34    5.10     6.16  5.50     6.20      5.64

Gumbel copula with a Kendall's tau of 0.75
User time:  104.3 sec
       mpl mpl.ave mpl.rand  itau itau.ave itau.rand
bias -1.24   -1.98   -10.67 -1.03     4.57     -5.84
std   2.92    3.16     2.37  3.20     3.31      3.04
rmse  3.05    3.48     8.34  3.28     4.75      5.32
```

Finally, we redo the previous simulations but this time with a ties-generating mechanism that affects only the first variable:

```
> ## An alternative definition of the discretizing function
> ## producing ties only in the first component sample
> discrFirst <- function(U, b)
      cbind(cut(U[,1], breaks = 0:b/b, labels = 0.5:(b - 0.5)/b), U[,2])
```

For the Clayton family, for instance, we obtain:

```
Clayton copula with a Kendall's tau of 0.25
User time:  43.3 sec
       mpl mpl.ave mpl.rand  itau itau.ave itau.rand
bias 3.29     9.26    -7.75 -0.99     5.92     -4.62
std  6.15     6.06     5.18  6.09     6.69      6.02
rmse 6.18     6.46     5.50  6.07     6.81      6.10

Clayton copula with a Kendall's tau of 0.5
User time:  50.9 sec
        mpl mpl.ave mpl.rand  itau itau.ave itau.rand
bias -0.16     0.41   -12.15 -0.65     6.66     -3.93
std   4.24     4.53     3.40  4.83     5.16      4.62
rmse  4.22     4.51     6.95  4.82     6.12      5.00

Clayton copula with a Kendall's tau of 0.75
User time:  47.4 sec
        mpl mpl.ave mpl.rand itau itau.ave itau.rand
bias -1.07    -2.64   -15.94 0.13     5.88     -4.62
std   2.58     3.16     2.18 2.88     2.85      2.54
rmse  2.69     3.71    12.15 2.87     5.24      4.29
```

It thus seems that, overall, copula parameter estimation in the presence of ties suffers from a larger bias than in the absence of ties. The variability of the estimation appears to remain roughly unaffected, though. The effect of the bias can become catastrophic in the case of strong dependence, particularly in the case of maximum pseudo-likelihood estimation based on multivariate scaled random ranks. Overall, maximum pseudo-likelihood estimation based on multivariate scaled average ranks seems to give the best results in terms of root mean squared error. As far as method-of-moments estimation based on Kendall's tau is concerned, the use of random ranks seems slightly preferable to the use of average ranks.

An alternative to maximum pseudo-likelihood estimation based on multivariate scaled average ranks is suggested in Li et al. (2016) who propose to treat tied observations as interval censored observations when constructing the pseudo-likelihood maximized in (4.11). A simulation study in the bivariate case seems to indicate that the method is unbiased regardless of the strength of dependence. Attractive as the method appears to be in the bivariate case, it is not much more than a proof of concept at the moment because implementing it in dimension d is challenging due to the required cross partial derivatives of order up to d of the copula under consideration. □

As we have seen in Chap. 5, parametric estimation is often accompanied by graphical diagnostics, formal tests, and model selection. When dealing with ties, data analysis is even more complicated due to the fact that the vast majority of inference procedures on C proposed in the literature are theoretically studied and implemented under the assumption of no ties in any of the d component series of the given data $\boldsymbol{X}_1, \ldots, \boldsymbol{X}_n$, that is, no ties in the component samples of the multivariate scaled ranks (4.10). A noticeable exception is the test of extreme-value dependence proposed by Cormier et al. (2014, Section 6) which can be adapted for discontinuous margins.

As far as the inference procedures described in Chap. 5 are concerned, unfortunately, not all provide meaningful results when applied to data sets containing ties (unless the proportion of ties is negligible). This is illustrated in the following example.

Example 6.1.4 (Tests Not Adapted for Ties) We consider samples generated from a Gumbel–Hougaard copula which are discretized as in Example 6.1.3 with the function `discrAll()`. The test of exchangeability based on (5.4), the test of extreme-value dependence based on (5.8), and the parametric bootstrap and multiplier goodness-of-fit tests based on (5.9) are then applied to each generated sample. The setup is as follows:

```
> ##' @title Auxiliary function for computing the empirical levels of tests
> ##'        of exchangeability and extreme-value dependence, and
> ##'        parametric bootstrap and multiplier goodness-of-fit tests
> ##' @param n sample size
> ##' @param cop copula object (from which the data is generated)
> ##' @param discretize discretization function
> ##' @param b number of bins
> ##' @return p-values (numeric(4))
> pvalTies <- function(n, cop, discretize, b)
  {
      U.ties <- discretize(rCopula(n, cop), b = b) # binned samples
      c(exch = exchTest(U.ties, ties = FALSE)$p.value,
        ev   = evTestC(U.ties)$p.value,
        pb   = gofCopula(cop, U.ties, optim.method = "BFGS",
                         ties = FALSE)$p.value,
        mult = gofCopula(cop, U.ties, optim.method = "BFGS",
                         sim = "mult")$p.value)
  }
```

The meaning of the argument `ties` in the call of `exchTest()` and in the first call of `gofCopula()` will be discussed in Example 6.1.5 below. Note that, in accordance with the empirical conclusions of Example 6.1.3, we base the goodness-of-fit tests on the default maximum pseudo-likelihood estimator. The auxiliary function `pvalTies()` is then executed 100 times with sample size `n = 100` and its run time is measured using `withTime()`:

```
> gc <- gumbelCopula(iTau(gumbelCopula(), tau = 0.5)) # Gumbel-Hougaard copula
> set.seed(4478)
> pv <- withTime(replicate(100, pvalTies(n = 100, cop = gc,
                                         discretize = discrAll, b = 10)))
```

The run time (measured as user time) is:

```
> pv$sys.time # user time
```

```
User time:   2254.1 sec
```

The empirical levels corresponding to the nominal levels 0.01, 0.05, and 0.1, respectively, are obtained as follows:

```
> alpha <- c(0.01, 0.05, 0.1) # nominal levels
> rbind(nom.level = alpha,
        emp.level.exch = ecdf(pv$value["exch",])(alpha),
        emp.level.ev   = ecdf(pv$value["ev",  ])(alpha),
        emp.level.pb   = ecdf(pv$value["pb",  ])(alpha),
        emp.level.mult = ecdf(pv$value["mult",])(alpha))
```

```
               [,1] [,2] [,3]
nom.level      0.01 0.05 0.10
emp.level.exch 1.00 1.00 1.00
emp.level.ev   1.00 1.00 1.00
emp.level.pb   0.44 0.60 0.79
emp.level.mult 0.00 0.00 0.00
```

The above results show that these four tests do not maintain their level in the presence of ties, as, otherwise, the p-values would have been approximately uniformly distributed over the unit interval since the samples are generated from a Gumbel–Hougaard copula (which is exchangeable and an extreme-value copula). While the first three tests are clearly too liberal (that is, they reject the null hypothesis too often when it is true), the last one is way too conservative (it will hardly ever reject the null hypothesis when it is true). Indeed:

```
> summary(pv$value["mult",])
```

```
  Min. 1st Qu.  Median    Mean 3rd Qu.    Max.
0.4680  0.8559  0.9281  0.8984  0.9648  0.9965
```

Note in passing that the test of extreme-value dependence based on (5.7) is also too liberal in the presence of ties; see Genest et al. (2011). □

Unfortunately, most of the statistical tests presented in Chap. 5 do not appear to provide meaningful results when applied to data sets containing ties (unless, as mentioned earlier, the proportion of ties is negligible). To carry out inference using these tests, there are thus two possible approaches: (i) transform the original data set so that the resulting multivariate scaled ranks are free of ties, or (ii) adapt the inference procedures so that they give meaningful results in the presence of ties.

Approach (i) could be achieved by deleting multivariate observations with at least one tied coordinate before carrying out the inference. This rather extreme approach is not advisable as discussed in Genest et al. (2011, Section 2) since it will not only result in a reduced sample size and thus in less efficient statistical procedures, but it may also alter the underlying dependence structure. A second way of carrying out Approach (i) consists of breaking ties at random, which is sometimes referred to as *jittering*, that is, adding independent realizations of a small continuous random variable to all univariate observations (so that the ranks of untied observations remain unaffected). As we have already seen in Example 6.1.2, computing multivariate scaled ranks from jittered data is equivalent to computing multivariate scaled ranks from the original data using `ties.method = "random"`. As we have hinted at in Example 6.1.3, because jittering depends on the seed, the analysis should be performed multiple times.

In the presence of a moderate amount of ties, inference on C based on multivariate scaled ranks computed using `ties.method = "random"` is considered in Kojadinovic and Yan (2010, Section 5.1) and Bücher and Kojadinovic (2015, Section 5.2). The proposed approach consists of carrying out the inference for many different seeds with the hope that the resulting sets of multivariate scaled ranks will not be too different from the pseudo-observations that would have been obtained if there were no ties in the data. For a given inference procedure, instead of looking at the average of the results over all seeds, the distribution of the results is considered. The motivation for this stems from the following heuristic: If the p-value of a given test is very small (respectively, large) for all seeds, then it is meaningful to reject (respectively, not reject) the corresponding null hypothesis for the original data set.

For data sets containing a large proportion of ties, such as the one generated in Example 6.1.2, the previous approach is not expected to work well given the huge number of possible sets of multivariate scaled ranks. For a given test, the set of p-values obtained from different seeds is thus likely to be large and to include both very small and large p-values, thereby making it impossible to draw a conclusion for the original data set according to the aforementioned heuristic.

An extension of the aforementioned method is proposed in Pappadà et al. (2017). In the context of a study focusing on structural risk, the authors suggest a jittering which consists of adding to each initial d-dimensional observation a realization of a d-dimensional random vector with copula J and univariate marginal distributions $\mathrm{U}(0, \Delta_1), \dots, \mathrm{U}(0, \Delta_d)$, where, for any $j \in \{1, \dots, d\}$, $\Delta_j > 0$ is smaller than or equal to the smallest absolute difference between any two distinct observations in the jth component sample $X_{1j}, \dots, X_{nj}$ of the data. Three choices are considered for J: the independence copula Π (which leads to the usual jittering illustrated in Example 6.1.2), the comonotone copula M, and a convex combination of the two, given by $C = \lambda M + (1 - \lambda)\Pi$ for $\lambda \in (0, 1)$. Note that, in the bivariate case, $\tau = \lambda(\lambda + 2)/3 \in [0, 1]$ so that $\lambda = -1 + \sqrt{1 + 3\tau}$, where τ is Kendall's tau of C. Assuming that the data at hand are positively dependent, the latter property is exploited in the suggested jittering when $d = 2$ by setting λ according to the sample value of Kendall's tau.

Let us now turn our attention to Approach (ii). Developing tie-adapted versions of the tests presented in Chap. 5 is difficult. This is mostly a consequence of the difficulty of extending certain key theoretical results (such as the weak convergence of an appropriate version of the empirical copula process, see Sect. 4.2.1) from the case of continuous margins to the case of discontinuous margins. Recent promising results in that direction are due to Genest et al. (2017) and might allow for a sound theoretical derivation of tie-adapted copula inference procedures in the near future.

A more empirical approach is considered in Kojadinovic (2017b). It consists of trying to modify the resampling schemes of existing tests so that they are not too liberal anymore in the presence of ties (as is the case, for instance, of some of the tests considered in Example 6.1.4). Such modifications are specifically proposed for the test of exchangeability based on (5.4), the test of radial symmetry based on (5.5), the test of extreme-value dependence based on (5.7), and the parametric bootstrap-based goodness-of-fit test relying on (5.9). By construction, the resulting tie-adapted tests should never be too liberal. They may, however, be too conservative (and thus lack power) when the sample size is small or the amount of ties is large. For the parametric bootstrap-based goodness-of-fit test relying on (5.9), for example, the key idea, initially appearing in Bücher and Kojadinovic (2015), consists of imposing on each parametric bootstrap sample the same "tie structure" as the one found in the original sample. Reassuringly, when there are no ties, the modified test leads exactly to the same computations as those in Algorithm 5.2.11; see Kojadinovic (2017b, Section 6) for more details. While conditions under which such a tie-adapted parametric bootstrap-based test is asymptotically valid under the null hypothesis remain to be established, the following example shows empirically that it seems to maintain its level.

Example 6.1.5 (Parametric Bootstrap-Based Goodness-of-Fit Test Adapted for Ties) We consider a similar setup as in Example 6.1.4, the crucial difference being that the parametric bootstrap-based goodness-of-fit test is now called with argument `ties = TRUE`, which results in the execution of its tie-adapted version:

```
> ##' @title Auxiliary function for computing the empirical levels of the
> ##'        parametric bootstrap-based goodness-of-fit test
> ##' @param n sample size
> ##' @param cop copula object (from which the data is generated)
> ##' @param discretize discretization function
> ##' @param b number of bins
> ##' @return p-value (numeric(1))
> pvalPB <- function(n, cop, discretize, b)
  {
      U <- rCopula(n, copula = cop)
      U.ties <- discretize(U, b = b)
      gofCopula(cop, x = U.ties, optim.method = "BFGS", ties = TRUE)$p.value
  }
```

Note that the default value of the argument `ties` is `NA`, which forces the presence of ties in the component samples to be checked for automatically: If ties are detected (respectively, not detected), `ties` is set to `TRUE` (respectively, `FALSE`) and the tie-adapted (respectively, standard) test is applied. For `simulation = "mult"`, however, the value of the argument `ties` is currently ignored.

The function `pvalPB()` is then executed 100 times on samples of size `n = 100`. This is computationally demanding because each call to `gofCopula()` generates (by default) `N = 1000` parametric bootstrap replications of the test statistic (see Algorithm 5.2.11), resulting in 100,000 calls to `rCopula()`, `fitCopula()`, and `gofTstat()`:

```
> set.seed(8848)
> pvAll <- withTime(replicate(100, pvalPB(n = 100, cop = gc,
                                          discretize = discrAll, b = 10)))
> pvAll$sys.time # run time

User time:  2180.9 sec
```

A similar experiment based on the alternative discretizing function `discrFirst()` (which produces ties only in the first component sample) is also carried out:

```
> set.seed(3298)
> pvFirst <- withTime(replicate(100, pvalPB(n = 100, cop = gc,
                                            discretize = discrFirst, b = 10)))
> pvFirst$sys.time # run time

User time:  2048.8 sec
```

Given the small number of replications, there is no evidence that the adapted goodness-of-fit test does not hold its level in the considered settings:

```
> alpha <- c(0.01, 0.05, 0.1) # nominal levels
> rbind(nom.level = alpha, emp.level.all = ecdf(pvAll$value)(alpha),
        emp.level.first = ecdf(pvFirst$value)(alpha))

                [,1] [,2] [,3]
nom.level       0.01 0.05 0.10
emp.level.all   0.00 0.07 0.13
emp.level.first 0.01 0.07 0.13
```

Empirical levels computed from 1000 replications are reported in Tables 10 and 11 of Kojadinovic (2017b) and empirically confirm that the tie-adapted parametric bootstrap maintains its level for both even and uneven, coarse and fine discretizations, as well as in the absence of ties. Provided the dependence is not strong, the behavior of the test seems to remain adequate even when the tie-adapted parametric bootstrap is based on inverting Kendall's tau. □

Because it essentially relies on parametric estimation, one of the few inference procedures on C whose results might remain meaningful in the presence of ties is cross-validation as implemented in the function `xvCopula()`; see Sect. 5.3.

Example 6.1.6 (Effect of Ties on Cross-Validation) In order to empirically investigate the effect of ties on the results of cross-validation, let us carry out a small experiment. It consists of applying the following function to samples generated from Clayton, Gumbel–Hougaard, or Frank copulas discretized as in the previous examples. The setup is as follows:

```
> ##' @title 10-fold cross-validation in the presence of ties
> ##' @param n sample size
> ##' @param cop copula object (from which the data is generated)
> ##' @param discretize discretization function
> ##' @param b number of bins
> ##' @param optim.method optimization method for MPLE
> ##' @return label of the copula with the highest xv score
> xv <- function(n, cop, discretize = discrAll, b, optim.method = "BFGS")
  {
      U <- rCopula(n, copula = cop)
      U.ties <- discretize(U, b = b)
      score <- c(xvCopula(claytonCopula(), x = U.ties, k = 10,
                          optim.method = optim.method),
                 xvCopula(gumbelCopula(), x = U.ties, k = 10,
                          optim.method = optim.method),
                 xvCopula(frankCopula(), x = U.ties, k = 10,
                          optim.method = optim.method))
      c("C", "GH", "F")[which(score == max(score))]
  }
```

Note that, following Example 6.1.3, we carry out cross-validation using the maximum pseudo-likelihood estimator based on multivariate scaled average ranks. We then execute the function `xv()` 100 times on samples of size `n = 100` generated from each of the three copulas and report the number of times `"C"`, `"GH"`, or `"F"` is returned. When sampling from the Clayton copula with a Kendall's tau of 0.5, we obtain:

```
> cc <- claytonCopula(iTau(claytonCopula(), tau = 0.5))
> set.seed(2885)
> withTime(table(replicate(100, xv(n = 100, cop = cc, b = 20))))

  C
100
User time:  41.2 sec
```

When sampling from the Gumbel–Hougaard copula with a Kendall's tau of 0.5, we get:

```
> set.seed(2885)
> withTime(table(replicate(100, xv(n = 100, cop = gc, b = 20))))

 C  F GH
 4 35 61
User time:  38.4 sec
```

Finally, when sampling from the Frank copula with a Kendall's tau of 0.5, we obtain:

```
> fc <- frankCopula(iTau(frankCopula(), tau = 0.5))
> set.seed(2885)
> withTime(table(replicate(100, xv(n = 100, cop = fc, b = 20))))

 C  F GH
 8 88  4
User time:  42 sec
```

Hence, overall, cross-validation seems to be doing a reasonable job at selecting the true copula.

Reproducing the previous computations with a smaller number of bins, say 10 (thus increasing the proportion of ties), we observe that, as expected, the selection of the true copula family becomes less likely. □

It thus appears from the previous example that the ranking induced by `xvCopula()` can be meaningful even in the presence of ties. However, one needs to be careful. As we have seen earlier, rounding or lack of measurement precision can hide crucial dependence features, in particular, in the tails. In the presence of ties, expert knowledge thus needs to be given even more weight than in the absence of ties. For instance, if the modeled random phenomenon is known to involve a copula with a certain type of tail dependence, one should not rely on the results of the inference if they do not point in the same direction.

We end this section with the analysis of a well-known data set.

Example 6.1.7 (Analysis of the `loss` *Insurance Data)* The Loss/ALAE insurance data are frequently used for illustration purposes in the copula literature; see, for instance, Frees and Valdez (1998), Ben Ghorbal et al. (2009), Kojadinovic and Yan (2010), and Bücher and Kojadinovic (2015). Available in the `copula` package, the data set consists of 1500 claims of an insurance company. Two variables are observed for each claim: `loss`, an indemnity payment, and `alae`, the corresponding allocated loss adjustment expense. Among the 1500 claims, 34 are censored. As is well known from survival analysis, ignoring censored observations is likely to lead to biased conclusions. Yet, because the inference procedures investigated in this section cannot account for censoring, we shall, for illustration purposes only, restrict ourselves to the 1466 uncensored claims here:

```
> data(loss)
> X <- as.matrix(subset(loss, censored == 0, select = c("loss", "alae")))
```

Ties affect mostly the first variable:

```
> ## Percentages of ties
> 100 * apply(X, 2, function(x) 1 - length(unique(x))/length(x))
```

```
     loss      alae
63.096862  4.433834
```

This can also be seen from the corresponding multivariate scaled average ranks and normal scores shown in Fig. 6.2:

```
> U <- pobs(X)
> plot(U, xlab = quote(U[1]~~"(Loss)"), ylab = quote(U[2]~~"(ALAE)"))
> Y <- qnorm(U)
> plot(Y, xlab = quote(Y[1]~~"(Loss)"), ylab = quote(Y[2]~~"(ALAE)"))
```

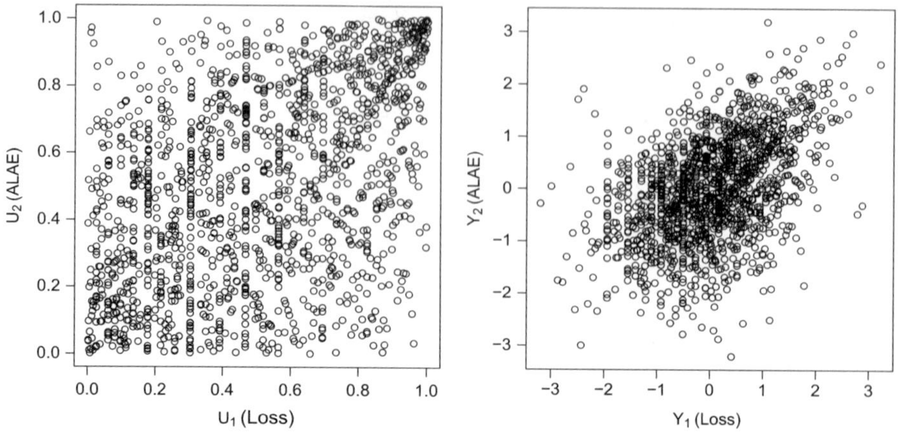

Fig. 6.2 Scatter plot of the multivariate scaled average ranks (left) and of the corresponding normal scores (right) of the Loss/ALAE data

We start by applying the tie-adapted tests of exchangeability, radial symmetry, and extreme-value dependence based on (5.4), (5.5), and (5.7), respectively:

```
> set.seed(3070)
> withTime(round(c(  exchTest(X, ties = TRUE)$p.value,
                    radSymTest(X, ties = TRUE)$p.value,
                       evTestK(X, ties = TRUE)$p.value), 4))

[1] 0.0534 0.0005 0.8375
User time:  206.5 sec
```

There is thus no strong evidence against exchangeability and extreme-value dependence in the data, while there is strong evidence against radial symmetry. These results are in accordance with the symmetry with respect to the main diagonal, asymmetry with respect to the secondary diagonal, and the upper tail dependence seemingly appearing in both plots in Fig. 6.2.

The above conclusions suggest to consider an exchangeable extreme-value copula as candidate model. Given the strong similarity among the existing bivariate parametric exchangeable extreme-value copula families, see Sect. 3.3 and, in particular, Example 3.3.4, the Gumbel–Hougaard copula family appears to be a natural choice.

For illustration purposes, besides the Gumbel–Hougaard family, we also carry out goodness-of-fit testing for the survival Clayton, Frank, Plackett, and normal families:

```
> ## Goodness-of-fit testing
> set.seed(4634)
> optim.method <- "BFGS" # the numerical optimization method for MPLE
> withTime(
      round(c(gofCopula(gumbelCopula(), x = X, ties = TRUE,
                        optim.method = optim.method)$p.value,
```

```
              gofCopula(rotCopula(claytonCopula()), x = X, ties = TRUE,
                        optim.method = optim.method)$p.value,
              gofCopula(frankCopula(), x = X, ties = TRUE,
                        optim.method = optim.method)$p.value,
              gofCopula(plackettCopula(), x = X, ties = TRUE,
                        optim.method = optim.method)$p.value,
              gofCopula(normalCopula(), x = X, ties = TRUE,
                        optim.method = optim.method)$p.value), 4)
  )
```

```
[1] 0.1683 0.0005 0.0005 0.0005 0.0005
User time:  790.1 sec
```

Finally, we perform model selection based on `xvCopula()`:

```
> ## Model selection
> set.seed(4807)
> k <- 50 # for k-fold cross-validation
> withTime(
      round(c(xvCopula(gumbelCopula(), x = X, k = k,
                       optim.method = optim.method),
              xvCopula(rotCopula(claytonCopula()), x = X, k = k,
                       optim.method = optim.method),
              xvCopula(frankCopula(), x = X, k = k,
                       optim.method = optim.method),
              xvCopula(plackettCopula(), x = X, k = k,
                       optim.method = optim.method),
              xvCopula(normalCopula(), x = X, k = k,
                       optim.method = optim.method)), 1)
  )
```

```
[1] 189.9 182.2 160.4 161.6 169.4
User time:  27.9 sec
```

As one can see, the Gumbel–Hougaard family is the only one not rejected at the, say, 1% level. This extreme-value family is also ranked first by the cross-validation criterion.

All of our inference procedures thus suggest to select the Gumbel–Hougaard family as a model for the dependence which is in accordance with the results obtained, for instance, in Chen and Fan (2005), Genest et al. (2006), and Kojadinovic and Yan (2010). Notice that, with `ties = FALSE`, the p-value returned by the test of exchangeability based on (5.4) is the smallest possible (that is, 0.5/1001) which, to an unwary user, would strongly suggest to discard the Gumbel–Hougaard family:

```
> withTime(exchTest(X, ties = FALSE))
```

```
        Test of exchangeability for bivariate copulas with argument
        'm' set to 0

data:  X
```

```
statistic = 0.14727, p-value = 0.0004995

User time:  2.4 sec
```

The latter result illustrates again the fact that the presence of ties should not be ignored when carrying out copula inference. □

6.2 Selected Copula Tests and Models for Time Series

Assume that one is given a d-dimensional sample $\boldsymbol{X}_1, \ldots, \boldsymbol{X}_n$ and considers applying some of the statistical procedures described in Chaps. 4 and 5. As mentioned therein, for these statistical techniques to provide fully valid results, $\boldsymbol{X}_1, \ldots, \boldsymbol{X}_n$ need to be iid and their common unknown df H needs to be continuous.

The iid assumption can fail in various ways. If $\boldsymbol{X}_1, \ldots, \boldsymbol{X}_n$ cannot be considered as a stretch from a stationary multivariate time series, the use of the inference procedures presented in Chap. 5 on $\boldsymbol{X}_1, \ldots, \boldsymbol{X}_n$ is meaningless. As we shall see, the assumption of stationarity can be formally tested by proceeding, for instance, as in Sect. 6.2.1. Clearly, even if $\boldsymbol{X}_1, \ldots, \boldsymbol{X}_n$ can be assumed to be a stretch from a stationary multivariate time series, $\boldsymbol{X}_1, \ldots, \boldsymbol{X}_n$ can still fail to be serially independent. The presence of mild serial dependence will not affect, asymptotically, the point-estimation of parameter values when fitting copula families to $\boldsymbol{X}_1, \ldots, \boldsymbol{X}_n$. However, it will result in the underestimation of the corresponding standard errors as well as in underestimated approximate p-values when applying the statistical tests of Chap. 5, which tend to be too liberal in this case. It is thus crucial to be able to assess whether $\boldsymbol{X}_1, \ldots, \boldsymbol{X}_n$ are serially independent. Several approaches to do so will be introduced in Sect. 6.2.2.

The fact that the statistical procedures of Chap. 5 should not be applied to $\boldsymbol{X}_1, \ldots, \boldsymbol{X}_n$ in the presence of serial dependence may actually not be an issue of strong practical importance. Indeed, when dealing with time series, the interest often lies in forecasting. To this end, one does not need to model the distribution of the random vector $\boldsymbol{X}$ of which $\boldsymbol{X}_1, \ldots, \boldsymbol{X}_n$ are serially dependent copies, but rather the conditional distribution of $\boldsymbol{X}_i$ given the information generated by the past values $\boldsymbol{X}_1, \ldots, \boldsymbol{X}_{i-1}$. A brief overview of the main copula-based models for this conditional distribution will be given in Sect. 6.2.3 with particular emphasis on applications in finance and econometrics.

6.2.1 *Tests of Stationarity*

The first step prior to any modeling is to assess whether $\boldsymbol{X}_1, \ldots, \boldsymbol{X}_n$ can be considered as a stretch from a stationary multivariate time series $(\boldsymbol{X}_i)_{i \in \mathbb{Z}}$. A time series $(\boldsymbol{X}_i)_{i \in \mathbb{Z}}$ is said to be *stationary* if, for any $k \in \mathbb{N}$ and $m \in \mathbb{Z}$, the random vector $(\boldsymbol{X}_1, \ldots, \boldsymbol{X}_k)$ and the random vector $(\boldsymbol{X}_{1+m}, \ldots, \boldsymbol{X}_{k+m})$ have the same distribution (such a property is sometimes referred to as strong stationarity in the literature). Non-stationarity such as the presence of a trend or a change in the variance could be detected to some extent by plotting the d component series. To detect or confirm more subtle departures from stationarity, the use of formal tests is recommended. Several genuine tests of stationarity for univariate time series are available in R: The tests of Priestley and Subba Rao (1969), Nason (2013), and Cardinali and Nason (2013) are implemented in the R packages `fractal` of Constantine and Percival (2016), `locits` of Nason (2016), and `costat` of Nason and Cardinali (2013), respectively. The finite-sample behavior of the first test is particularly disappointing. Overall, from the extensive simulations carried out in Bücher et al. (forthcoming), none of the three tests appear to maintain their levels in general, especially when applied to non-Gaussian time series.

As an imperfect alternative to testing in full generality whether $\boldsymbol{X}_1, \ldots, \boldsymbol{X}_n$ is a stretch from a stationary time series, one could consider tests particularly sensitive to departures from the hypothesis

$$\mathcal{H}_0 : \text{There exists a df } H \text{ such that } \boldsymbol{X}_1, \ldots, \boldsymbol{X}_n \text{ have df } H. \tag{6.1}$$

Such statistical procedures are commonly referred to as *tests for change-point detection*; see, for instance, Csörgö and Horváth (1997) and Aue and Horváth (2013) for an overview of the literature.

We shall now present a nonparametric test with good sensitivity to many alternatives to $\mathcal{H}_0$. For any $1 \le k \le l \le n$, consider the subsample $\boldsymbol{X}_k, \ldots, \boldsymbol{X}_l$ obtained from $\boldsymbol{X}_1, \ldots, \boldsymbol{X}_n$ and let

$$H_{k:l}(\boldsymbol{x}) = \frac{1}{l-k+1} \sum_{i=k}^{l} \mathbf{1}(\boldsymbol{X}_i \le \boldsymbol{x}), \quad \boldsymbol{x} \in \mathbb{R}^d,$$

be the empirical df computed from $\boldsymbol{X}_k, \ldots, \boldsymbol{X}_l$ with the convention that $H_{k:l} = 0$ if $l < k$. Following Csörgö and Horváth (1997) and Gombay and Horváth (1999), a broad class of nonparametric tests of stationarity particularly sensitive to departures from $\mathcal{H}_0$ can be derived from the empirical process

$$\mathbb{D}_n^H(t, \boldsymbol{x}) = \sqrt{n}\lambda_n(0,t)\lambda_n(t,1)\big(H_{1:\lfloor nt \rfloor}(\boldsymbol{x}) - H_{(\lfloor nt \rfloor+1):n}(\boldsymbol{x})\big), \quad (t, \boldsymbol{x}) \in [0,1] \times \mathbb{R}^d, \tag{6.2}$$

where $\lambda_n(t, t') = (\lfloor nt' \rfloor - \lfloor nt \rfloor)/n$ for $0 \le t \le t' \le 1$ and where $\lfloor \cdot \rfloor$ denotes the floor function. The normalizing term $\lambda_n(0,t)\lambda_n(t,1)$ is typical of cumulative sum change-point procedures to which the above class of tests belongs. A meaningful test statistic is then

$$S_n^H = \sup_{t \in [0,1]} \int_{\mathbb{R}^d} \big(\mathbb{D}_n^H(t, \boldsymbol{x})\big)^2 \mathrm{d}H_{1:n}(\boldsymbol{x}) = \max_{1 \le k \le n-1} S_{n,k}^H, \tag{6.3}$$

where

$$S_{n,k}^H = \frac{1}{n} \sum_{i=1}^{n} \left(\mathbb{D}_n^H(k/n, \boldsymbol{X}_i)\right)^2, \tag{6.4}$$

and was, for instance, used in Holmes et al. (2013) for independent $\boldsymbol{X}_1, \ldots, \boldsymbol{X}_n$. Under stationarity, the difference between $H_{1:k}$ and $H_{(k+1):n}$ should be small for all $k \in \{1, \ldots, n-1\}$, which implies that S_n^H should be small. Large values of S_n^H provide evidence against stationarity.

The computation of an approximate p-value for the test based on S_n^H relies on a resampling procedure taking into account the strength of the serial dependence in $\boldsymbol{X}_1, \ldots, \boldsymbol{X}_n$. Its asymptotic validity under stationarity and when H in (6.1) has continuous margins can be shown using results from Bücher and Kojadinovic (2016) for short range serial dependence; see also Bücher et al. (forthcoming, Section 2.2).

Example 6.2.1 (Test of Stationarity Based on S_n^H) We begin by testing the stationarity of the trivariate time series of financial log-returns `rdj` considered initially in Example 4.1.3. To plot the component series against time, we first build an `xts` object representing the trivariate time series with the R package `xts` of Ryan and Ulrich (2017):

```
> data(rdj)
> library(xts)
> Xrdj <- xts(rdj[,-1], order.by = rdj[,1])
```

Plots of the three component series against time are shown in Fig. 6.3 based on `plot.zoo()`:

```
> plot.zoo(Xrdj, main = "", xlab = "", mar = c(0, 7, 0, 2.1))
```

The graphs seem to reveal an increased volatility towards the end of the observation period. Whether stationarity should be rejected based on these plots is, however, unclear.

The test based on S_n^H is implemented in the function `cpDist()` of the R package `npcp` of Kojadinovic (2017a). The computation of an approximate p-value for S_n^H requires the knowledge of a bandwidth parameter corresponding to the argument `b` in `cpDist()`. To have `b` estimated from the available (possibly serially dependent) data, we set `b` to `NULL`. As the test is based on resampling, the seed is set to ensure reproducibility:

```
> library(npcp)
> set.seed(981)
> (res <- withTime(cpDist(Xrdj, b = NULL)))
```

```
        Test for change-point detection sensitive to changes in the
        distribution function with 'method'="nonseq"

data:  Xrdj
cvmmax = 325.12, p-value = 0.002498

User time:  43.8 sec
```

The small p-value provides evidence against stationarity. Assuming that stationarity does not hold only because $\mathcal{H}_0$ in (6.1) does not hold, and under the hypothesis of at most one change in the df of the $\boldsymbol{X}_i$, the date of change can be estimated as $\mathrm{argmax}_{1 \le k \le n-1} S_{n,k}^H$, which corresponds to June 15, 1998:

```
> out <- res$value # the output of the test (object of class 'htest')
> rdj[which(out$cvm == out$statistic), 1]

[1] "1998-06-15"
```

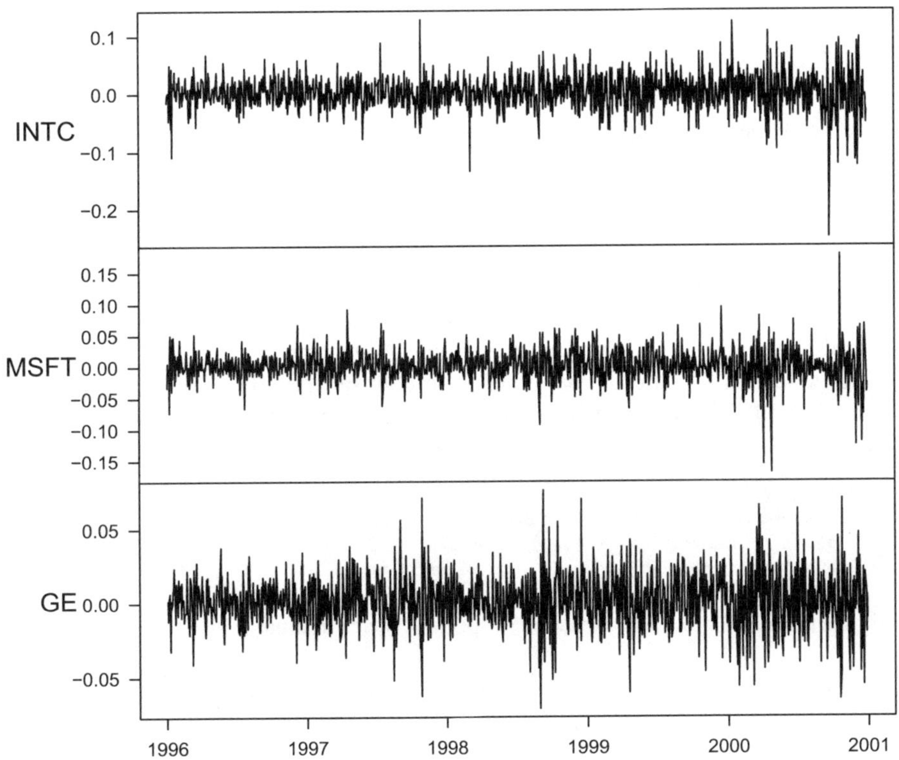

Fig. 6.3 Daily log-returns for the Intel (INTC), Microsoft (MSFT), and General Electric (GE) stocks for the period 1996–2000

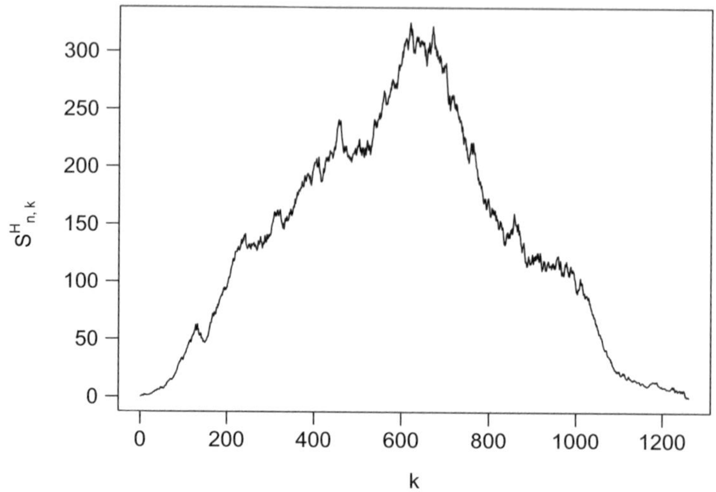

Fig. 6.4 The intermediate test statistics $S^H_{n,k}$ given in (6.4) for $k \in \{1, \dots, n-1\}$ for the `rdj` data set. The values $k = 1$ and $k = n - 1$ correspond to January 3, 1996 and December 28, 2000, respectively

It can also be informative to plot the intermediate test statistics $S^H_{n,1}, \dots, S^H_{n,n-1}$ given by (6.4):

```
> plot(out$cvm, type = "l", xlab = "k", ylab = quote({S^H}[list(n,k)]))
```

The resulting graph is given in Fig. 6.4. Note that, under the aforementioned additional assumptions, the results obtained so far do not provide any information regarding the strength of the departure from stationarity.

As a second illustration, we consider a data set initially analyzed in Grégoire et al. (2008). It consists of bivariate daily log-returns computed from 3 years of daily prices (for the period July 2003 to July 2006) of crude oil and natural gas:

```
> data(gasoil) # oil and gas prices
> library(qrmtools) # for returns()
> Rgasoil <- returns(gasoil[,-1]) # bivariate daily log-returns
> Xgasoil <- xts(Rgasoil, order.by = gasoil[-1, 1]) # corresponding xts object
```

Figure 6.5 shows plots of the two component series against time:

```
> plot.zoo(Xgasoil, main = "", xlab = "", mar = c(0, 7, 0, 2.1))
```

We can now apply the test based on S^H_n as done previously:

```
> set.seed(292)
> withTime(cpDist(Xgasoil, b = NULL))
```

```
        Test for change-point detection sensitive to changes in the
        distribution function with 'method'="nonseq"

data:  Xgasoil
cvmmax = 60.212, p-value = 0.3362

User time:   4.6 sec
```

The large p-value provides no evidence against stationarity as measured by S_n^H. □

The test based on S_n^H is a nonparametric, general-purpose test which seems to be powerful against alternatives involving changes in the marginal dfs of the observations. However, it turns out to be rather insensitive to changes which only affect the underlying copula; see Holmes et al. (2013).

Assume that $\mathcal{H}_0$ holds and that, additionally, the common marginal dfs $F_1, \ldots, F_d$ of $\boldsymbol{X}_1, \ldots, \boldsymbol{X}_n$ are continuous. By Sklar's Theorem, see Theorem 2.3.1, the common multivariate df H of the observations can be written as

$$H(\boldsymbol{x}) = C\big(F_1(x_1), \ldots, F_d(x_d)\big), \quad \boldsymbol{x} \in \mathbb{R}^d,$$

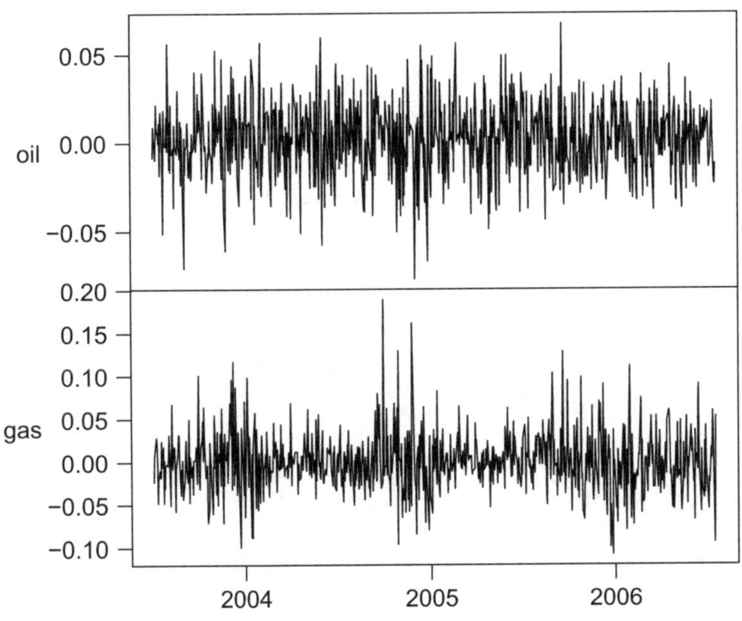

Fig. 6.5 Daily log-returns computed from 3 years of daily prices (from July 2003 to July 2006) of crude oil and natural gas

where C is the unique copula associated with H. It follows that $\mathcal{H}_0$ can be decomposed as $\mathcal{H}_{0,m} \wedge \mathcal{H}_{0,c}$, where

$$\begin{aligned}\mathcal{H}_{0,m} &: \text{There exist } F_1, \ldots, F_d \text{ such that } \boldsymbol{X}_1, \ldots, \boldsymbol{X}_n \text{ have marginal dfs } F_1, \ldots, F_d,\\ \mathcal{H}_{0,c} &: \text{There exists a copula } C \text{ such that } \boldsymbol{X}_1, \ldots, \boldsymbol{X}_n \text{ have copula } C.\end{aligned} \tag{6.5}$$

Roughly speaking, the multivariate df changes if and only if the marginal dfs change or the copula changes.

Let $\neg\mathcal{H}_{0,c}$ denote the negation of $\mathcal{H}_{0,c}$. Several nonparametric tests of stationarity designed to be particularly sensitive to certain alternatives under $\mathcal{H}_{0,m} \wedge (\neg\mathcal{H}_{0,c})$ are proposed in the literature. For instance, tests particularly sensitive to changes in Kendall's tau are investigated by Gombay and Horváth (1999), Quessy et al. (2013), and Dehling et al. (2017). To derive tests particularly sensitive to changes in the copula, Bücher et al. (2014) consider the following variant of the process $\mathbb{D}_n^H$ in (6.2):

$$\mathbb{D}_n^C(t, \boldsymbol{u}) = \sqrt{n}\lambda_n(0, t)\lambda_n(t, 1)\big(C_{1:\lfloor nt\rfloor}(\boldsymbol{u}) - C_{(\lfloor nt\rfloor+1):n}(\boldsymbol{u})\big), \quad (t, \boldsymbol{u}) \in [0, 1]^{d+1}, \tag{6.6}$$

where, for any $1 \le k \le l \le n$,

$$C_{k:l}(\boldsymbol{u}) = \frac{1}{l-k+1}\sum_{i=k}^{l} \mathbf{1}(\boldsymbol{U}_i^{k:l} \le \boldsymbol{u}), \quad \boldsymbol{u} \in [0, 1]^d, \tag{6.7}$$

is the empirical copula of $\boldsymbol{X}_k, \ldots, \boldsymbol{X}_l$, see Sect. 4.2.1, with the convention that $C_{k:l} = 0$ if $l < k$. In (6.7), the sample of pseudo-observations $\boldsymbol{U}_k^{k:l}, \ldots, \boldsymbol{U}_l^{k:l}$ is defined by

$$\boldsymbol{U}_i^{k:l} = (F_{k:l,1}(X_{i1}), \ldots, F_{k:l,d}(X_{id}))\frac{l-k+1}{l-k+2}, \quad i \in \{k, \ldots, l\},$$

where $F_{k:l,j}$ is the empirical df of $X_{kj}, \ldots, X_{lj}$. Similar to (6.3) and (6.4), a natural test statistic is then

$$S_n^C = \sup_{t\in[0,1]} \int_{[0,1]^d} \big(\mathbb{D}_n^C(t, \boldsymbol{u})\big)^2 \,\mathrm{d}C_{1:n}(\boldsymbol{u}) = \max_{1\le k\le n-1} \frac{1}{n}\sum_{i=1}^{n}\big(\mathbb{D}_n^C(k/n, \boldsymbol{U}_i^{1:n})\big)^2.$$

The test based on S_n^C also relies on resampling. Conditions under which it is asymptotically valid under stationarity are stated in Bücher et al. (2014).

Example 6.2.2 (Test of Stationarity Based on S_n^C) We first apply the test based on S_n^C to the `rdj` data set:

```
> set.seed(314)
> withTime(cpCopula(Xrdj, b = NULL, method = "nonseq"))
```

```
        Test for change-point detection sensitive to changes in the
        copula with 'method'="nonseq"

data:  Xrdj
cvmmax = 57.28, p-value = 0.01049

User time:  51.5 sec
```

The default value for the argument `method` is `"seq"`, which leads to better finite-sample behavior for smaller samples. We chose `"nonseq"` here to speed-up the computations; more details can be obtained from `?cpCopula`. Similar to the result of S_n^H, the small p-value again provides evidence against stationarity for the `rdj` data set.

For the bivariate daily log-returns from the `gasoil` data set, we obtain:

```
> set.seed(137)
> withTime(cpCopula(Xgasoil, b = NULL, method = "nonseq"))

        Test for change-point detection sensitive to changes in the
        copula with 'method'="nonseq"

data:  Xgasoil
cvmmax = 6.2836, p-value = 0.6429

User time:  5.9 sec
```

As measured by S_n^H and S_n^C, there is thus no evidence against stationarity in the bivariate log-returns computed from `gasoil`. □

Non-stationarity can also be a consequence of changes in the serial dependence of a multivariate time series. By construction, the tests based on S_n^H and S_n^C described previously are not designed to be sensitive to such changes. As demonstrated in Bücher et al. (forthcoming), the ideas underlying the test based on S_n^C can, however, be easily adapted to derive a test sensitive to changes in the serial dependence of a univariate time series. Given a sequence of random variables $X_1, \dots, X_{n'}$ and an integer $p > 1$, the approach first consists of forming the $n = n' - p + 1$ p-dimensional vectors of observations

$$\boldsymbol{Y}_i = (X_i, \dots, X_{i+p-1}), \quad i \in \{1, \dots, n\}, \tag{6.8}$$

where $p - 1$ is the maximum lag under investigation and p is known as the *embedding dimension*. If $X_1, \dots, X_{n'}$ is a stretch from a continuous stationary time series, all the $\boldsymbol{Y}_i$ have the same df, denoted by H^s. By Sklar's Theorem, see Theorem 2.3.1, H^s can be uniquely represented using a p-dimensional copula C^s such that

$$H^s(\boldsymbol{x}) = C^s\big(G(x_1), \dots, G(x_p)\big), \quad \boldsymbol{x} \in \mathbb{R}^p, \tag{6.9}$$

where G is the common df of each X_i. The copula C^s controls the dependence between the components of the $\boldsymbol{Y}_i$. Equivalently, it controls the serial dependence up to lag $p-1$ in the time series, which is why it is sometimes called the lag $p-1$ *autocopula* or *serial copula*.

To derive tests of stationarity for continuous univariate time series particularly sensitive to changes in the serial dependence, Bücher et al. (forthcoming) consider the following variant of the process $\mathbb{D}_n^C$ in (6.6):

$$\mathbb{D}_n^{C^s}(t,\boldsymbol{u}) = \sqrt{n}\lambda_n(0,t)\lambda_n(t,1)\big(C^s_{1:\lfloor nt\rfloor}(\boldsymbol{u}) - C^s_{(\lfloor nt\rfloor+1):n}(\boldsymbol{u})\big), \qquad (t,\boldsymbol{u}) \in [0,1]^{p+1},$$

where, for any $1 \le k \le l \le n$,

$$C^s_{k:l}(\boldsymbol{u}) = \frac{1}{l-k+1}\sum_{i=k}^{l}\prod_{j=1}^{p} \mathbf{1}(G_{k:l}(X_{i+j-1}) \le u_j), \qquad \boldsymbol{u} \in [0,1]^p, \tag{6.10}$$

and

$$G_{k:l}(x) = \frac{1}{l+p-k}\sum_{j=k}^{l+p-1} \mathbf{1}(X_j \le x), \qquad x \in \mathbb{R}. \tag{6.11}$$

The quantity $C^s_{k:l}$ is a nonparametric estimator of C^s based on $\boldsymbol{Y}_k, \dots, \boldsymbol{Y}_l$ that could be called the lag $p-1$ *empirical autocopula*. The latter was used, for instance, in Genest and Rémillard (2004) for testing serial independence as we shall see in Sect. 6.2.2. It can be verified that it is a straightforward transposition (to the serial context under consideration) of the definition adopted in (6.7). A natural test statistic is then

$$S_n^{C^s} = \sup_{t\in[0,1]} \int_{[0,1]^p} \big(\mathbb{D}_n^{C^s}(t,\boldsymbol{u})\big)^2 \mathrm{d}C^s_{1:n}(\boldsymbol{u}) = \max_{1\le k\le n-1} \int_{[0,1]^p} \big(\mathbb{D}_n^{C^s}(k/n,\boldsymbol{u})\big)^2 \mathrm{d}C^s_{1:n}(\boldsymbol{u}).$$

Similar to the tests based on S_n^H and S_n^C, the computation of an approximate p-value for the test based on $S_n^{C^s}$ relies on resampling. The asymptotic validity of the resampling scheme under stationarity is shown in Bücher et al. (forthcoming).

Prior to applying the test based on $S_n^{C^s}$ to a univariate time series, it is necessary to choose the embedding dimension p or, equivalently, the maximum lag $p-1$. This aspect is discussed in Bücher et al. (forthcoming, Section 3.4) where a first rough recommendation is to consider $p \in \{2, 3, 4\}$.

Multivariate extensions of the test based on $S_n^{C^s}$ are mentioned in Bücher et al. (forthcoming) but seem to need more investigation, in particular, from the point of view of their finite-sample properties. In the presence of a multivariate time series, an imperfect alternative is to apply the test based on $S_n^{C^s}$ to each component series.

Example 6.2.3 (Test of Stationarity Based on $S_n^{C^s}$) We first apply the test based on $S_n^{C^s}$ to the first component series of the `rdj` data set with $p-1=1$:

```
> set.seed(3355)
> withTime(cpAutocop(Xrdj[,1], lag = 1))

        Test of change-point detection sensitive to changes in the
        2-dimensional autocopula

data:  Xrdj[, 1]
cvmmax = 37.553, p-value = 0.03447

User time:  18.9 sec
```

The p-value provides weak evidence against stationarity for the `rdj` data set. If the maximum lag to be investigated is set to 2, we obtain:

```
> set.seed(3355)
> withTime(cpAutocop(Xrdj[,1], lag = 2))

        Test of change-point detection sensitive to changes in the
        3-dimensional autocopula

data:  Xrdj[, 1]
cvmmax = 35.028, p-value = 0.04645

User time:  18.5 sec
```

Finally, with $p-1=3$, we get:

```
> set.seed(3355)
> withTime(cpAutocop(Xrdj[,1], lag = 3))

        Test of change-point detection sensitive to changes in the
        4-dimensional autocopula

data:  Xrdj[, 1]
cvmmax = 23.419, p-value = 0.08242

User time:  23.9 sec
```

The increase in the p-value as p is increased is due to the fact that, as discussed in Bücher et al. (forthcoming), in many cases the power of the test based on $S_n^{C^s}$ decreases as p is increased. This is partly a consequence of the fact that the underlying empirical serial copulas are of dimension p.

For the bivariate daily log-returns obtained from the `gasoil` data set, we obtain, for the first component series with $p-1=1$:

```
> set.seed(3105)
> withTime(cpAutocop(Xgasoil[,1], lag = 1))
```

```
        Test of change-point detection sensitive to changes in the
        2-dimensional autocopula

data:  Xgasoil[, 1]
cvmmax = 11.309, p-value = 0.2303

User time:  4.4 sec
```

For the second component series, we have with $p - 1 = 1$:

```
> set.seed(2895)
> withTime(cpAutocop(Xgasoil[,2], lag = 1))

        Test of change-point detection sensitive to changes in the
        2-dimensional autocopula

data:  Xgasoil[, 2]
cvmmax = 13.243, p-value = 0.1643

User time:  4.4 sec
```

The results are qualitatively similar when larger values of p are considered. Overall, there is thus no evidence against stationarity in the bivariate log-returns computed from `gasoil`. □

It is important to keep in mind that although it is designed to be particularly sensitive to changes in the serial dependence, the test based on $S_n^{C^s}$ is a test of stationarity. Consequently, a small p-value should not be used to conclude that there is a change in the serial dependence unless one has reasons to believe that, for the data at hand, non-stationarity can only be a consequence of changes in the serial dependence. Similarly, a small p-value for the test based on S_n^H (respectively, S_n^C) should not be used to conclude that $\mathcal{H}_0$ in (6.1) (respectively, $\mathcal{H}_{0,c}$ in (6.5)) does not hold unless one is ready to believe that, for the data at hand, non-stationarity can only be a consequence of departures from $\mathcal{H}_0$ (respectively, $\mathcal{H}_{0,c}$).

6.2.2 *Tests of Serial Independence*

If there is no evidence against the hypothesis that $\boldsymbol{X}_1, \ldots, \boldsymbol{X}_n$ is a stretch from a stationary multivariate time series, a next natural question is whether $\boldsymbol{X}_1, \ldots, \boldsymbol{X}_n$ are serially independent.

A classical approach to assess whether serial dependence is present in the data at hand is to estimate the autocorrelation and cross-correlation functions of the multivariate time series and complement this by Ljung–Box-type portmanteau tests.

Example 6.2.4 (Correlogram and Ljung–Box Test of Serial Independence) We consider again the bivariate log-returns computed from the `gasoil` data set initially described in Example 6.2.1. When investigating the presence of serial dependence in financial log-returns, it is customary to consider the time series of squared (or absolute) log-returns. This is due to a stylized fact known as *volatility clustering*, that is, the tendency for large changes to be followed by large changes, although not necessarily with the same sign; see McNeil et al. (2015, Section 3.1.1). The correlogram representing the estimated autocorrelation and cross-correlation functions of the squared log-returns can be displayed as follows:

```
> colnames(Xgasoil) <- c("Oil", "Gas")
> acf(Xgasoil^2, ci.col = 1)
```

The resulting plots are shown in Fig. 6.6.

The correlogram of the second squared component series appears to display significant autocorrelations. To formally test for serial independence, we apply Ljung–Box tests (via `Box.test()`) to both squared component series. We first

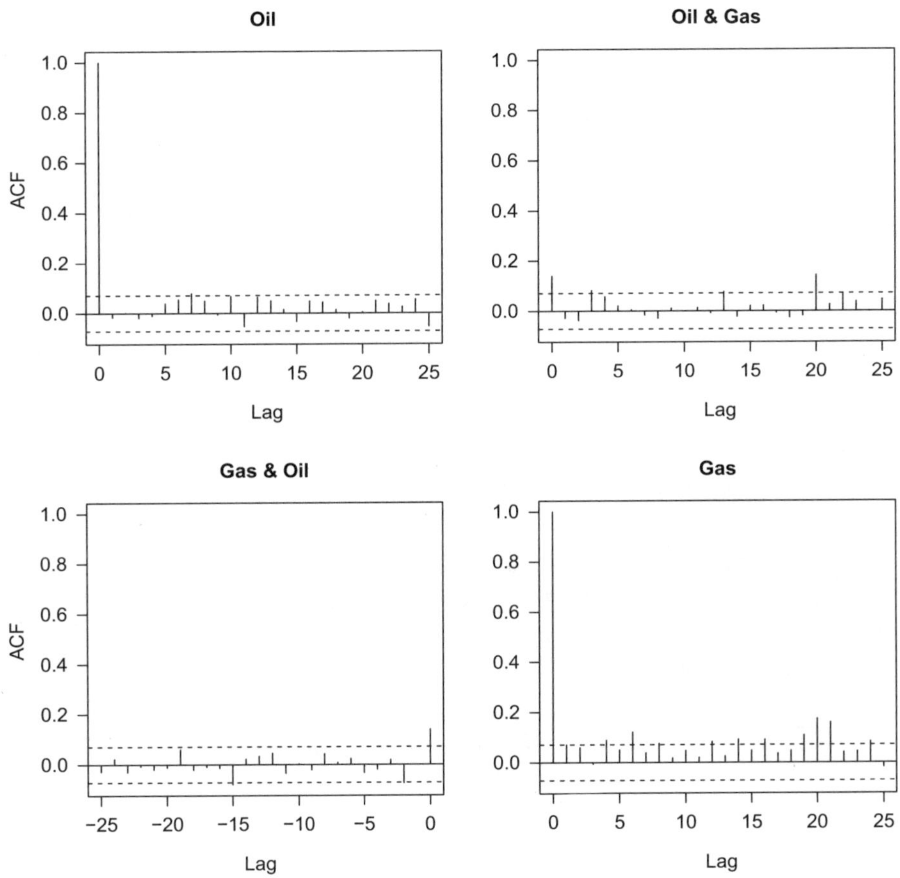

Fig. 6.6 Correlogram of the bivariate log-returns of the `gasoil` data

test for significant autocorrelations up to lag 5, and then, up to lag20:

```
> Box.test(Xgasoil[,1]^2, lag =  5, type = "Ljung-Box")

        Box-Ljung test

data:  Xgasoil[, 1]^2
X-squared = 1.7878, df = 5, p-value = 0.8777

> Box.test(Xgasoil[,2]^2, lag =  5, type = "Ljung-Box")

        Box-Ljung test

data:  Xgasoil[, 2]^2
X-squared = 14.752, df = 5, p-value = 0.01148

> Box.test(Xgasoil[,1]^2, lag = 20, type = "Ljung-Box")

        Box-Ljung test

data:  Xgasoil[, 1]^2
X-squared = 26.712, df = 20, p-value = 0.1435

> Box.test(Xgasoil[,2]^2, lag = 20, type = "Ljung-Box")

        Box-Ljung test

data:  Xgasoil[, 2]^2
X-squared = 90.924, df = 20, p-value = 5.108e-11
```

The tests provide strong evidence against serial independence in the second squared component series, that is, the squared log-returns of gas prices. □

Portmanteau tests (such as the Ljung–Box test) applied on squared observations are, however, known to be often too liberal when the observations are heavy-tailed. This is illustrated in the following example.

Example 6.2.5 (Ljung–Box Tests Can Be Too Liberal) We apply Ljung–Box tests a large number of times to samples of squared independent realizations from a univariate t distribution with 4 degrees of freedom by means of the following function:

```
> pvalBox <- function(n)
  {
      x2 <- rt(n, df = 4)^2
      c(lag5  = Box.test(x2, type = "Ljung-Box", lag =  5)$p.value,
        lag20 = Box.test(x2, type = "Ljung-Box", lag = 20)$p.value)
  }
```

The function `pvalBox()` is executed 10,000 times on samples of size 500 and empirical levels are computed:

```
> set.seed(3298)
> pv <- replicate(10000, pvalBox(500))
> alpha <- c(0.01, 0.05, 0.1) # nominal levels
> rbind(nom.level = alpha,
        emp.level.lag5  = ecdf(pv["lag5", ])(alpha),
        emp.level.lag20 = ecdf(pv["lag20",])(alpha))
```

```
                 [,1]   [,2]   [,3]
nom.level       0.010 0.0500 0.1000
emp.level.lag5  0.032 0.0579 0.0805
emp.level.lag20 0.045 0.0766 0.1029
```

As one can see, the lower tails of the distributions of the p-values do not appear to be uniform. □

More robust tests of serial independence, based on ranks, for instance, might be preferable. One possible such test is the one based on the serial empirical copula proposed by Genest and Rémillard (2004). It is a serial extension of the multivariate test of independence initially proposed by Deheuvels (1981) and briefly presented in Sect. 5.2.1. Given a stationary sequence of continuous random variables $X_1, \dots, X_{n'}$ and an embedding dimension $p > 1$, the approach of Genest and Rémillard (2004) consists of first forming the $n = n' - p + 1$ p-dimensional random vectors $\boldsymbol{Y}_i$, $i \in \{1, \dots, n\}$, defined by (6.8). It follows from the stationarity assumption that all the $\boldsymbol{Y}_i$ have the same df H^s which, as mentioned in the previous section, can be decomposed as in (6.9) in terms of the serial copula C^s and the common df G of each X_i. Starting from the fact that, under serial independence of $X_1, \dots, X_{n'}$, C^s is the independence copula Π, Genest and Rémillard (2004) consider test statistics derived from the empirical process

$$\sqrt{n}\big(C^s_{1:n}(\boldsymbol{u}) - \Pi(\boldsymbol{u})\big), \quad \boldsymbol{u} \in [0,1]^p,$$

where $C^s_{1:n}$ is the serial empirical copula defined by (6.10). This amounts to working with the ranks $R_1, \dots, R_{n'}$ associated with $X_1, \dots, X_{n'}$ as $R_i = n' G_{1:n}(X_i)$, $i \in \{1, \dots, n'\}$, where $G_{1:n}$ is defined by (6.11). A sensible test statistic is then

$$S_n^{\Pi^s} = \int_{[0,1]^p} n(C^s_{1:n}(\boldsymbol{u}) - \Pi(\boldsymbol{u}))^2 \mathrm{d}\boldsymbol{u}.$$

Its realizations under serial independence can be obtained by simulation, thereby allowing one to compute an approximate p-value for $S_n^{\Pi^s}$ empirically. The previous construction was extended to multivariate time series by Kojadinovic and Yan (2011) who used a permutation principle to compute approximate p-values.

Example 6.2.6 (Tests of Serial Independence Based on $S_n^{\Pi^s}$) We consider again the bivariate log-returns of the `gasoil` data, which, according to Examples 6.2.1, 6.2.2, and 6.2.3, might well be a stretch from a stationary time series. We choose to apply the test based on $S_n^{\Pi^s}$ with $p = 6$, which corresponds to a maximum lag of 5. To compute an approximate p-value for $S_n^{\Pi^s}$, it is first necessary to generate a large number, say 1000, of realizations of the test statistic under serial independence for sample size n and embedding dimension p:

```
> set.seed(137)
> sI.d <- withTime(serialIndepTestSim(nrow(Xgasoil), lag.max = 5))
> sI.d$sys.time # the run time

User time:   321.8 sec
```

The generated realizations of $S_n^{\Pi^s}$, stored in the output `sI.d$value` of the function `serialIndepTestSim()`, can then be used to carry out the test on the two squared component series:

```
> serialIndepTest(Xgasoil[,1]^2, d = sI.d$value)

Global Cramer-von Mises statistic: 0.03519808 with p-value 0.02247752
Combined p-values from the Mobius decomposition:
   0.2862138  from Fisher's rule,
   0.3361638  from Tippett's rule.

> serialIndepTest(Xgasoil[,2]^2, d = sI.d$value)

Global Cramer-von Mises statistic: 0.07246049 with p-value 0.002497502
Combined p-values from the Mobius decomposition:
   0.08841159  from Fisher's rule,
   0.0004995005  from Tippett's rule.
```

The first p-values in the outputs are those corresponding to the test statistic $S_n^{\Pi^s}$. The definitions of the test statistics corresponding to the remaining two p-values can be found in Genest and Rémillard (2004). The results are in line with those obtained from the Ljung–Box test in Example 6.2.4 in that there is strong evidence against serial independence in the second squared component series. □

6.2.3 *Models for Multivariate Time Series Based on Conditional Copulas*

Suppose that there is no evidence against stationarity but evidence against serial independence of $\boldsymbol{X}_1, \dots, \boldsymbol{X}_n$, implying that $\boldsymbol{X}_1, \dots, \boldsymbol{X}_n$ can be regarded as a stretch from a stationary d-dimensional time series. As explained at the beginning of Sect. 6.2, in such a time series context and with forecasting in mind, the interest often does not lie in the (unconditional) distribution of the random vector $\boldsymbol{X}$ of which $\boldsymbol{X}_1, \dots, \boldsymbol{X}_n$ are serially dependent copies, but rather in the conditional distribution of $\boldsymbol{X}_i$ given the sigma-field $\mathcal{G}_{i-1}$ generated by $\boldsymbol{X}_1, \dots, \boldsymbol{X}_{i-1}$, that is, the information available at time $i-1$. Under additional assumptions, the conditional df of $\boldsymbol{X}_i$ given $\mathcal{G}_{i-1}$ can be decomposed as in Sklar's Theorem which leads to the concept of *conditional copula* initially introduced in Patton (2006); see also Fermanian and

Wegkamp (2012). The aim of this section is to provide a brief overview of time series models based on this notion.

To introduce the concept of conditional copula, we first consider a simpler setting: let $(\boldsymbol{X}, \boldsymbol{Z})$ be a $(d+q)$-dimensional random vector, let $\operatorname{ran} \boldsymbol{Z} = \{\boldsymbol{z} \in \mathbb{R}^q : \mathbb{P}(\boldsymbol{Z} \in (\boldsymbol{z} - \boldsymbol{h}, \boldsymbol{z}]) > 0 \text{ for all } \boldsymbol{h} > \boldsymbol{0}\}$ be the *range* of the random vector $\boldsymbol{Z}$, and, given $\boldsymbol{z} \in \operatorname{ran} \boldsymbol{Z}$, let us focus on the conditional df of $\boldsymbol{X}$ given $\boldsymbol{Z} = \boldsymbol{z}$, that is, on the conditional df

$$H_{\boldsymbol{z}}(\boldsymbol{x}) = \mathbb{P}(X_1 \le x_1, \dots, X_d \le x_d \mid \boldsymbol{Z} = \boldsymbol{z}), \quad \boldsymbol{x} \in \mathbb{R}^d. \tag{6.12}$$

Assuming that $H_{\boldsymbol{z}}$ is continuous, we immediately obtain from Sklar's Theorem, see Theorem 2.3.1, that

$$H_{\boldsymbol{z}}(\boldsymbol{x}) = C_{\boldsymbol{z}}\big(F_{\boldsymbol{z},1}(x_1), \dots, F_{\boldsymbol{z},d}(x_d)\big), \quad \boldsymbol{x} \in \mathbb{R}^d, \tag{6.13}$$

where $F_{\boldsymbol{z},j}(x) = \mathbb{P}(X_j \le x \mid \boldsymbol{Z} = \boldsymbol{z})$, $j \in \{1, \dots, d\}$, are the univariate margins of $H_{\boldsymbol{z}}$ and $C_{\boldsymbol{z}}$ is the copula of $H_{\boldsymbol{z}}$. Equivalently,

$$C_{\boldsymbol{z}}(\boldsymbol{u}) = \mathbb{P}\big(F_{\boldsymbol{z},1}(X_1) \le u_1, \dots, F_{\boldsymbol{z},d}(X_d) \le u_d \mid \boldsymbol{Z} = \boldsymbol{z}\big), \quad \boldsymbol{u} \in [0,1]^d. \tag{6.14}$$

Assuming that $H_{\boldsymbol{z}}$ is continuous for all $\boldsymbol{z} \in \operatorname{ran} \boldsymbol{Z}$ (which implies that (6.13) holds for all $\boldsymbol{z} \in \operatorname{ran} \boldsymbol{Z}$), the underlying conditional copula is the random function $C_{\boldsymbol{Z}}$ obtained by replacing $\boldsymbol{z}$ by $\boldsymbol{Z}$ in (6.14).

Providing a rigorous definition of the notion of conditional copula in the time series setting considered in this section is mathematically more involved; see Fermanian and Wegkamp (2012). We shall only attempt to convey the main intuitions. Recall that we are interested in modeling the conditional df of $\boldsymbol{X}_i$ given the information $\mathcal{G}_{i-1}$ at time $i-1$, that is, the conditional df

$$H_{\mathcal{G}_{i-1}}(\boldsymbol{x}) = \mathbb{P}(X_{i1} \le x_1, \dots, X_{id} \le x_d \mid \mathcal{G}_{i-1}), \quad \boldsymbol{x} \in \mathbb{R}^d. \tag{6.15}$$

Furthermore, let $F_{\mathcal{G}_{i-1},j}(x) = \mathbb{P}(X_{ij} \le x \mid \mathcal{G}_{i-1})$, $x \in \mathbb{R}$, $j \in \{1, \dots, d\}$, denote the corresponding margins. Then, under continuity and measurability assumptions, there exists an almost surely unique random function $C_{\mathcal{G}_{i-1}}$, referred to as a *conditional copula*, such that

$$H_{\mathcal{G}_{i-1}}(\boldsymbol{x}) = C_{\mathcal{G}_{i-1}}\big(F_{\mathcal{G}_{i-1},1}(x_1), \dots, F_{\mathcal{G}_{i-1},d}(x_d)\big), \quad \boldsymbol{x} \in \mathbb{R}^d.$$

Starting from the above decomposition, a first natural approach to model $H_{\mathcal{G}_{i-1}}$ is to consider parametric families for the marginal dfs $F_{\mathcal{G}_{i-1},j}$, $j \in \{1, \dots, d\}$, and the conditional copula $C_{\mathcal{G}_{i-1}}$, and to allow the corresponding parameter vectors to change according to the information $\mathcal{G}_{i-1}$ available at time $i-1$. The latter thus incurs an additional modeling difficulty compared to, for instance, the (unconditional) setting considered in Sect. 4.1.1 since it is necessary to specify how the marginal and the copula parameters change with $\mathcal{G}_{i-1}$.

When dealing with financial time series, the most frequently encountered model in this setting, see Patton (2006, 2012, 2013) and the references therein, is as follows:

- For any $j \in \{1, \ldots, d\}$, it is assumed that

$$X_{ij} = \mu_{ij}(\boldsymbol{\beta}_j) + \sigma_{ij}(\boldsymbol{\beta}_j)\varepsilon_{ij}, \tag{6.16}$$

  where $\mu_{ij}(\boldsymbol{\beta}_j) = \mathbb{E}(X_{ij} \mid \mathcal{G}_{i-1})$ is the conditional mean of X_{ij} given $\mathcal{G}_{i-1}$, $\sigma_{ij}^2(\boldsymbol{\beta}_j) = \mathrm{Var}(X_{ij} \mid \mathcal{G}_{i-1})$ is the conditional variance of X_{ij} given $\mathcal{G}_{i-1}$ and the conditional distribution of the innovations ε_{ij} given $\mathcal{G}_{i-1}$ does not depend on $\mathcal{G}_{i-1}$. In addition, the conditional mean and variance in (6.16) are defined up to the finite-dimensional parameter vector $\boldsymbol{\beta}_j$.
- For identifiability, it is further assumed that, for any $j \in \{1, \ldots, d\}$, the (nonrandom) conditional distribution of ε_{ij} given $\mathcal{G}_{i-1}$ has mean zero and variance one and that its df belongs to an absolutely continuous parametric family of univariate dfs $\mathcal{F}_j = \{F_{j,\boldsymbol{\gamma}_j} : \boldsymbol{\gamma}_j \in \Gamma_j\}$, where Γ_j is the parameter space. Note that the conditional distribution of $\boldsymbol{\varepsilon}_i = (\varepsilon_{i1}, \ldots, \varepsilon_{id})$ given $\mathcal{G}_{i-1}$ may depend on $\mathcal{G}_{i-1}$. In other words, although the univariate margins of the conditional df of $\boldsymbol{\varepsilon}_i$ given $\mathcal{G}_{i-1}$ are not random (they do not depend on $\mathcal{G}_{i-1}$), the conditional copula of $\boldsymbol{\varepsilon}_i$ given $\mathcal{G}_{i-1}$ may depend on $\mathcal{G}_{i-1}$.
- It is additionally assumed that there exists an absolutely continuous parametric family of copulas $\mathcal{C} = \{C_{\boldsymbol{\theta}} : \boldsymbol{\theta} \in \Theta\}$, where Θ is the parameter space, and a parametric *copula calibration function* $\boldsymbol{\varphi}$ that transforms $\mathcal{G}_{i-1}$ into $\boldsymbol{\theta}_{i-1} \in \Theta$ so that the copula of the conditional distribution of $\boldsymbol{\varepsilon}_i$ given $\mathcal{G}_{i-1}$ is $C_{\boldsymbol{\varphi}(\mathcal{G}_{i-1})} = C_{\boldsymbol{\theta}_{i-1}}$. The calibration function $\boldsymbol{\varphi}$ governs the temporal dynamics of the parameter vector of the copula: It determines how the copula parameter vector changes with the information $\mathcal{G}_{i-1}$ available at time $i-1$.
- The copula calibration function $\boldsymbol{\varphi}$ is further assumed to be completely defined up to a finite-dimensional parameter vector $\boldsymbol{\beta}$ (which is not to be confused with the finite-dimensional parameter vectors $\boldsymbol{\beta}_1, \ldots, \boldsymbol{\beta}_d$ defining the conditional means and variances in (6.16)).

As a consequence, under this model, the conditional df of $\boldsymbol{\varepsilon}_i$ given $\mathcal{G}_{i-1}$ can be written as $C_{\boldsymbol{\varphi}(\mathcal{G}_{i-1})}\big(F_{1,\boldsymbol{\gamma}_1}(\cdot), \ldots, F_{d,\boldsymbol{\gamma}_d}(\cdot)\big)$. It follows from (6.16) that the conditional df $H_{\mathcal{G}_{i-1}}$ of $\boldsymbol{X}_i$ given $\mathcal{G}_{i-1}$, see (6.15), can be expressed as

$$H_{\mathcal{G}_{i-1}}(\boldsymbol{x}) = C_{\boldsymbol{\varphi}(\mathcal{G}_{i-1})}\left(F_{1,\boldsymbol{\gamma}_1}\left(\frac{x_1 - \mu_{i1}(\boldsymbol{\beta}_1)}{\sigma_{i1}(\boldsymbol{\beta}_1)}\right), \ldots, F_{d,\boldsymbol{\gamma}_d}\left(\frac{x_d - \mu_{id}(\boldsymbol{\beta}_d)}{\sigma_{id}(\boldsymbol{\beta}_d)}\right)\right), \quad \boldsymbol{x} \in \mathbb{R}^d.$$

Denoting by $c_{\boldsymbol{\theta}}$ the density of $C_{\boldsymbol{\theta}}$ and by $f_{j,\boldsymbol{\gamma}_j}$ the density of $F_{j,\boldsymbol{\gamma}_j}$, $j \in \{1, \ldots, d\}$, we immediately obtain that the conditional density $h_{\mathcal{G}_{i-1}}$ of $\boldsymbol{X}_i$ given $\mathcal{G}_{i-1}$ can be written as

$$h_{\mathcal{G}_{i-1}}(\boldsymbol{x}) = c_{\boldsymbol{\varphi}(\mathcal{G}_{i-1})}\left(F_{1,\boldsymbol{\gamma}_1}\left(\frac{x_1 - \mu_{i1}(\boldsymbol{\beta}_1)}{\sigma_{i1}(\boldsymbol{\beta}_1)}\right), \dots, F_{d,\boldsymbol{\gamma}_d}\left(\frac{x_d - \mu_{id}(\boldsymbol{\beta}_d)}{\sigma_{id}(\boldsymbol{\beta}_d)}\right)\right)$$
$$\cdot \prod_{j=1}^{d} \frac{1}{\sigma_{ij}(\boldsymbol{\beta}_j)} f_{j,\boldsymbol{\gamma}_j}\left(\frac{x_j - \mu_{ij}(\boldsymbol{\beta}_j)}{\sigma_{ij}(\boldsymbol{\beta}_j)}\right).$$

The log-likelihood of the model is then

$$\ell_n(\boldsymbol{\beta}, \boldsymbol{\beta}_1, \dots, \boldsymbol{\beta}_d, \boldsymbol{\gamma}_1, \dots, \boldsymbol{\gamma}_d) = \log \prod_{i=1}^{n} h_{\mathcal{G}_{i-1}}(\boldsymbol{X}_i) = \sum_{i=1}^{n} \log h_{\mathcal{G}_{i-1}}(\boldsymbol{X}_i).$$

The latter is often maximized in stages in the spirit of the inference functions for margins estimator presented in Sect. 4.1.1. The marginal parameters $\boldsymbol{\beta}_j$ and $\boldsymbol{\gamma}_j$, $j \in \{1, \dots, d\}$, are estimated first in d separate steps, each involving one of the d component series. Denoting by $\boldsymbol{\beta}_{j,n}$ and $\boldsymbol{\gamma}_{j,n}$ the estimators of $\boldsymbol{\beta}_j$ and $\boldsymbol{\gamma}_j$, respectively, the estimated standardized residuals are given by

$$\varepsilon_{ij,n} = \frac{X_{ij} - \mu_{ij}(\boldsymbol{\beta}_{j,n})}{\sigma_{ij}(\boldsymbol{\beta}_{j,n})}, \quad i \in \{1, \dots, n\}, \; j \in \{1, \dots, d\}. \tag{6.17}$$

The parameter $\boldsymbol{\beta}$ specifying the copula calibration function can then be estimated by

$$\operatorname*{argsup}_{\boldsymbol{\beta}} \sum_{i=1}^{n} \log c_{\boldsymbol{\varphi}(\mathcal{G}_{i-1})}\big(F_{1,\boldsymbol{\gamma}_{1,n}}(\varepsilon_{i1,n}), \dots, F_{d,\boldsymbol{\gamma}_{d,n}}(\varepsilon_{id,n})\big).$$

A simplified version of the above model is investigated in Chen and Fan (2006) and Rémillard (2017), among others: The main difference is that the conditional copula of $\boldsymbol{\varepsilon}_i$ given $\mathcal{G}_{i-1}$ is also assumed not to depend on $\mathcal{G}_{i-1}$, implying that the conditional distribution of $\boldsymbol{\varepsilon}_i$ given $\mathcal{G}_{i-1}$ does not depend on $\mathcal{G}_{i-1}$, or equivalently, that the copula calibration function $\boldsymbol{\varphi}$ is constant and equals $\boldsymbol{\theta}$. As a consequence, its parameter vector $\boldsymbol{\beta}$ can be identified with the unknown parameter vector $\boldsymbol{\theta}$ of the underlying non-time-varying copula and $\boldsymbol{\varepsilon}_1, \dots, \boldsymbol{\varepsilon}_n$ are iid. Chen and Fan (2006) then propose to estimate $\boldsymbol{\beta}$ semi-parametrically in the spirit of the maximum pseudo-likelihood estimator presented in Sect. 4.1.2. The resulting estimator $\boldsymbol{\beta}_n$ of $\boldsymbol{\beta} = \boldsymbol{\theta}$ is given by

$$\boldsymbol{\beta}_n = \operatorname*{argsup}_{\boldsymbol{\theta} \in \Theta} \sum_{i=1}^{n} \log c_{\boldsymbol{\theta}}\big(F_{1,n}(\varepsilon_{i1,n}), \dots, F_{d,n}(\varepsilon_{id,n})\big), \tag{6.18}$$

where, for any $j \in \{1, \ldots, d\}$,

$$F_{j,n}(x) = \frac{1}{n+1} \sum_{i=1}^{n} \mathbf{1}(\varepsilon_{ij,n} \le x), \quad x \in \mathbb{R}. \tag{6.19}$$

The estimator $\boldsymbol{\beta}_n$ is thus nothing else but the maximum pseudo-likelihood estimator of Sect. 4.1.2 computed from the estimated standardized residuals given in (6.17).

Under correct marginal specifications, Chen and Fan (2006) show the somewhat surprising result that the above estimator behaves asymptotically as if it were based on the unobservable innovations $\boldsymbol{\varepsilon}_i = (\varepsilon_{i1}, \ldots, \varepsilon_{id})$, $i \in \{1, \ldots, n\}$, and not on the estimated standardized residuals $\boldsymbol{\varepsilon}_{i,n} = (\varepsilon_{i1,n}, \ldots, \varepsilon_{id,n})$, $i \in \{1, \ldots, n\}$, defined by (6.17). Under similar assumptions, Rémillard (2017) shows that the empirical copula process (see Sect. 4.2.1) based on the estimated standardized residuals has the same asymptotics as if it were based on the unobservable innovations instead. In other words, the aforementioned asymptotics do not depend on the conditional mean and variance parameters. Hence, as explained in Rémillard (2017), provided the assumptions of the aforementioned simplified model hold, one can apply the rank-based copula inference procedures of Chaps. 4 and 5 on the estimated standardized residuals as if they were the innovations.

In applications, frequent choices for the marginal models are ARMA models for the conditional means $\mu_{ij}(\boldsymbol{\beta}_j)$, $j \in \{1, \ldots, d\}$, and GARCH models for the conditional variances $\sigma_{ij}^2(\boldsymbol{\beta}_j)$, $j \in \{1, \ldots, d\}$. In a related way, Chen and Fan (2006) and Jondeau and Rockinger (2006) seem to be among the first to popularize the expression *copula–GARCH* to refer to certain models of this type.

Example 6.2.7 (Conditional Modeling Based on ARMA–GARCH Marginal Models) From Examples 6.2.1, 6.2.2, and 6.2.3, we have no reasons to believe that the bivariate log-returns computed from the `gasoil` data set do not form a sample from a stationary multivariate time series. However, as seen in Example 6.2.6, the log-returns cannot be considered iid. To apply the aforementioned modeling steps, we consider ARMA–GARCH models for the margins using their implementation in the R package `rugarch` of Ghalanos (2017). For the sake of simplicity, we do not attempt to find the best fitting ARMA–GARCH models but simply consider the popular ARMA(1,1)–GARCH(1,1) model with t innovations (scaled to have variance one) for both component series. Furthermore, after finding no evidence of time varying dependence using the test based on S_n^C presented in Sect. 6.2.1, we decide to link the marginal innovations with a t copula. Note that a more adequate modeling of these data is carried out in Grégoire et al. (2008). For illustration purposes, we also simulate 1000 paths of size 260 (the latter are roughly the number of trading days in 1 year) from the fitted bivariate copula–GARCH time-series model for the log-returns. We then use these paths to estimate the 1-year ahead bivariate predictive density of oil and gas prices.

Model specification is carried out as follows based on the function `ugarchspec()` of `rugarch`:

```
> library(rugarch)
> ## Specify ARMA(1,1)-GARCH(1,1) model with Student t innovations
> meanModel <- list(armaOrder = c(1,1)) # ARMA(1,1)
> varModel  <- list(model = "sGARCH", garchOrder = c(1,1)) # GARCH(1,1)
> uspec <- ugarchspec(varModel, mean.model = meanModel,
                      distribution.model = "std") # scaled Student t
```

Next, we estimate the component models with the function `ugarchfit()` and extract the estimated standardized residuals in (6.17):

```
> ## Fit marginal ARMA-GARCH models
> fit <- apply(Xgasoil, 2, function(x) ugarchfit(uspec, data = x))
> ## Extract the estimated standardized residuals
> eps <- sapply(fit, residuals, standardize = TRUE) # standardized residuals
> (nus <- sapply(fit, function(x) x@fit$coef[["shape"]])) # fitted d.o.f.
```

```
      Oil       Gas
13.298668  5.821641
```

Assuming correct marginal specifications, the function `cpCopula()` of the R package `npcp` can be used to test the hypothesis that the conditional copula does not depend on $\mathcal{G}_{i-1}$, or, equivalently, that it is not time-varying. Recall from Sect. 6.2.1 that the computation of an approximate p-value for the underlying test statistic S_n^C requires the knowledge of a bandwidth parameter corresponding to the argument `b`. Since the estimated standardized residuals should be approximately serially independent, we change the default `b = NULL` (forcing the estimation of the bandwidth parameter) to `b = 1` (see `?cpCopula` for more details):

```
> set.seed(2013)
> withTime(cpCopula(eps, b = 1, method = "nonseq"))
```

```
        Test for change-point detection sensitive to changes in the
        copula with 'method'="nonseq"

data:  eps
cvmmax = 6.6343, p-value = 0.5839

User time:  5.5 sec
```

The large p-value provides no evidence against a constant, that is, non-time-varying, conditional copula. We then form the corresponding pseudo-observations based on (6.19):

```
> U <- pobs(eps) # pseudo-observations from the residuals eps
> plot(U, xlab = expression(U[1]), ylab = expression(U[2]))
```

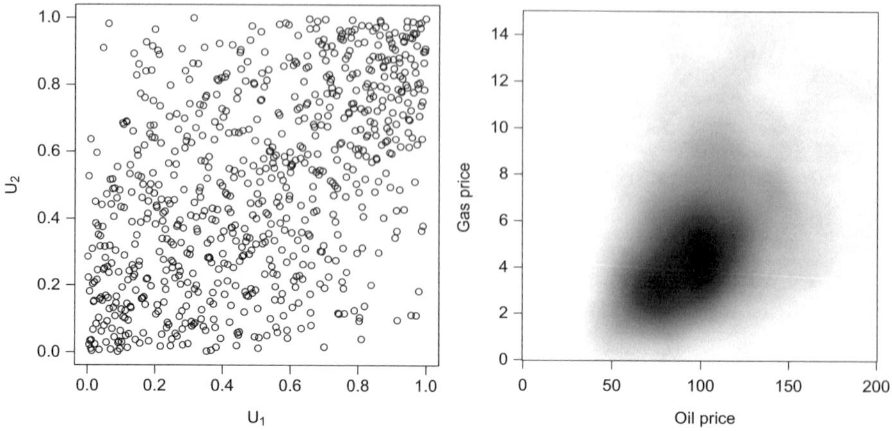

Fig. 6.7 Pseudo-observations of the residuals obtained after ARMA–GARCH modeling of the oil and gas log-returns (left) and estimated predictive density for oil and gas prices 1 year ahead (right)

The left-hand side of Fig. 6.7 shows a scatter plot of the pseudo-observations of the estimated standardized residuals. Next, we fit a t copula to these pseudo-observations by means of (6.18):

```
> ## Fit a t copula to the estimated standardized residuals
> fitcop <- fitCopula(tCopula(), data = U, method = "mpl")
> fitcop@estimate # estimated correlation parameter rho and d.o.f. nu
```

```
    rho.1         df
0.5019151 45.5276670
```

```
> cop <- fitcop@copula # fitted t copula
```

Note the large degrees of freedom, so we could have equally well chosen a normal copula here.

We are now in the position to simulate the cross-sectionally dependent innovations $\boldsymbol{\varepsilon}_i = (\varepsilon_{i1}, \varepsilon_{i2})$ of the two component time-series models. Specifically, to estimate the bivariate predictive density of oil and gas prices 1 year ahead, we simulate 1000 paths of size 260 of oil and gas prices via `ugarchsim()`:

```
> ## Simulate from the bivariate model
> ## 1) Simulate from the fitted copula
> set.seed(271) # set seed
> n.sim <- 260 # sample size
> m.sim <- 1000 # number of paths
> U. <- rCopula(n.sim * m.sim, cop) # simulate from the fitted copula
> ## 2) Quantile-transform the corresponding innovations
> ##    Note: eps have to be standardized (mean 0, variance 1) for ugarchsim()
> eps. <- sapply(1:2, function(j)
      sqrt((nus[j]-2)/nus[j]) * qt(U.[,j], df = nus[j]))
> ## 3) Feed the (cross-sec. dependent) innovations to the marginal ARMA-GARCH
> ##    models and simulate from them
> sim <- lapply(1:2, function(j)
      ugarchsim(fit[[j]], # fitted marginal ARMA-GARCH model
                n.sim = n.sim, # sample size
                m.sim = m.sim, # number of trajectories/paths
                custom.dist = list(name = "sample", # our innovations
                                   distfit = matrix(eps.[,j], ncol = m.sim))))
> ## 4) Extract the simulated (cross-sec. dependent) series X_t and build the
> ##    corresponding (predicted/simulated) oil and gas prices
> X. <- lapply(sim, function(x) fitted(x)) # equal to seriesSim in x@simulation
> S.t <- as.numeric(tail(gasoil, n = 1)[,2:3]) # last available prices S_t
> library(qrmtools) # for returns()
> S. <- lapply(1:2, function(j) # predicted prices for each stock
      returns(X.[[j]], inverse = TRUE, start = rep(S.t[j], m.sim)))
> S.T <- sapply(1:2, function(j) tail(S.[[j]], n = 1)) # pick out prices at T
```

Based on these 1000 sample paths, we can then compute an estimate of the bivariate predictive density for the daily oil and gas prices for the following year:

```
> library(MASS)
> pred.dens <- kde2d(S.T[,1], S.T[,2], n = 300, lims = c(0, 200, 0, 15))
> image(pred.dens, xlab = "Oil price", ylab = "Gas price",
        col = gray(seq(1, 0, length.out = 100)))
```

An estimate of the predictive density is displayed on the right-hand side of Fig. 6.7; note that the larger the density, the darker the shade of gray.

□

Additional examples of marginal conditional modeling based on ARMA–GARCH models prior to copula inference can be found, for instance, in the documentation of the function `fit_ARMA_GARCH()` of the R package `qrmtools`; see also the vignette `copula_GARCH` of `copula`.

Other types of time series models based on copulas are, for instance, discussed in Darsow et al. (1992), Ibragimov (2009), Beare (2010), and Rémillard et al. (2012); see also the references therein.

6.3 Regression

As already mentioned in Sect. 6.2.3, in many applications, besides the random vector $\boldsymbol{X}$ of interest, one also observes a q-variate random vector $\boldsymbol{Z}$ of covariates or explanatory variables. The focus then typically shifts from modeling the unconditional distribution of $\boldsymbol{X}$ to modeling the conditional distribution of $\boldsymbol{X}$ given the covariates $\boldsymbol{Z}$. Under continuity assumptions, the latter can be based, again, on the concept of conditional copula already introduced in Sect. 6.2.3. The aim of this section is to present a brief overview of the underlying statistical challenges.

In the setting considered above, conditional copulas capture the dependence among the components of $\boldsymbol{X}$ conditionally on $\boldsymbol{Z}$. They are used, for example, to estimate the relationship between diastolic and systolic blood pressure given the cholesterol level, see Lambert (2007), or to relate life expectancies of males and females for different categories of countries, see Veraverbeke et al. (2011). A more involved application on longitudinal data is presented in Sun et al. (2008). Last but not least, conditional copulas are at the heart of the actively investigated field of vine copulas not addressed in this book.

The formal setting is similar to the one considered at the beginning of Sect. 6.2.3. The aim is to model the random df $H_{\boldsymbol{Z}}$, where, for any $\boldsymbol{z}$ in ran $\boldsymbol{Z} \subseteq \mathbb{R}^q$, $H_{\boldsymbol{z}}$ defined by (6.12) is the conditional df of $\boldsymbol{X}$ given $\boldsymbol{Z} = \boldsymbol{z}$. As explained in Sect. 6.2.3, under continuity assumptions, it holds that, almost surely,

$$H_{\boldsymbol{Z}}(\boldsymbol{x}) = C_{\boldsymbol{Z}}\big(F_{\boldsymbol{Z},1}(x_1), \dots, F_{\boldsymbol{Z},d}(x_d)\big), \quad \boldsymbol{x} \in \mathbb{R}^d, \tag{6.20}$$

where, for any $\boldsymbol{z} \in$ ran $\boldsymbol{Z}$, $F_{\boldsymbol{z},j}(x) = \mathbb{P}(X_j \le x \mid \boldsymbol{Z} = \boldsymbol{z})$, $j \in \{1, \dots, d\}$, and $C_{\boldsymbol{Z}}$ is the conditional copula of $\boldsymbol{X}$ given $\boldsymbol{Z}$ defined through (6.14).

A key assumption frequently made in certain types of applications is that, for any $\boldsymbol{z} \in$ ran $\boldsymbol{Z}$, the copula $C_{\boldsymbol{z}}$ does not depend on $\boldsymbol{z}$, or equivalently, that the conditional copula $C_{\boldsymbol{Z}}$ is nonrandom. This assumption is referred to as the *simplifying assumption* in the literature. As an example, assume that $X_1, \dots, X_d$ are conditionally independent given $\boldsymbol{Z}$. The latter implies that, almost surely,

$$H_{\boldsymbol{Z}}(\boldsymbol{x}) = F_{\boldsymbol{Z},1}(x_1) \cdot \ldots \cdot F_{\boldsymbol{Z},d}(x_d), \quad \boldsymbol{x} \in \mathbb{R}^d,$$

and thus, in view of (6.20), that $C_{\boldsymbol{Z}}$ is the independence copula Π, almost surely. In other words, the simplifying assumption holds. Of course, the latter does not imply that the (unconditional) copula of $\boldsymbol{X}$ is Π since $X_1, \dots, X_d$ are not necessarily (unconditionally) independent.

To grasp an idea of the statistical challenges related to the use of conditional copulas in the current regression context, assume that we have at hand a random sample $(\boldsymbol{X}_1, \boldsymbol{Z}_1), \dots, (\boldsymbol{X}_n, \boldsymbol{Z}_n)$ from $(\boldsymbol{X}, \boldsymbol{Z})$ and recall that the overall aim is to estimate the random df $H_{\boldsymbol{Z}}$ through the decomposition (6.20).

Very naturally, and similar to what was already presented in Sect. 6.2.3 in a time series context, the early literature on the subject suggests to use parametric families for the marginal dfs $F_{Z,j}$, $j \in \{1, \ldots, d\}$, and the copula C_Z, and to allow the underlying parameter vectors to change according to the value of $\boldsymbol{Z}$. In the spirit of generalized linear models (GLMs), see, for instance, Lindsey (1997), the marginal and copula parameter vectors are often first transformed through a link function and then linearly regressed on $\boldsymbol{Z}$. The fitting of the model thus amounts to estimating the various regression coefficients. More formally, the fully parametric setting under consideration can be described as follows:

- For any $j \in \{1, \ldots, d\}$, there exists an absolutely continuous parametric family of univariate dfs $\mathcal{F}_j = \{F_{j,\boldsymbol{\gamma}_j} : \boldsymbol{\gamma}_j \in \Gamma_j\}$, where Γ_j is the parameter space.
- For any $j \in \{1, \ldots, d\}$, there exists a map $\boldsymbol{\varphi}_j : \operatorname{ran} \boldsymbol{Z} \to \Gamma_j$ that transforms any realization of the vector of covariates $\boldsymbol{Z}$ into a parameter vector of $\mathcal{F}_j$ such that, for any $z \in \operatorname{ran} \boldsymbol{Z}$, $\boldsymbol{\varphi}_j(z) = \boldsymbol{\gamma}_{j,z} \in \Gamma_j$ and $F_{z,j}(x) = F_{j,\boldsymbol{\gamma}_{j,z}}(x)$ for all $x \in \mathbb{R}$; in other words, for every $z \in \operatorname{ran} \boldsymbol{Z}$, there exists a parameter $\boldsymbol{\gamma}_{j,z} \in \Gamma_j$ such that the unknown df $F_{z,j}$ is equal to the parametric df $F_{j,\boldsymbol{\gamma}_{j,z}} \in \mathcal{F}_j$.
- For any $j \in \{1, \ldots, d\}$, the *marginal calibration function* $\boldsymbol{\varphi}_j$ is completely defined up to a finite-dimensional parameter vector $\boldsymbol{\beta}_j$ of regression coefficients.
- There exists an absolutely continuous parametric family of copulas $\mathcal{C} = \{C_{\boldsymbol{\theta}} : \boldsymbol{\theta} \in \Theta\}$, where Θ is the parameter space.
- There exists a map $\boldsymbol{\varphi} : \operatorname{ran} \boldsymbol{Z} \to \Theta$ that transforms any realization of the vector of covariates $\boldsymbol{Z}$ into a parameter vector of the copula family $\mathcal{C}$ such that, for any $z \in \operatorname{ran} \boldsymbol{Z}$, $\boldsymbol{\varphi}(z) = \boldsymbol{\theta}_z$ and $C_z(\boldsymbol{u}) = C_{\boldsymbol{\theta}_z}(\boldsymbol{u})$ for all $\boldsymbol{u} \in [0, 1]^d$, where C_z is given by (6.14); in other words, for every $z \in \operatorname{ran} \boldsymbol{Z}$, there exists a parameter $\boldsymbol{\theta}_z \in \Theta$ such that the unknown copula C_z is equal to the parametric copula $C_{\boldsymbol{\theta}_z} \in \mathcal{C}$.
- The *copula calibration function* $\boldsymbol{\varphi}$ is completely defined up to a finite-dimensional parameter vector $\boldsymbol{\beta}$ of regression coefficients (which is not to be confused with the finite-dimensional parameter vectors $\boldsymbol{\beta}_1, \ldots, \boldsymbol{\beta}_d$ defining the marginal calibration functions).

Denoting by $c_{\boldsymbol{\theta}}$ the density of $C_{\boldsymbol{\theta}}$ and by $f_{j,\boldsymbol{\gamma}_j}$ the density of $F_{j,\boldsymbol{\gamma}_j}$, $j \in \{1, \ldots, d\}$, we immediately obtain that, under the above parametric assumptions, the conditional density h_z of $\boldsymbol{X}$ given $\boldsymbol{Z} = z$, that is, the density associated with H_z in (6.13), is given by

$$\begin{aligned} h_z(\boldsymbol{x}) &= c_{\boldsymbol{\theta}_z}\big(F_{1,\boldsymbol{\gamma}_{1,z}}(x_1), \ldots, F_{d,\boldsymbol{\gamma}_{d,z}}(x_d)\big) \prod_{j=1}^{d} f_{j,\boldsymbol{\gamma}_{j,z}}(x_j) \\ &= c_{\boldsymbol{\varphi}(z)}\big(F_{1,\boldsymbol{\varphi}_1(z)}(x_1), \ldots, F_{d,\boldsymbol{\varphi}_d(z)}(x_d)\big) \prod_{j=1}^{d} f_{j,\boldsymbol{\varphi}_j(z)}(x_j). \end{aligned}$$

Mimicking classical regression, the conditional log-likelihood function of the model is

$$\ell_n(\boldsymbol{\beta}, \boldsymbol{\beta}_1, \ldots, \boldsymbol{\beta}_d) = \sum_{i=1}^{n} \log c_{\boldsymbol{\varphi}(\mathbf{Z}_i)}\big(F_{1,\varphi_1(\mathbf{Z}_i)}(X_{i1}), \ldots, F_{d,\varphi_d(\mathbf{Z}_i)}(X_{id})\big)$$

$$+ \sum_{j=1}^{d}\sum_{i=1}^{n} \log f_{j,\varphi_j(\mathbf{Z}_i)}(X_{ij}). \tag{6.21}$$

The latter can be maximized to estimate $(\boldsymbol{\beta}, \boldsymbol{\beta}_1, \ldots, \boldsymbol{\beta}_d)$. The maximizer will be denoted by $(\boldsymbol{\beta}_n, \boldsymbol{\beta}_{1,n}, \ldots, \boldsymbol{\beta}_{d,n})$, and the corresponding estimated copula and marginal calibration functions by $\boldsymbol{\varphi}_n$ and $\boldsymbol{\varphi}_{1,n}, \ldots, \boldsymbol{\varphi}_{d,n}$, respectively. The resulting estimator of $H_{\mathbf{Z}}$ is then $C_{\boldsymbol{\varphi}_n(\mathbf{Z})}\big(F_{1,\varphi_{1,n}(\mathbf{Z})}(\cdot), \ldots, F_{d,\varphi_{d,n}(\mathbf{Z})}(\cdot)\big)$.

Several remarks are in order:

- The optimization task becomes simpler if the simplifying assumption is satisfied, that is, if $C_{\mathbf{z}}$ in (6.14) does not depend on $\mathbf{z}$. Deriving procedures for testing whether this crucial hypothesis holds is currently an active area of research; see, for instance, Acar et al. (2013), Derumigny and Fermanian (2017), and the references therein.
- Since the copula and marginal models do not have parameters in common, in the spirit of the inference functions for margins estimator presented in Sect. 4.1.1, the maximization of (6.21) can be broken into two main stages. First, estimate the parameter vector $\boldsymbol{\beta}_j$ of the jth marginal model from the jth component sample for all $j \in \{1, \ldots, d\}$. Then, plug the estimated marginal parameters $\boldsymbol{\beta}_{1,n}, \ldots, \boldsymbol{\beta}_{d,n}$ into the first term on the right-hand side of (6.21) and maximize the resulting logarithmic pseudo-likelihood with respect to $\boldsymbol{\beta}$.
- The previous setup is fully parametric. A slightly more flexible specification based on a nonparametric form of the calibration function $\boldsymbol{\varphi}$ is considered in Acar et al. (2011) and tests of the functional form of $\boldsymbol{\varphi}$ are proposed in Acar et al. (2013).
- Nonparametric estimation of the conditional copula is studied in Veraverbeke et al. (2011) and, under the simplifying assumption, in Gijbels et al. (2015) and Portier and Segers (2018). Semi-parametric approaches extending maximum pseudo-likelihood estimation, see Sect. 4.1.2, are considered, for instance, in Abegaz et al. (2012) and Fermanian and Lopez (2018).

A simplified version of the aforementioned fully parametric model is studied in Gijbels et al. (2015). Assuming the simplifying assumption to hold, the authors propose to consider location-scale models for the components. Specifically, for any $j \in \{1, \ldots, d\}$, it is supposed that

$$X_{ij} = \mu_j(\mathbf{Z}_i) + \sigma_j(\mathbf{Z}_i)\varepsilon_{ij}, \quad i \in \{1, \ldots, n\}, \tag{6.22}$$

where the functions μ_j and σ_j are known up to finite-dimensional parameter vectors, and $\boldsymbol{\varepsilon}_i = (\varepsilon_{i1}, \ldots, \varepsilon_{id})$, $i \in \{1, \ldots, n\}$, is a random sample, independent of $\mathbf{Z}_1, \ldots, \mathbf{Z}_n$, from an unknown continuous distribution with copula C. Some thought then reveals that, since the simplifying assumption holds, the conditional copula (6.14) coincides with C. Gijbels et al. (2015) provide conditions under which the empirical copula process based on the unobservable random sample $\boldsymbol{\varepsilon}_i = (\varepsilon_{i1}, \ldots, \varepsilon_{id})$, $i \in \{1, \ldots, n\}$, is asymptotically equivalent to the empirical copula process based on the estimated residuals

$$\boldsymbol{\varepsilon}_{i,n} = \left(\frac{X_{i1} - \mu_{1,n}(\mathbf{Z}_i)}{\sigma_{1,n}(\mathbf{Z}_i)}, \ldots, \frac{X_{id} - \mu_{d,n}(\mathbf{Z}_i)}{\sigma_{d,n}(\mathbf{Z}_i)} \right), \quad i \in \{1, \ldots, n\}, \tag{6.23}$$

where $\mu_{j,n}$ and $\sigma_{j,n}$ are estimators of the functions μ_j and σ_j, respectively. Interestingly enough, the obtained result is of the same nature as the one of Rémillard (2017) for the simplified multivariate time series model considered in Sect. 6.2.3. Consequently, provided the underlying assumptions are satisfied, inference on the (conditional) copula C could be carried out by applying the rank-based procedures of Chaps. 4 and 5 to the estimated residuals (6.23).

Example 6.3.1 (Conditional Modeling Based on Marginal Gamma GLMs) We consider an example from Frees (2010) based on data from the US National Education Longitudinal Study (NELS) of 1988. The mathematics (`Math`), science (`Science`) and reading (`Reading`) scores of 14,707 students were assessed along with eight explanatory variables: Five were binary and indicate whether the student comes from an ethnic minority group (`Minority`), is a female (`Female`), whether the school is publicly funded (`Public`), and whether it is located in an urban area (`Urban`) or a rural area (`Rural`) (note in passing that the sum of `Urban` and `Rural` is at most, but not necessarily equal to one); the three remaining covariates report the socioeconomic status of the student and its family (`SES`), the student's school identifier (`ID`) and the size of the student's school (`Size`). Following Frees (2010), we make the assumption that the way the three student scores are related to the seven explanatory variables `Minority`, `SES`, `Female`, `Public`, `Size`, `Urban`, and `Rural` does not depend on the student's school (and thus on `ID`) and work with a randomly selected sample of $n = 1000$ students which is available in the R package `copulaData` of Hofert et al. (2017a):

```
> data(NELS88, package = "copulaData")
> nels <- subset(NELS88, select = -ID) # remove school ID
```

As we continue, the $d = 3$ scores are to be regarded as the components of the random vector of interest $\boldsymbol{X}$, while the $q = 7$ remaining variables (taken in the aforementioned order; this is also the order in which they appear in the data frame `nels`) correspond to the components of the random vector $\boldsymbol{Z}$ of covariates. We further assume that the rows of `nels` can be seen as $n = 1000$ independent realizations of the random vector $(\boldsymbol{X}, \boldsymbol{Z})$. An inspection of the estimated standard deviations of the components of $\boldsymbol{Z}$ reveals that `Size` is substantially more dispersed than the other covariates. To ensure that all explanatory variables have standard deviations of the same order of magnitude, `Size` is scaled using `scale()`:

```
> nels$Size <- scale(nels$Size) # mean 0, standard deviation 1
```

Following Frees (2010), for every $j \in \{1, 2, 3\}$, the family of univariate parametric dfs $\mathcal{F}_j$ for modeling X_j is taken to consist of gamma dfs. Their densities of the form $x \mapsto x^{a_j-1}e^{-x/s_j}/(s^{a_j}\Gamma(a_j))$, $x > 0$, for a shape parameter $a_j > 0$ and a scale parameter $s_j > 0$, are reparametrized in terms of the expectation $\mu_j = a_j s_j$ and the shape a_j. The next step consists of defining, for every $j \in \{1, 2, 3\}$, the marginal calibration map $\boldsymbol{\varphi}_j$ transforming every realization z of $\boldsymbol{Z}$ into the corresponding parameter vector $(\mu_{j,z}, a_{j,z})$ of the gamma distribution. Following the aforementioned order of the covariates and their observed ranges, $\boldsymbol{\varphi}_j$ can be regarded as a map from $\{0, 1\} \times \mathbb{R} \times \{0, 1\}^2 \times \mathbb{R} \times \{0, 1\}^2$ to $(0, \infty)^2$ and will be parametrized by $\boldsymbol{\beta}_j = (\beta_{j,1}, \ldots, \beta_{j,9}) \in \mathbb{R}^8 \times (0, \infty)$ such that the first and second coordinate maps of $\boldsymbol{\varphi}_j$ are defined by $z \mapsto \mu_{j,z} = \exp(\beta_{j,1} + \sum_{l=2}^{8} \beta_{j,l} z_j)$ and $z \mapsto a_{j,z} = 1/\beta_{j,9}$, respectively. In other words, for every $j \in \{1, 2, 3\}$, the logarithm of the expectation of the gamma distribution used to model X_j is taken to depend linearly on the covariates while the (inverse of the) shape parameter does not depend on the covariates.

The marginal conditional negative log-likelihood for X_j (that is, the jth of the d inner sums in the second term on the right-hand side of (6.21)) can be implemented as follows:

```
> ##' @title Marginal conditional negative log-likelihood
> ##' @param beta.m parameter vector defining the marginal calibration map
> ##' @param x vector of values of one of the three scores
> ##' @param z design matrix
> ##' @param pobs logical indicating whether, additionally, the parametric
> ##'        pseudo-observations shall be computed and returned
> ##' @return -log-likelihood and, possibly, the parametric pseudo-observations
> nmLL <- function(beta.m, x, z, pobs = FALSE)
  {
      p <- ncol(z) + 1 # number of parameters
      mu.z <- exp(z %*% beta.m[1:(p-1)])
      a.z <- 1 / beta.m[p] # shape
      s.z <- mu.z * beta.m[p] # scale
      nLL <- -sum(dgamma(x, shape = a.z, scale = s.z, log = TRUE))
      if (!pobs) nLL else
          list(nLL = nLL, U = pgamma(x, shape = a.z, scale = s.z))
  }
```

For the actual fitting, we also need the design matrix:

```
> ## Build the design matrix
> z <- model.matrix(~ Minority + SES + Female + Public + Size +
                    Urban + Rural, data = nels)
> p <- ncol(z) + 1 # number of parameters per margin
```

For the mathematics score X_1, maximizing this log-likelihood, or, equivalently, minimizing the negative log-likelihood, could be done by executing `res <- optim(init, nmLL, x = nels[,"Math"], z = z, method = "BFGS")`, where the argument `init` is a vector of initial values for $\boldsymbol{\beta}_1$ which would need to be provided. In principle, the same could be done for the science score X_2 and the reading score X_3. However, there is a more convenient way of fitting the marginal models in our setup here. It can be verified that the above specification for the distribution of X_j, $j \in \{1, 2, 3\}$, is actually equivalent to modeling X_j using a gamma GLM with logarithmic link (in particular, $\beta_{j,9}$ is then known as the *dispersion parameter*). For the mathematics score X_1, the GLM can be fitted as follows:

```
> math.glm <- glm(Math ~ Minority + SES + Female + Public + Size +
                  Urban + Rural, data = nels, family = Gamma(link = "log"))
> (math.summary <- summary(math.glm))

Call:
glm(formula = Math ~ Minority + SES + Female + Public + Size +
    Urban + Rural, family = Gamma(link = "log"), data = nels)

Deviance Residuals:
     Min        1Q     Median        3Q       Max

-0.76596  -0.25062  -0.03988   0.18564   0.94128

Coefficients:
             Estimate Std. Error t value Pr(>|t|)
(Intercept)  3.611152   0.025491 141.664  < 2e-16 ***
Minority1   -0.074292   0.023017  -3.228  0.00129 **
SES          0.174095   0.013412  12.980  < 2e-16 ***
Female1      0.014506   0.018943   0.766  0.44400
Public1     -0.048937   0.026531  -1.845  0.06540 .
Size         0.003419   0.011240   0.304  0.76107
Urban1       0.014709   0.023741   0.620  0.53569
Rural1       0.016608   0.025388   0.654  0.51317
---
Signif. codes:  0 '***' 0.001 '**' 0.01 '*' 0.05 '.' 0.1 ' ' 1

(Dispersion parameter for Gamma family taken to be 0.08912198)

    Null deviance: 114.850  on 999  degrees of freedom
Residual deviance:  87.612  on 992  degrees of freedom
AIC: 7477.3

Number of Fisher Scoring iterations: 4
```

The previous output suggests that not all $q = 7$ explanatory variables need to be included in the model for X_1; additional diagnostics could be obtained by calling, for instance, `glm.diag.plots()` from the `R` package `boot` of Canty and Ripley (2017). Note that the line starting with `(Dispersion parameter ...)` provides the estimate of the parameter $\beta_{j,9}$. As we continue, we shall assume that the fit of the previous GLM for X_1 is decent enough and we shall not attempt to optimize it in terms of explanatory variables.

Next, we gather the estimates of $\boldsymbol{\beta}_1$ in a vector (for future use):

```
> ## The estimates of (beta_{1,1}, ..., beta_{1,9})
> (ts.math <- c(math.glm$coefficients, disp = math.summary$dispersion))

 (Intercept)   Minority1         SES     Female1     Public1
 3.611151573 -0.074291617  0.174094924  0.014505741 -0.048937308
        Size      Urban1      Rural1        disp
 0.003418825  0.014708865  0.016607564  0.089121984
```

Notice that the reason for naming the vector of estimates `ts.math` is that the previous computations can be seen as part of the first stage of a two-stage estimation of the full model (whose conditional copula part yet remains to be specified).

Reassuringly, if the previous estimates are used as starting values when minimizing the negative marginal conditional log-likelihood implemented in `nmLL()`, the results are roughly the same:

```
> ## Minimizing the marginal conditional negative log-likelihood
> ## using the previously obtained estimates as initial values
> res <- optim(ts.math, nmLL, x = nels[,"Math"], z = z, method = "BFGS")

> ## Compare GLM and ML estimates: change is small
> stopifnot(all.equal(ts.math, res$par, tolerance = 1e-3))
```

However, additional experiments reveal that the previous minimization is very much affected by the choice of the initial values. This does not come as a surprise given that the fitting steps in `glm.fit()` are actually substantially more involved than minimizing the negative marginal conditional log-likelihood presented above; see, for instance, `?glm.fit` and the references therein.

The corresponding gamma GLMs for the science score X_2 and the reading score X_3 can be fitted as follows:

```
> ## Science score
> sci.glm <- glm(Science ~ Minority + SES + Female + Public + Size +
                 Urban + Rural, data = nels, family = Gamma(link = "log"))
> ## The estimates of (beta_{2,1}, ..., beta_{2,9})
> (ts.sci <- c(sci.glm$coefficients, disp = summary(sci.glm)$dispersion))

 (Intercept)   Minority1         SES     Female1     Public1
 2.982860943 -0.103505731  0.112851160 -0.023818260 -0.040889382
        Size      Urban1      Rural1        disp
-0.001163258 -0.010660049  0.006951770  0.057958226

> ## Reading score
> read.glm <- glm(Reading ~ Minority + SES + Female + Public + Size +
                  Urban + Rural, data = nels, family = Gamma(link = "log"))
> ## The estimates of (beta_{3,1}, ..., beta_{3,9})
> (ts.read <- c(read.glm$coefficients, disp = summary(read.glm)$dispersion))

  (Intercept)     Minority1           SES       Female1       Public1
 3.3288096565 -0.0783444150  0.1405291605  0.0796672851 -0.1000690298
         Size        Urban1        Rural1          disp
 0.0002235484 -0.0158321791  0.0164005009  0.0855469523
```

To complete the specification of the model, it remains to specify the parametric copula family C for the conditional copula and the corresponding copula calibration function $\boldsymbol{\varphi}$ parametrized by the vector $\boldsymbol{\beta}$. For the sake of simplicity, we assume that the simplifying assumption holds, that is, that the conditional copula does not depend on the covariates. Under this assumption and the assumption of correct marginal specifications, parametric pseudo-observations from the underlying copula can be formed as follows:

```
> ## Parametric pseudo-observations from the underlying trivariate conditional
> ## copula under the parametric assumptions and the simplifying assumption
> V <- cbind("V[1]" = nmLL(ts.math, nels[,"Math"],    z, pobs = TRUE)$U,
             "V[2]" = nmLL(ts.sci,  nels[,"Science"], z, pobs = TRUE)$U,
             "V[3]" = nmLL(ts.read, nels[,"Reading"], z, pobs = TRUE)$U)
```

The latter can be visualized using the following code (see Fig. 6.8 for the plots):

```
> splom2(V, cex = 0.3, col.mat = "black") # scatter-plot matrix
> library(lattice)
> cloud2(V) # 3d cloud plot based on lattice's cloud()
```

Having in mind the simplified model of Gijbels et al. (2015), see (6.22), the related theoretical results and their practical implications, one may hope that, provided the marginal assumptions and the simplifying assumption hold, inference on the unknown conditional copula in the current setting might be safely based on the multivariate scaled ranks computed from the sample of componentwise residuals obtained from the three marginal models; see also Côté et al. (2016). For GLMs, several definitions of residuals exist; see, for instance, `?residuals.glm`

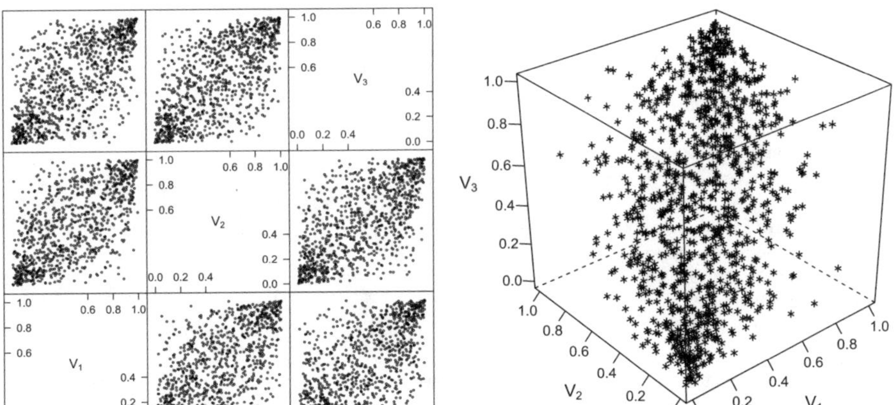

Fig. 6.8 Scatter-plot matrix (left) and cloud plot (right) of $n = 1000$ parametric pseudo-observations from the unknown conditional trivariate copula under the hypothesis of correct marginal specifications and the simplifying assumption

and the references therein. Below we consider multivariate scaled ranks computed from the so-called deviance residuals:

```
> U <- pobs(cbind(residuals(math.glm), residuals(sci.glm),
                  residuals(read.glm)))
```

The latter actually coincide with the multivariate scaled ranks computed from V:

```
> stopifnot(all.equal(U, pobs(V), check.attributes = FALSE))
```

Hence, provided the marginal assumptions and the simplifying assumption hold, the rows of U could also be regarded as pseudo-observations from the unknown trivariate conditional copula.

After inspecting the scatter-plot matrix of U and the corresponding normal scores (simply replace U by qnorm(U) in the plotting code), we decide to consider three trivariate parametric copula families for C: the Frank family, the homogeneous (exchangeable) normal family, and the inhomogeneous (non-exchangeable; three parameter) normal family. For the sake of illustration, we also include the Gumbel–Hougaard family. These four families can be fitted as follows:

```
> summary(ts.g <- fitCopula(gumbelCopula(dim = 3), data = U))

Call: fitCopula(copula, data = data)
Fit based on "maximum pseudo-likelihood" and 1000 3-dimensional observations.
Gumbel copula, dim. d = 3
      Estimate Std. Error
alpha    1.713      0.036
The maximized loglikelihood is 589.4
Optimization converged
Number of loglikelihood evaluations:
function gradient
       7        7

> summary(ts.f <- fitCopula(frankCopula (dim = 3), data = U))

Call: fitCopula(copula, data = data)
Fit based on "maximum pseudo-likelihood" and 1000 3-dimensional observations.
Frank copula, dim. d = 3
      Estimate Std. Error
alpha    4.791      0.188
The maximized loglikelihood is 594
Optimization converged
Number of loglikelihood evaluations:
function gradient
       5        5

> summary(ts.n.ex <- fitCopula(normalCopula(dim = 3, dispstr = "ex"), data=U))
```

```
Call: fitCopula(copula, data = data)
Fit based on "maximum pseudo-likelihood" and 1000 3-dimensional observations.
Normal copula, dim. d = 3
      Estimate Std. Error
rho.1   0.6603      0.013
The maximized loglikelihood is 650.3
Optimization converged
Number of loglikelihood evaluations:
function gradient
       8        8
```

```
> summary(ts.n.un <- fitCopula(normalCopula(dim = 3, dispstr = "un"), data=U))
```

```
Call: fitCopula(copula, data = data)
Fit based on "maximum pseudo-likelihood" and 1000 3-dimensional observations.
Normal copula, dim. d = 3

      Estimate Std. Error
rho.1   0.6628      0.017
rho.2   0.6548      0.016
rho.3   0.6632      0.017
The maximized loglikelihood is 650.4
Optimization converged
Number of loglikelihood evaluations:
function gradient
      30        6
```

To choose among the four families, we perform model selection based on `xvCopula()`:

```
> k <- 100 # for k-fold cross-validation
> optim.method <- "BFGS" # the numerical optimization method for MPLE
> set.seed(3090)
> withTime(
      round(c(xvCopula(gumbelCopula(dim = 3),
                       x = V, k = k, optim.method = optim.method),
              xvCopula(frankCopula(dim = 3),
                       x = V, k = k, optim.method = optim.method),
              xvCopula(normalCopula(dim = 3), # homogeneous (exchangeable)
                       x = V, k = k, optim.method = optim.method),
              xvCopula(normalCopula(dim = 3, dispstr = "un"), # unstruct. cor.
                       x = V, k = k, optim.method = optim.method)), 1)
  )
```

```
[1] 586.4 592.0 648.1 646.1
User time:  59.5 sec
```

As one can see, the homogeneous normal family is ranked first. Next, we assess its fit using the multiplier goodness-of-fit test based on (5.9):

```
> set.seed(2723)
> withTime(gofCopula(normalCopula(dim = 3), x = V, simulation = "mult"))
```

```
        Multiplier bootstrap-based goodness-of-fit test of Normal
        copula, dim. d = 3, with 'method'="Sn", 'estim.method'="mpl":

data:  x
statistic = 0.025396, parameter = 0.66025, p-value = 0.473

User time:  3.1 sec
```

Hence, the homogeneous normal family is not rejected.

The two-stage estimation of H_Z defined by (6.20) could thus be regarded as complete: Gamma GLMs seem to be sensible marginal models and a homogeneous normal copula with parameter $\rho \approx 0.66$ might be an adequate choice for the conditional copula under correct marginal models and the simplifying assumption. The resulting conditional df H_Z can be used, for instance, to estimate the conditional probability of various events related to the student scores (given specific values of the covariates); see the end of this example.

Let us now turn to the maximization of the full conditional log-likelihood in (6.21), or, equivalently, to the minimization of the negative log-likelihood. Assuming, for the sake of simplicity, that the conditional copula belongs to a one-parameter trivariate copula family, the full conditional negative log-likelihood can be implemented as follows:

```
> ##' @title Full conditional negative log-likelihood function
> ##' @param par param. vector defining the marg. and copula calibration maps
> ##' @param x matrix of values of the responses
> ##' @param z design matrix
> ##' @param copula a trivariate one-parameter copula object
> ##' @return -log-likelihood
> nfLL <- function(par, x, z, copula)
  {
      beta <- par[1] # copula parameter
      tc <- tryCatch(copula <- setTheta(copula, beta), # try to set parameters
                     error = function(e) NULL)
      if (is.null(tc)) return(-Inf) # in case of failure, return -Inf
      p <- ncol(z) + 1 # number of parameters per margin
      beta.1 <- par[1 + 1:p] # parameters of the first marginal model
      beta.2 <- par[p+1 + 1:p] # parameters of the second marginal model
      beta.3 <- par[2*p+1 + 1:p] # parameters of the third marginal model
      ## Marginal log-likelihood evaluation and computing the
      ## corresponding parametric pseudo-observations
      nmLL.1 <- nmLL(beta.1, x[,1], z, pobs = TRUE)
      nmLL.2 <- nmLL(beta.2, x[,2], z, pobs = TRUE)
      nmLL.3 <- nmLL(beta.3, x[,3], z, pobs = TRUE)
      ## In case of invalid evaluation of the marg. likelihoods, return -Inf
      if (any(is.na(c(nmLL.1$nLL, nmLL.2$nLL, nmLL.3$nLL)))) return(-Inf)
      ## Parametric pseudo-observations as a matrix
      U <- cbind(nmLL.1$U, nmLL.2$U, nmLL.3$U)
      ## -log-likelihood
      -sum(dCopula(u = U, copula = copula, log = TRUE)) +
          nmLL.1$nLL + nmLL.2$nLL + nmLL.3$nLL
  }
```

We first consider the model based on the trivariate Gumbel–Hougaard copula family:

```
> res.g <- optim(c(coef(ts.g), ts.math, ts.sci, ts.read), nfLL,
                 x = nels, z = z, copula = gumbelCopula(dim = 3),
                 method = "BFGS", control = list(maxit = 1e6), hessian = TRUE)
> stopifnot(res.g$convergence == 0) # the optimization has converged
> -res.g$value # the maximized likelihood

[1] -9441.34
```

For the Frank family, we obtain:

```
> ## Maximization of the full conditional likelihood function using
> ## as initial values the estimates obtained with the two-stage approach
> res.f <- optim(c(coef(ts.f), ts.math, ts.sci, ts.read), nfLL,
                 x = nels, z = z, copula = frankCopula(dim = 3),
                 method = "BFGS", control = list(maxit = 1e6), hessian = TRUE)
> stopifnot(res.f$convergence == 0) # the optimization has converged
> -res.f$value # the maximized likelihood

[1] -9455.478
```

Finally, for the homogeneous normal family, we get:

```
> res.n <- optim(c(coef(ts.n.ex), ts.math, ts.sci, ts.read), nfLL,
                 x = nels, z = z, copula = normalCopula(dim = 3),
                 method = "BFGS", control = list(maxit = 1e6), hessian = TRUE)
> stopifnot(res.n$convergence == 0) # the optimization has converged
> -res.n$value # the maximized likelihood

[1] -9385.596
```

Hence, the largest maximized likelihood (or, equivalently, the largest AIC, see (5.10), since all models under consideration have the same number of parameters) is obtained for the homogeneous normal family. Let us extract the parameter estimates of the copula and the three marginal models:

```
> full.n.ex <- res.n$par[1] # parameter estimate of the normal copula
> full.math <- res.n$par[1 + 1:p] # param. est. of the 1st marginal model
> full.sci  <- res.n$par[p+1 + 1:p] # param. est. of the 2nd marginal model
> full.read <- res.n$par[2*p+1 + 1:p] # param. est. of the 3rd marg. model
```

Reassuringly, the estimates appear to be reasonably close to those obtained with the two-stage approach:

```
> ## Comparison of the estimated copula param. (two-stage vs full likelihood)
> c(ts.n.ex@estimate, full.n.ex)

0.6602536 0.6590429

> ## Comparison of the estimated parameters of the three marginal models
> round(cbind(ts.math, full.math, ts.sci, full.sci, ts.read, full.read), 3)
```

```
            ts.math full.math ts.sci full.sci ts.read full.read
(Intercept)   3.611     3.615  2.983    2.984   3.329     3.331
Minority1    -0.074    -0.076 -0.104   -0.103  -0.078    -0.078
SES           0.174     0.172  0.113    0.111   0.141     0.139
Female1       0.015     0.005 -0.024   -0.030   0.080     0.072
Public1      -0.049    -0.048 -0.041   -0.040  -0.100    -0.098
Size          0.003     0.004 -0.001   -0.001   0.000    -0.001
Urban1        0.015     0.016 -0.011   -0.009  -0.016    -0.012
Rural1        0.017     0.016  0.007    0.007   0.016     0.015
disp          0.089     0.086  0.058    0.057   0.086     0.087
```

With inference in mind, it is also of interest to compute the standard errors of the estimates. When these are computed by maximization of the full conditional log-likelihood in (6.21), standard errors can be obtained from the Hessian matrix estimated during the optimization. Specifically, the covariance matrix of the vector of estimates can be obtained as follows:

```
> cov.fLL <- solve(res.n$hessian)
```

The standard errors of the estimates can then be computed from the diagonal of the latter. For instance, for the copula parameter, we obtain:

```
> sqrt(cov.fLL[1,1])

[1] 0.01443763
```

The previous value can be compared with the standard error obtained in the two-stage approach by calling `vcov()` on the fitted object `ts.n.ex`:

```
> sqrt(vcov(ts.n.ex))

           rho.1
rho.1 0.01308801
```

For the marginal parameters, a comparison of the standard errors is provided by the following code:

```
> ## Standard errors of marginal parameters (two-stage vs full likelihood)
> ## Note that, when called on the fitted objects returned by 'glm()',
> ## 'diag(vcov())' does not provide the variance of the dispersion parameter
> full.SE <- sqrt(diag(cov.fLL))
> all.SE <- cbind(ts.math   = c(sqrt(diag(vcov(math.glm))), NA),
                  full.math = full.SE[1 + 1:p],
                  ts.sci    = c(sqrt(diag(vcov(sci.glm))), NA),
                  full.sci  = full.SE[p+1 + 1:p],
                  ts.read   = c(sqrt(diag(vcov(read.glm))), NA),
                  full.read = full.SE[2*p+1 + 1:p])
> rownames(all.SE)[p] <- "disp"
> round(all.SE, 4)
```

```
             ts.math full.math ts.sci full.sci ts.read full.read
(Intercept)   0.0255    0.0248 0.0206   0.0200  0.0250    0.0248
Minority1     0.0230    0.0226 0.0186   0.0182  0.0226    0.0227
SES           0.0134    0.0130 0.0108   0.0106  0.0131    0.0130
Female1       0.0189    0.0185 0.0153   0.0150  0.0186    0.0186
Public1       0.0265    0.0259 0.0214   0.0210  0.0260    0.0261
Size          0.0112    0.0111 0.0091   0.0090  0.0110    0.0111
Urban1        0.0237    0.0234 0.0191   0.0188  0.0233    0.0234
Rural1        0.0254    0.0250 0.0205   0.0202  0.0249    0.0251
disp              NA    0.0038     NA   0.0025      NA    0.0038
```

We end this example by illustrating the practical usefulness of the obtained (say, full likelihood-based) estimate of the conditional distribution H_Z given by (6.20). We focus on the probability that four fictitious students have all three scores below their 10% quantiles:

```
> q.math <- quantile(nels$Math,    probs = 0.1)
> q.sci  <- quantile(nels$Science, probs = 0.1)
> q.read <- quantile(nels$Reading, probs = 0.1)
```

The four students are assumed to be males from public schools of average size in urban areas. They only differ in their value for `Minority` and `SES` (socioeconomic status). Their descriptions in terms of the covariates are stored in the following data frame:

```
> probs <- c(0.25, 0.75) # quantile orders of SES
> stdnts <- data.frame(Minority = c(0, 1, 0, 1),
                       SES = rep(quantile(nels$SES, probs = probs), each = 2),
                       Female = 0, Public = 1, Size = 0, Urban = 1, Rural = 0)
```

For each type of score and each student, the probability that the student's score is below its 10% quantile can be computed as follows:

```
> newz <- model.matrix(~ Minority + SES + Female + Public + Size +
                         Urban + Rural, data = stdnts) # design matrix
> ## The four marginal probabilities for the math, science and reading score
> prob.math <- nmLL(full.math, q.math, newz, pobs = TRUE)$U
> prob.sci  <- nmLL(full.sci,  q.sci,  newz, pobs = TRUE)$U
> prob.read <- nmLL(full.read, q.read, newz, pobs = TRUE)$U
```

The corresponding four joint probabilities that all three scores are below their 10% quantiles are then simply:

```
> u <- cbind(prob.math, prob.sci, prob.read)
> joint.prob.full <- pCopula(u, copula = normalCopula(full.n.ex, dim = 3))
> data.frame(Minority = stdnts[,1] == 1, SES.q.order = rep(probs, each = 2),
             joint.prob = round(joint.prob.full, 6),
             joint.prob.ind = round(pCopula(u, copula = indepCopula(3)), 6))
```

```
  Minority SES.q.order joint.prob joint.prob.ind
1    FALSE        0.25   0.026183       0.001006
2     TRUE        0.25   0.052111       0.004072
3    FALSE        0.75   0.006447       0.000055
4     TRUE        0.75   0.014345       0.000291
```

As could have been expected, students from minorities and from low socioeconomic background are the most disadvantaged (which can be inferred here from the larger estimated probability of having simultaneously low scores). Furthermore, as can be seen from the last column of the output, ignoring the conditional dependence between the scores would have led to a strong underestimation of the four joint probabilities. □

References

Abegaz, F., Gijbels, I., & Veraverbeke, N. (2012). Semiparametric estimation of conditional copulas. *Journal of Multivariate Analysis, 110*, 43–73.

Acar, E. F., Craiu, R. V., & Yao, Y. (2011). Dependence calibration in conditional copulas: A nonparametric approach. *Biometrics, 67*, 445–453.

Acar, E. F., Craiu, R. V., & Yao, Y. (2013). Statistical testing of covariate effects in conditional copula models. *Electronic Journal of Statistics, 7*, 2822–2850.

Agresti, A. (2010). *Analysis of ordinal categorical data* (2nd ed.). New York: Wiley.

Aue, A., & Horváth, L. (2013). Structural breaks in time series. *Journal of Time Series Analysis, 34*(1), 1–16.

Beare, B. K. (2010). Copulas and temporal dependence. *Econometrica, 78*(1), 395–410.

Ben Ghorbal, M., Genest, C., & Nešlehová, J. G. (2009). On the test of Ghoudi, Khoudraji, and Rivest for extreme-value dependence. *The Canadian Journal of Statistics, 37*(4), 534–552.

Bücher, A., Fermanian, J.-D., & Kojadinovic, I. (forthcoming). Combining cumulative sum change-point detection tests for assessing the stationarity of continuous univariate time series. *Journal of Time Series Analysis*.

Bücher, A., & Kojadinovic, I. (2015). An overview of nonparametric tests of extreme-value dependence and of some related statistical procedures. In D. Dey & J. Yan (Eds.), *Extreme value modeling and risk analysis: Methods and applications* (pp. 377–398). Boca Baton, FL: Chapman & Hall/CRC.

Bücher, A., & Kojadinovic, I. (2016). A dependent multiplier bootstrap for the sequential empirical copula process under strong mixing. *Bernoulli, 22*(2), 927–968.

Bücher, A., Kojadinovic, I., Rohmer, T., & Segers, J. (2014). Detecting changes in cross-sectional dependence in multivariate time series. *Journal of Multivariate Analysis, 132*, 111–128.

Canty, A., & Ripley, B. D. (2017). boot: Bootstrap R (S-Plus) functions, R package version 1.3-20. https://CRAN.R-project.org/package=boot

Cardinali, A., & Nason, G. P. (2013). Costationarity of locally stationary time series using costat. *Journal of Statistical Software, 55*(1), 1–22.

Chen, X., & Fan, Y. (2005). Pseudo-likelihood ratio tests for semiparametric multivariate copula model selection. *The Canadian Journal of Statistics, 33*, 389–414.

Chen, X., & Fan, Y. (2006). Estimation and model selection of semiparametric copula-based multivariate dynamic models under copula misspecifcation. *Journal of Econometrics, 135*, 125–154.

Constantine, W., & Percival, D. (2016). fractal: Fractal time series modeling and analysis, R package version 2.0-1. https://CRAN.R-project.org/package=fractal

Cormier, E., Genest, C., & Nešlehová, J. G. (2014). Using B-splines for nonparametric inference on bivariate extreme-value copulas. *Extremes, 17*, 633–659.

Côté, M.-P., Genest, C., & Abdallah, A. (2016). Rank-based methods for modeling dependence between loss triangles. *European Actuarial Journal, 6*, 377–408.

Csörgö, M., & Horváth, L. (1997). *Limit theorems in change-point analysis*. Chichester: Wiley.

Darsow, W. W. F., Nguyen, B., & Olsen, E. T. (1992). Copulas and Markov processes. *Illinois Journal of Mathematics, 36*, 600–642.

Deheuvels, P. (1981). A non parametric test for independence. *Publications de l'Institut de Statistique de l'Université de Paris, 26*, 29–50.

Dehling, H., Vogel, D., Wendler, M., & Wied, D. (2017). Testing for changes in Kendall's tau. *Econometric Theory, 33*(6), 1352–1386.

Derumigny, A., & Fermanian, J.-D. (2017). About tests of the "simplifying" assumption for conditional copulas. *Dependence Modeling, 5*(1), 154–197.

Fermanian, J.-D., & Lopez, O. (2018). Single-index copulae. *Journal of Multivariate Analysis, 165*, 27–55.

Fermanian, J.-D., & Wegkamp, M. (2012). Time-dependent copulas. *Journal of Multivariate Analysis, 110*, 19–29.

Frees, E. W. (2010). Multivariate regression using copulas. University Lecture.

Frees, E. W., & Valdez, E. A. (1998). Understanding relationships using copulas. *North American Actuarial Journal, 2*, 1–25.

Genest, C., Nešlehová, J. G., & Rémillard, B. (2014). On the empirical multilinear copula process for count data. *Bernoulli, 20*, 1344–1371.

Genest, C., Nešlehová, J. G., & Rémillard, B. (2017). Asymptotic behavior of the empirical multilinear copula process under broad conditions. *Journal of Multivariate Analysis, 20*, 82–110.

Genest, C., Nešlehová, J. G., & Ruppert, M. (2011). Comment on the paper by S. Haug, C. Klüppelberg and L. Peng entitled "Statistical models and methods for dependence in insurance data". *Journal of the Korean Statistical Society, 40*, 141–148.

Genest, C., & Rémillard, B. (2004). Tests of independence and randomness based on the empirical copula process. *Test, 13*(2), 335–369.

Genest, C., Quessy, J.-F., & Rémillard, B. (2006). Goodness-of-fit procedures for copula models based on the probability integral transformation. *Scandinavian Journal of Statistics, 33*, 337–366.

Ghalanos, A. (2017). rugarch: Univariate GARCH models. R package version 1.3-8. https://CRAN.R-project.org/package=rugarch

Gijbels, I., Omelka, M., & Veraverbeke, N. (2015). Estimation of a copula when a covariate affects only marginal distributions. *Scandinavian Journal of Statistics, 42*(4), 1109–1126.

Gombay, E., & Horváth, L. (1999). Change-points and bootstrap, *Environmetrics, 10*(6), 725–736.

Grégoire, V., Genest, C., & Gendron, M. (2008). Using copulas to model price dependence in energy markets. *Energy Risk, 5*(5), 58–64.

Hofert, M., Kojadinovic, I., Mächler, M., & Yan, J. (2017a). copulaData: Data sets for copula modeling. R package version 0.0-1. https://CRAN.R-project.org/package=copulaData

Holmes, M., Kojadinovic, I., & Quessy, J.-F. (2013). Nonparametric tests for change-point detection à la Gombay and Horváth. *Journal of Multivariate Analysis, 115*, 16–32.

Ibragimov, R. (2009). Copula based characterizations for higher-order Markov processes. *Econometric Theory, 25*, 819–846.

Joe, H. (2014). *Dependence modeling with copulas*. Boca Raton, FL: Chapman & Hall/CRC.

Jondeau, E., & Rockinger, M. (2006). The Copula–GARCH model of conditional dependencies: An international stock market application. *Journal of International Money and Finance, 25*, 827–853.

Kendall, M. G. (1945). The treatment of ties in rank problems. *Biometrika, 3*, 239–251.

Kojadinovic, I. (2017a). npcp: Some nonparametric tests for change-point detection in possibly multivariate observations. R package version 0.1-9. https://CRAN.Rproject.org/package=npcp

Kojadinovic, I. (2017b). Some copula inference procedures adapted to the presence of ties. *Computational Statistics and Data Analysis, 142*, 24–41.

Kojadinovic, I., & Yan, J. (2010). Modeling multivariate distributions with continuous margins using the `copula` R package. *Journal of Statistical Software, 34*(9), 1–20.

Kojadinovic, I., & Yan, J. (2011). Tests of serial independence for continuous multivariate time series based on a Möbius decomposition of the independence empirical copula process. *Annals of the Institute of Statistical Mathematics, 63*, 347–373.

Lambert, P. (2007). Archimedean copula estimation using Bayesian splines smoothing techniques. *Computational Statistics and Data Analysis, 51*(12), 6307–6320.

Li, Y., Li, Y., Qin, Y., & Yan, J. (2016). Copula modeling for data with ties. Tech. Rep., ArXiv E-prints. arxiv:1612.06968.

Lindsey, J. K. (1997). *Applying generalized linear models*. New York: Springer.

McNeil, A. J., Frey, R., & Embrechts, P. (2015). *Quantitative risk management: Concepts, techniques and tools* (2nd ed.). Princeton, NJ: Princeton University Press.

Nason, G. (2016). locits: Tests of stationarity and localized autocovariance. R package version 1.7.3. https://CRAN.R-project.org/package=locits

Nason, G., & Cardinali, A. (2013). costat: Time series costationarity determination. R package version 2.3. http://CRAN.R-project.org/package=costat

Nason, G. P. (2013). A test for second-order stationarity and approximate confidence intervals for localized autocovariances for locally stationary time series. *Journal of the Royal Statistical Society Series B, 75*, 879–904.

Pappadà, R., Durante, F., & Salvadori, G. (2017). Quantification of the environmental structural risk with spoiling ties: Is randomization worthwhile? *Stochastic Environmental Research and Risk Assessment, 31*, 2483–2497.

Patton, A. J. (2006). Modelling asymmetric exchange rate dependence. *International Economic Review, 47*(2), 527–556.

Patton, A. J. (2012). A review of copula models for economic time series. *Journal of Multivariate Analysis, 110*, 4–18.

Patton, A. J. (2013). Copula methods for forecasting multivariate time series. In G. Elliott & A. Timmermann (Eds.), *Handbook of economic forecasting* (pp. 899–960). New York: Springer.

Portier, F., & Segers, J. (2018). On the weak convergence of the empirical conditional copula under a simplifying assumption. *Journal of Multivariate Analysis, 166*, 160–181.

Priestley, M. B., & Subba Rao, T. (1969). A test for stationarity of time series. *Journal of the Royal Statistical Society Series B, 31*, 140–149.

Quessy, J.-F., Saïd, M., & Favre, A.-C. (2013). Multivariate Kendall's tau for change-point detection in copulas. *The Canadian Journal of Statistics, 41*, 65–82.

Rémillard, B. (2017). Goodness-of-fit tests for copulas of multivariate time series. *Econometrics, 5*(13), 1–23.

Rémillard, B., Papageorgiou, N., & Soustra, F. (2012). Copula-based semiparametric models for multivariate time series. *Journal of Multivariate Analysis, 110*, 30–42.

Ryan, J. A., & Ulrich, J. M. (2017). xts: eXtensible Time Series. R package version 0.10-0. http://CRAN.R-project.org/package=xts

Sun, J., Frees, E. W., & Rosenberg, M. A. (2008). Heavy-tailed longitudinal data modeling using copulas. *Insurance: Mathematics and Economics, 42*, 817–830.

Veraverbeke, N., Omelka, M., & Gijbels, I. (2011). Estimation of a conditional copula and association measures. *Scandinavian Journal of Statistics, 38*, 766–780.

Appendix A
R and Package Versions

In this appendix, we list detailed information about the R session and the packages used to produce this book including figures and other R results.

```
> sfsmisc::sessionInfoX(pkgs = "copula")

Extended  sessionInfo():
-----------------------
specific  packageDescription()s:
$copula
Package: copula
Version: 0.999-19
VersionNote: Last CRAN: 0.999-18 on 2017-08-31
Date: 2017-11-11
Title: Multivariate Dependence with Copulas
Authors@R: c(person("Marius", "Hofert", role = "aut", email =
        "marius.hofert@uwaterloo.ca") , person("Ivan",
        "Kojadinovic", role = "aut", email =
        "ivan.kojadinovic@univ-pau.fr") ,
        person("Martin","Maechler", role=c("aut","cre"), email
        = "maechler@stat.math.ethz.ch") , person("Jun", "Yan",
        rol = "aut", email = "jun.yan@uconn.edu") ,
        person(c("Johanna", "G."), "Nešlehová", role = "ctb",
        comment = "evTestK()"))
Depends: R (>= 3.1.0)
Imports: stats, graphics, methods, stats4, Matrix, lattice,
        colorspace, gsl, ADGofTest, stabledist (>= 0.6-4),
        mvtnorm, pcaPP, pspline, numDeriv
Suggests: MASS, KernSmooth, sfsmisc, scatterplot3d, Rmpfr,
        bbmle, knitr, parallel, gridExtra, lcopula, mvnormtest,
        partitions, polynom, qrng, randtoolbox, rugarch,
        Runuran, tseries, VGAM, VineCopula, zoo
SuggestsNote: the last lines' packages {parallel, ..., zoo}
        are only used in vignettes, demos and few tests.
Enhances: nor1mix, copulaData
```


M. Hofert et al., *Elements of Copula Modeling with R*, Use R!,
https://doi.org/10.1007/978-3-319-89635-9

```
Description: Classes (S4) of commonly used elliptical,
       Archimedean, extreme-value and other copula families,
       as well as their rotations, mixtures and
       asymmetrizations. Nested Archimedean copulas, related
       tools and special functions. Methods for density,
       distribution, random number generation, bivariate
       dependence measures, Rosenblatt transform, Kendall
       distribution function, perspective and contour plots.
       Fitting of copula models with potentially partly fixed
       parameters, including standard errors. Serial
       independence tests, copula specification tests
       (independence, exchangeability, radial symmetry,
       extreme-value dependence, goodness-of-fit) and model
       selection based on cross-validation. Empirical copula,
       smoothed versions, and non-parametric estimators of the
       Pickands dependence function.
License: GPL (>= 3) | file LICENCE
ByteCompile: yes
VignetteBuilder: knitr
Collate: AllClass.R Classes.R AllGeneric.R Auxiliaries.R .....
Encoding: UTF-8
URL: http://copula.r-forge.r-project.org/
NeedsCompilation: yes
Packaged: 2017-11-21 22:35:00 UTC; mhofert
Author: Marius Hofert [aut], Ivan Kojadinovic [aut], Martin
       Maechler [aut, cre], Jun Yan [aut], Johanna G.
       Nešlehová [ctb] (evTestK())
Maintainer: Martin Maechler <maechler@stat.math.ethz.ch>
Built: R 3.4.2; x86_64-apple-darwin16.7.0; 2017-11-21 22:36:13
       UTC; unix

-- File: /Users/mhofert/R/copula/pkg/copula.Rcheck/copula/Meta/package.rds

~~~~~~~~~~~~~~~~~~~~~~~~~~~~~~~~


Capabilities:
       jpeg         png        tiff       tcltk          X11
          X           X           X           X            -
       aqua    http/ftp     sockets      libxml         fifo
          X           X           X           X            X
     cledit       iconv         NLS     profmem        cairo
          -           X           X           -            X
        ICU long.double     libcurl
          X           X           X
Sys.info:
nodename                sklar.home
user                    mhofert

LAPACK version: 3.7.1
External software (versions):
zlib                    1.2.11
bzlib                   1.0.6, 6-Sept-2010
xz                      5.2.3
PCRE                    8.41 2017-07-05
ICU                     59.1
TRE                     TRE 0.8.0 R_fixes (BSD)
iconv                   GNU libiconv 1.11
```

```
readline              EditLine wrapper
BLAS                  /usr/local/R/R-rc_build/lib/libRblas.dylib

R_LIBS:
libPath [.libPaths()] contents in addition to R_LIBS and .Library:
[1] "/usr/local/R/library"
Main R env. variables (for more, inspect the 'xR.env' component):
                [,1]
R_ENVIRON       ""
R_PROFILE       ""
R_CHECK_ENVIRON ""
---------------- standard sessionInfo():
R version 3.4.3 RC (2017-11-23 r73782)
Platform: x86_64-apple-darwin17.2.0 (64-bit)
Running under: macOS High Sierra 10.13.2

Matrix products: default
BLAS: /usr/local/R/R-rc_build/lib/libRblas.dylib
LAPACK: /usr/local/R/R-rc_build/lib/libRlapack.dylib

locale:
[1] en_US.UTF-8/en_US.UTF-8/en_US.UTF-8/C/en_US.UTF-8/en_US.UTF-8

attached base packages:
[1] parallel  stats4    stats     graphics  grDevices utils
[7] datasets  methods   base

other attached packages:
 [1] copulaData_0.0-1    rugarch_1.3-6       npcp_0.1-9
 [4] xts_0.10-0          zoo_1.8-0           latticeExtra_0.6-28
 [7] lattice_0.20-35     lcopula_1.0         bbmle_1.0.19
[10] MASS_7.3-47         qrng_0.0-4          qrmtools_0.0-7
[13] nor1mix_1.2-3       mvtnorm_1.0-6       sfsmisc_1.1-1
[16] RColorBrewer_1.1-2  copula_0.999-19

loaded via a namespace (and not attached):
 [1] multicool_0.1-10            ks_1.10.7
 [3] pcaPP_1.9-72                expm_0.999-2
 [5] htmltools_0.3.6             SkewHyperbolic_0.3-2
 [7] nloptr_1.0.4                TTR_0.23-2
 [9] quantmod_0.4-10             Quandl_2.8.0
[11] pspline_1.0-18              htmlwidgets_0.9
[13] codetools_0.2-15            misc3d_0.8-4
[15] knitr_1.16                  httpuv_1.3.5
[17] ADGofTest_0.3               curl_2.8.1
[19] Rcpp_0.12.12                KernSmooth_2.23-15
[21] xtable_1.8-2                DistributionUtils_0.5-1
[23] jsonlite_1.5                truncnorm_1.0-7
[25] FNN_1.1                     mime_0.5
[27] spd_2.0-1                   digest_0.6.12
[29] shiny_1.0.3                 gsl_1.9-10.3
[31] numDeriv_2016.8-1           grid_3.4.3
[33] stabledist_0.7-1            tools_3.4.3
[35] magrittr_1.5                rgl_0.98.1
[37] Rsolnp_1.16                 GeneralizedHyperbolic_0.8-1
[39] Matrix_1.2-12               httr_1.2.1
[41] R6_2.2.2                    compiler_3.4.3
```

Index

Symbols

M. Hofert et al., *Elements of Copula Modeling with R*, Use R!,
https://doi.org/10.1007/978-3-319-89635-9

Made in the USA
Monee, IL
13 September 2026

62975066R00155